Energy Principles in
STRUCTURAL
MECHANICS

McGRAW-HILL
BOOK COMPANY
New York
St. Louis
San Francisco
Düsseldorf
Johannesburg
Kuala Lumpur
London
Mexico
Montreal
New Delhi
Panama
Paris
São Paulo
Singapore
Sydney
Tokyo
Toronto

THEODORE R. TAUCHERT

Department of Engineering Mechanics
University of Kentucky

Energy Principles in STRUCTURAL MECHANICS

This book was set in Times New Roman.
The editors were B. J. Clark and M. E. Margolies;
the cover was designed by Anne Canevari Green;
and the production supervisor was Bill Greenwood.
The drawings were done by John Cordes, J & R Technical Services, Inc.
Kingsport Press, Inc., was printer and binder.

Library of Congress Cataloging in Publication Data

Tauchert, T   R   date
  Energy principles in structural mechanics.

  Includes bibliographies.
  1. Structures, Theory of. I. Title.
TA654.T37   624.17   73–17205
ISBN 0–07–062925–0

Energy Principles in
STRUCTURAL MECHANICS

1 2 3 4 5 6 7 8 9 0 KPKP 7 9 8 7 6 5 4

CONTENTS

PREFACE

As engineering structures and their environments become more diverse and complex, it is not enough that the engineer be adept at applying the classical methods of structural analysis. More importantly, he must be aware of the limitations of the underlying theories and be able to make intelligent judgments about the validity of the basic assumptions. It is hoped that, by starting with a discussion of the classical theory of elasticity, this text will make clear the applicability and limitations of linear structural mechanics.

The emphasis of the book is on the development and applications of work and energy methods. The principles of virtual work, complementary virtual work, and various energy theorems derived therefrom are used to study the behavior of linearly elastic structures. While no attempt is made to cover the many ad hoc techniques which are appropriate for special types of structures, the basic force and displacement approaches treated herein have a wide range of application and are particularly adaptable to machine computation.

This book was developed from class notes used in teaching a two-term introductory course in structural mechanics at Princeton University. Portions of the notes have also been used in advanced strength-of-materials and mechanical vibration courses at the University of Kentucky. Those enrolled in the courses include juniors, seniors, and beginning graduate students from the departments of aerospace, mechanical, and civil engineering, and engineering mechanics. It is presumed that the students have had the normal undergraduate courses in engineering mechanics and have been exposed to ordinary differential equations.

Following an introductory chapter, the book is divided into three parts. Part I, comprising Chapters 2 to 5, is concerned with the foundations of solid mechanics. The concepts of stress, strain, and material behavior are reviewed in Chapters 2, 3, and 4. Virtual work principles are developed in Chapter 5 and are used to derive reciprocal theorems and minimum energy principles. Exact and approximate solutions are shown for the stress and deformation distributions in several structural elements.

Part II contains four chapters dealing with the behavior of structures under stationary loads. Relatively simple, statically indeterminate beams, trusses, and frames are analyzed in Chapter 6. The conjugate force and displacement methods are formulated in matrix notation in Chapter 7, and are applied to more complicated framed and stiffened structures. The basic equations governing the nonisothermal behavior of elastic bodies are developed in Chapter 8, and the response of structures to combined thermal and mechanical loadings are examined. Chapter 9 provides an introduction to elastic stability.

Part III of the text is concerned with the behavior of structures subject to dynamic loads. Structures which can be idealized as discrete-mass systems are considered in Chapters 10 and 11. Chapter 12 deals with the dynamic response of distributed-mass systems.

For readers who are unfamiliar with cartesian tensors, matrix algebra, or the calculus of variations, these topics are discussed in sufficient detail in Appendixes A, B, and C.

I am indebted to many students and colleagues for their valuable criticisms and suggestions. In particular I wish to acknowledge several inspiring discussions with Professor S. M. Vogel on the subject of energy principles. I also wish to thank Miss Elizabeth Thompson for her care and cheerfulness in typing and retyping the manuscript.

Finally, I am most grateful to Ann for her patience, and to Amy, Charles, Sarah, Rebecca, and Macy, who have tried to learn the art of being silent.

THEODORE R. TAUCHERT

Energy Principles in
STRUCTURAL
MECHANICS

1

INTRODUCTION

1.1 THE SUBJECT OF STRUCTURAL MECHANICS

The design of a modern engineering structure encompasses many unique and technically challenging problems. Design considerations differ markedly, depending upon the structure's function and its environment. In some cases the design is governed by the presence of stationary loads, while in other instances the effects of moving loads and temperature variations are of importance. In the design of a tall building, for example, it is necessary to consider wind loads and the possibility of earthquake motions in addition to the more obvious types of loads. The designer of a nuclear reactor, on the other hand, must worry about the effects of heat generation upon the structure. And in designing a missile or spacecraft, the influence of solar heating, bombardment by meteoroids, and various other phenomena which are difficult to forecast must be considered.

Owing to dissimilarities in the functions and environments of structures, the relative importance of factors such as strength, stiffness, and weight will vary. Obtaining optimum designs in different situations naturally requires the use of different structural configurations and materials. In short, each particular design project unveils a totally new set of problems.

The role of *structural mechanics* in the design process is to provide a description of the states of stress and deformation throughout the structure. An accurate knowledge of these response quantities is one of the essential ingredients in any design. Other factors include cost, appearance, fabrication techniques, and operational considerations. Thus structural mechanics represents but one of several aspects of an engineering design. However, it is often the most interesting feature and is generally the most technically demanding one. The structural analyst must be constantly aware of the fact that his errors and oversights can result in catastrophic failures.

1.2   CLASSIFICATIONS OF STRUCTURES

Engineering structures may be classified in several ways. The most common classification is based upon a combination of the geometric configuration and the loading characteristics of the structure. From a geometrical point of view the simplest structure is a *bar*, i.e., a single member having one dimension much larger than the other two. Depending upon whether the loading is tensile, compressive, torsional, or flexural, a straight bar is often referred to as a *rod*, *column*, *shaft*, or *beam*, respectively (Fig. 1.1a to d). A member which experiences simultaneous tension and bending is called a *tension beam* (Fig. 1.1e); one which undergoes both bending and compression is called a *beam column* (Fig. 1.1f). Curved members subject to bending are referred to as *curved beams* (Fig. 1.1g). If a curved bar is supported at its ends and is loaded in such a manner that it acts primarily in direct compression, it is called an *arch* (Fig. 1.1h). The term *ring* refers to a curved member which is closed (Fig. 1.1i).

Structures which consist of an assemblage of two or more bars are referred to as *framed structures*. If the bars are attached by frictionless hinges, and each member is subject to axial forces only, the system is called a *truss* (Fig. 1.1j). When the elements are rigidly attached and are subject to bending, shear, and axial loads, the structure is called a *frame* (Fig. 1.1k).

Another structural configuration consists of a body having two dimensions large in comparison with the third. If such a body is subject to in-plane loads (loads which act in a direction tangent to the two large surfaces), the structure is called a *panel* (Fig. 1.1l). When the body is acted on by out-of-plane loads (loads perpendicular to the large surfaces), it is referred to as a *plate* (Fig. 1.1m) or a *shell* (Fig. 1.1n). The surfaces of a plate are assumed to be flat, while those of a shell are curved. Panels, plates, or shells whose edges are reinforced with bars are said to be *stiffened*.

Structures may also be classified as *linear* or *nonlinear*, depending upon whether the equations which govern the structure's behavior are linear or nonlinear.

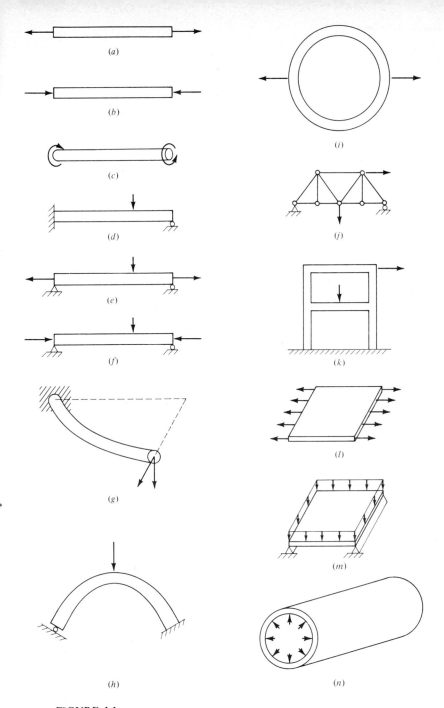

FIGURE 1.1
(a) Rod; (b) column; (c) shaft; (d) beam; (e) tension beam; (f) beam column;
(g) curved beam; (h) arch; (i) ring; (j) truss; (k) frame; (l) panel; (m) plate;
(n) shell.

When the equations are linear, the magnitude of any response quantity (stress, displacement, etc.) is proportional to the magnitude of the applied loads. The *principle of superposition* is then valid, and the response due to several loads may be obtained by summing the response produced by each load separately.

We shall see that a structure is linear only when (1) the material behavior is governed by Hooke's law and (2) the displacements are everywhere small (corresponding to infinitesimal strains and rotations). Although these conditions are never satisfied exactly for real structures, they are satisfied approximately in many instances, and they form the basis for most structural analyses.

Structures may also be classified as *determinate or indeterminate*. A structure is said to be *statically determinate* if the stress distribution or the unknown forces can be computed in terms of the applied loads using statics alone. If the number of unknowns exceeds the number of independent equations of statics, the system is termed *statically indeterminate*.

Similarly a structure is *kinematically determinate* if its displacement field can be related to a set of prescribed displacements from a consideration of kinematics alone; otherwise the structure is *kinematically indeterminate*.

1.3 CLASSIFICATIONS OF LOADS

The student is aware of the fact that the application of a force will produce stresses and deformations throughout a structure. The idea that distortion and stress may also be induced by temperature changes is usually less well understood. It is a fact, however, that in some instances the *thermal stresses* and *thermal displacements* are significantly larger than those resulting from the mechanically applied forces. Hence loads are sometimes classified according to their source as either *thermal* or *mechanical*.

Loads are also classified as either *distributed* or *discrete*. Although any real load is distributed over a finite region, the concept of a discrete or concentrated load is often very useful. If the region of application is small, the response to the distributed load and the response to a statically equivalent concentrated load will be very nearly the same everywhere in the structure except in the vicinity of the loading.[1]

One other common classification of loads involves the rate at which the loads are exerted on the structure. If they are applied so slowly that the system can be

---

[1]This conclusion was reached by Barre' de Saint-Venant in 1855 and is called Saint-Venant's principle. Various statements and mathematical proofs of the principle are discussed in: Y. C. Fung, "Foundations of Solid Mechanics," Prentice-Hall, Englewood Cliffs, N.J., 1965, pp. 300–309

considered to be in equilibrium at every instant of time, and if once applied they remain constant, then the loads are considered to be *static*. On the other hand *dynamic* loads refer to rapidly applied or continuously varying actions. The latter induce vibrations in the structure and generally result in stresses and deformations which are significantly greater than those produced by static loads of the same magnitude.

1.4 SCOPE OF THE TEXT

The emphasis of this text is on the development and applications of energy methods of structural analysis. After a review of the concepts of stress and strain, the principles of *virtual work* and *complementary virtual work* are introduced (Part I). By direct applications of these principles, or by using energy theorems derived therefrom, the response of structures to static (Part II) and dynamic loads (Part II) is studied.

Various types of structures are considered including bars, beams, beam columns, trusses, frames, and stiffened panels. The behavior of indeterminate as well as determinate systems is investigated. Although the text concentrates on linearly elastic structures, a few examples are included which illustrate applications of the methods in cases of more general structural behavior.

The response of structures to both thermal and mechanical loadings is examined. Problems involving elastic stability are also introduced. In each case the development of the theory is based upon fundamental principles of work and energy.

It is assumed that the reader has an understanding of elementary calculus and mechanics (statics, dynamics, and strength of materials). While a previous course in elementary structural analysis would be beneficial, it is not necessary.

The student is encouraged to work each of the problems given below as a review of the prerequisite material.

REVIEW PROBLEMS

Prepare a brief outline of the derivation of the basic formulas in Probs. 1.1 to 1.3 from the mechanics of deformable solids. Make sure that you understand the physical meaning of the various terms in each expression, as well as the assumptions used in the derivation.

1.1 (a) The tensile stress in a prismatic rod of cross-sectional area A, subject to an axial load P: $\sigma_{11} = P/A$.

(b) The normal strains in the rod, assuming that the material is elastic and has a Young's modulus E and Poisson's ratio ν: $e_{11} = P/EA$, $e_{22} = e_{33} = -\nu P/EA$.

(c) The axial displacement u_1 at the end $x_1 = L$ of the elastic rod shown: $u_1(L) = PL/EA$.

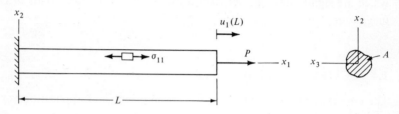

PROBLEM 1.1

1.2   (a) The shear stress in a circular shaft of polar moment of inertia J and shear modulus G, subject to a torque $\mathcal{T}$, referred to cylindrical coordinates (r,ϕ,x_1): $\tau = \mathcal{T}r/J$.

(b) The shear stress components referred to rectangular coordinates (x_1,x_2,x_3): $\sigma_{12} = -\mathcal{T}x_3/J$, $\sigma_{13} = \mathcal{T}x_2/J$.

(c) The angle of twist θ at the end $x_1 = L$ of the shaft shown: $\theta(L) = \mathcal{T}L/GJ$.

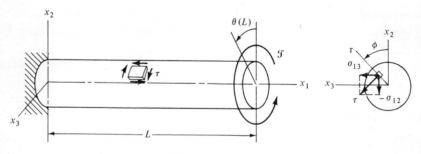

PROBLEM 1.2

1.3   (a) The stresses in a symmetric beam of cross-sectional area A and Young's modulus E: $\sigma_{11} = -Mx_2/I$, $\sigma_{12} = VQ/Ib$ where

$$I = \int_A x_2{}^2 \, dA \quad \text{and} \quad Q = \int_{A^*} x_2 \, dA$$

(b) The differential equations relating the bending moment $M(x_1)$, the shear force $V(x_1)$, and the distributed load $w(x_1)$ to the transverse deflection $u_2(x_1)$ (assumed positive upwards):

$$M = EI \frac{d^2 u_2}{dx_1^2}, \qquad V = -\frac{d}{dx_1}\left(EI \frac{d^2 u_2}{dx_1^2}\right), \qquad w = \frac{d^2}{dx_1^2}\left(EI \frac{d^2 u_2}{dx_1^2}\right)$$

PROBLEM 1.3

1.4 Determine the axial force N, the shear force V, and the bending moment M at the cross section a–a of the curved bar shown.

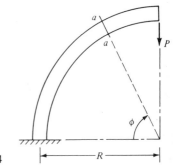

PROBLEM 1.4

1.5 A beam is subject to a distributed load (force per unit length) w_0 and a couple of magnitude $M_0 = 2w_0 L^2$, as shown. Plot the bending moment $M(x_1)$ and shear force $V(x_1)$ diagrams.

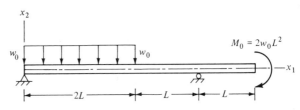

PROBLEM 1.5

1.6 A symmetric elastic beam of bending stiffness EI is acted on by the linearly varying distributed load shown; w_0 represents the intensity of the force per unit length at the end $x_1 = L$. Determine the maximum (absolute) value of the transverse displacement u_2 in the beam.

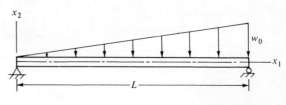

PROBLEM 1.6

PART ONE

Mechanics of Deformable Solids

ANALYSIS OF STRESS

2.1 INTRODUCTION

A thorough understanding of the physical and mathematical concepts of stress, strain, and the relationship of one to the other is essential in the study of structural mechanics. In this chapter we shall formalize some of the common notions of stress. A precise definition of stress at a point will be given, and the equations which govern the variation of stress throughout a body will then be derived.

2.2 STRESS AT A POINT

Consider a deformable body which is in equilibrium under a system of externally applied loads (Fig. 2.1). Some of the loads may be distributed over the surface of the body, such as those arising from contact with another body; these are called *surface forces*. Others may be distributed over the volume of the body, as for example the force of gravity, and these are referred to as *body forces*.

We next consider an element of surface $\Delta \mathscr{S}$ situated either within the body or on its boundary. The orientation of element $\Delta \mathscr{S}$ is specified by the unit normal

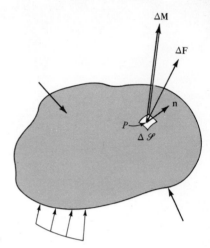

FIGURE 2.1
Deformable body in equilibrium.

vector **n**. We shall denote the resultant force and couple acting on $\Delta\mathscr{S}$ by $\Delta\mathbf{F}$ and $\Delta\mathbf{M}$, respectively. The concept of stress is obtained by letting the area $\Delta\mathscr{S}$ approach zero. It is assumed that the ratio $\Delta\mathbf{F}/\Delta\mathscr{S}$ then approaches a definite limit, and that $\Delta\mathbf{M}/\Delta\mathscr{S}$ vanishes;[1] that is,

$$\lim_{\Delta\mathscr{S}\to 0}\frac{\Delta\mathbf{F}}{\Delta\mathscr{S}} = \frac{d\mathbf{F}}{d\mathscr{S}} = \mathbf{T}$$

$$\lim_{\Delta\mathscr{S}\to 0}\frac{\Delta\mathbf{M}}{\Delta\mathscr{S}} = 0$$

(2.2.1)

The vector **T** in Eq. (2.2.1) is called the *stress vector*, and it represents the force per unit area at point P acting on an infinitesimal element having an orientation specified by **n**; it gives no information for elements at the same point which have other orientations, and hence it does not provide a complete description of the state of stress at the point P. In fact, we will show that a complete description requires a knowledge of the stress vectors acting on three elements having different orientations.

Now consider an infinitesimal rectangular parallelepiped at the point in question, and let the stress vectors $\mathbf{T}_1$, $\mathbf{T}_2$, and $\mathbf{T}_3$ represent the stress vectors on the faces perpendicular to the coordinate axes x_1, x_2, and x_3, respectively (Fig. 2.2). The components of $\mathbf{T}_1$ are denoted by σ_{11}, σ_{12}, and σ_{13}; the components of $\mathbf{T}_2$ by σ_{21}, σ_{22}, and σ_{23}; and the components of $\mathbf{T}_3$ by σ_{31}, σ_{32}, and σ_{33}.

---

[1] Theories have been proposed in which $\lim_{\Delta\mathscr{S}\to 0}(\Delta\mathbf{M}/\Delta\mathscr{S}) = \mathbf{m}$ represents a finite quantity known as the *couple-stress vector*. Such theories are far more complex than the conventional theory, and their use in structural mechanics is generally unwarranted.

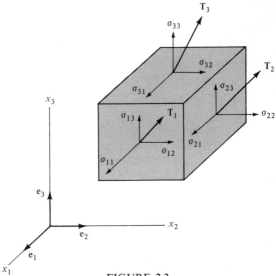

FIGURE 2.2
Stress vectors and their components.

In other words, σ_{ij} represents the projection of the stress vector $\mathbf{T}_i$ on the coordinate axis x_j. Hence we can write

$$\mathbf{T}_1 = \sigma_{11}\mathbf{e}_1 + \sigma_{12}\mathbf{e}_2 + \sigma_{13}\mathbf{e}_3$$
$$\mathbf{T}_2 = \sigma_{21}\mathbf{e}_1 + \sigma_{22}\mathbf{e}_2 + \sigma_{23}\mathbf{e}_3 \qquad (2.2.2)$$
$$\mathbf{T}_3 = \sigma_{31}\mathbf{e}_1 + \sigma_{32}\mathbf{e}_2 + \sigma_{33}\mathbf{e}_3$$

where $\mathbf{e}_1$, $\mathbf{e}_2$, and $\mathbf{e}_3$ are unit vectors. Employing the *summation convention* defined in Appendix *A*, Eqs. (2.2.2) may be written in the compact form[1]

$$\mathbf{T}_i = \sigma_{ij}\mathbf{e}_j \qquad (2.2.3)$$

The nine quantities σ_{ij} collectively are called the *stress tensor*. (We have yet to prove that σ_{ij} is a tensor quantity; i.e., in order to call it a tensor, we must first demonstrate that its components obey the transformation law for a second-order tensor.[2])

The sign conventions to be used for the stress components σ_{ij} are summarized in Fig. 2.3.[3] Note that the *normal stress* components σ_{11}, σ_{22}, and σ_{33} are

---

[1] Henceforth, the summation convention will apply to all repeated indices.
[2] See Eq. (A.5.1) of Appendix A.
[3] In general, the stress components acting on opposite faces of a volume element will not be equal in magnitude (as indicated in Fig. 2.3), but will differ by a small amount. For example, if the distance separating the right and left faces of the element in Fig. 2.3 is dx_2, then according to Taylor's theorem the normal stresses on these faces differ by the amount $d\sigma_{22} = (\partial\sigma_{22}/\partial x_2)\, dx_2$.

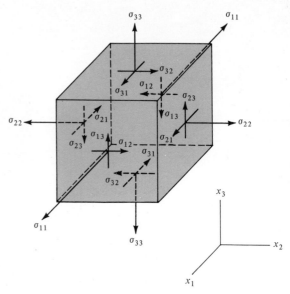

FIGURE 2.3
Sign convention for stress components.

taken to be positive when they produce tension, and negative when they produce compression. For example, a structural bar oriented parallel to the x_2 axis and subjected to a compressive end load of 2,000 $lb_f/in.^2$ would have a stress of $\sigma_{22} = -2,000 \ lb_f/in.^2$. The *shear stress* component σ_{12} is taken to be positive when it is directed in the positive x_2 direction on the face having outward unit normal $\mathbf{e}_1$ (and directed in the negative x_2 direction on the face whose outward normal is $-\mathbf{e}_1$). Similarly, shear stress σ_{13} is taken to be positive when it acts in the positive x_3 direction on the face having outward normal $\mathbf{e}_1$. Equivalent sign conventions apply to the other shear stress components.

We shall now prove that the state of stress at a point of the medium is described completely by the specification of the nine stress components. To do this, we will show that the stress vector $\mathbf{T}$ acting on an arbitrary element of surface may be found once the values of σ_{ij} are known. Consider a small tetrahedron with its apex at point P, as shown in Fig. 2.4. The unknown stress vector $\mathbf{T}$ acts on the face ABC, located at a small distance h from P. (Eventually the height h of the tetrahedron will be allowed to approach zero, so that the area ABC will pass through the point P.) Consistent with the notation introduced previously, the stress vectors $-\mathbf{T}_1$, $-\mathbf{T}_2$, and $-\mathbf{T}_3$ act on the faces having outward normals $-\mathbf{e}_1$, $-\mathbf{e}_2$, and $-\mathbf{e}_3$, respectively. The body force per unit volume is denoted by $\mathbf{f}$.

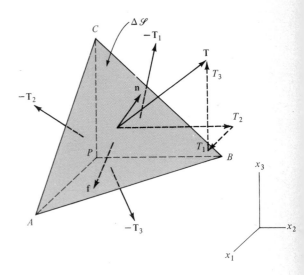

FIGURE 2.4
Stress vectors acting on an infinitesimal tetrahedron.

Since the tetrahedron is in equilibrium, the resultant of all forces acting on it vanishes; thus

$$\mathbf{T}\,\Delta\mathscr{S} - \mathbf{T}_1 n_1\,\Delta\mathscr{S} - \mathbf{T}_2 n_2\,\Delta\mathscr{S} - \mathbf{T}_3 n_3\,\Delta\mathscr{S} + \mathbf{f}(\tfrac{1}{3}h\,\Delta\mathscr{S}) = 0 \qquad (2.2.4)$$

where $\Delta\mathscr{S}$ is the area of face ABC, $n_1\,\Delta\mathscr{S}$ is the area of face PBC,[†] $n_2\,\Delta\mathscr{S}$ is the area of PAC, $n_3\,\Delta\mathscr{S}$ is the area of PAB, and $\tfrac{1}{3}$ h $\Delta\mathscr{S}$ represents the volume of the tetrahedron. Resolving the vector $\mathbf{T}$ into components along the coordinate axes $(\mathbf{T} = T_i \mathbf{e}_i)$, using Eqs. (2.2.2), and taking the limit $h \to 0$ gives, in index notation,

$$T_i = \sigma_{ji} n_j \qquad (2.2.5)$$

Hence, the components of the stress vector T_i at point P may be calculated for any direction n_j from a knowledge of the nine stress components σ_{ji}. [We shall prove shortly that the stress tensor is symmetric $(\sigma_{ij} = \sigma_{ji})$ and has, therefore, only six independent components.] Since T_i and n_j are the components of vectors, or tensors of order 1, it follows from Eq. (2.2.5) that σ_{ij} is indeed a tensor of order 2.[‡] This means that if the stress components at point P are known in one

[†] It is easily proved (see Ref. 2.2) that if the area of the face of the tetrahedron with outward normal **n** is $\Delta\mathscr{S}$, then the face normal to the x_i axis will have an area of $n_i\,\Delta\mathscr{S}$.
[‡] See Prob. A.14 of Appendix A.

FIGURE 2.5
Normal and shear stress components.

coordinate system, they may be calculated with respect to any other coordinate system by using the tensor transformation law; that is.,

$$\sigma'_{ij} = \alpha_{ik}\alpha_{jl}\sigma_{kl} \qquad (2.2.6)$$

where σ_{ij} are the stress components in the x_i coordinate system, σ'_{ij} are the components in the x'_i system, and α_{ij} are the direction cosines

$$\alpha_{ij} = \cos(x'_i, x_j) \qquad (2.2.7)$$

It is sometimes desirable to resolve the stress vector **T** into components directed normal and tangential to the surface element on which **T** acts (Fig. 2.5). The normal component T_n along **n** is given by

$$T_n = \mathbf{T} \cdot \mathbf{n} = T_i n_i \qquad (2.2.8)$$

or, using Eq. (2.2.5),

$$T_n = \sigma_{ji} n_j n_i \qquad (2.2.9)$$

The magnitude of the tangential or shear component T_s can then be computed from the relation

$$T_s^2 = T_i T_i - T_n^2 \qquad (2.2.10)$$

2.3 PRINCIPAL STRESSES

In the preceding section a stress vector was defined as the intensity of the resultant force acting upon an infinitesimal element of surface having a particular orientation. Surface elements at the same point having different orientations were shown to have different stress vectors. We shall now consider orientations for which the stress vector $\mathbf{T}$ is parallel to the outward normal vector $\mathbf{n}$, in which case the tangential or shearing stress components vanish. For such an orientation we have $\mathbf{T} = \sigma\mathbf{n}$, or

$$T_i = \sigma n_i \qquad (2.3.1)$$

where σ is the magnitude of the vector $\mathbf{T}$. Since Eq. (2.2.5) is valid for an element of arbitrary orientation, we may also write

$$T_i = \sigma_{ji} n_j \qquad (2.3.2)$$

From Eqs. (2.3.1) and (2.3.2) it follows that

$$(\sigma_{ji} - \sigma\delta_{ji})n_j = 0 \qquad (2.3.3)$$

Relation (2.3.3) represents a set of three homogeneous algebraic equations which are linear in the components n_j. An equation of this form constitutes what is known in mathematics as an *eigenvalue problem* (see Appendix B). A nonzero solution for the *eigenvalues* σ exists if and only if the determinant of the coefficients vanishes. Thus we require that

$$|\sigma_{ji} - \sigma\delta_{ji}| = 0 \qquad (2.3.4)$$

or, in expanded notation,

$$\begin{vmatrix} \sigma_{11} - \sigma & \sigma_{12} & \sigma_{13} \\ \sigma_{21} & \sigma_{22} - \sigma & \sigma_{23} \\ \sigma_{31} & \sigma_{32} & \sigma_{33} - \sigma \end{vmatrix} = 0 \qquad (2.3.5)$$

The expansion of this determinant yields a cubic equation in σ, the three roots of which are the desired eigenvalues, say $\sigma^{(I)}$, $\sigma^{(II)}$, and $\sigma^{(III)}$.[1] Consequently, when the stress tensor σ_{ij} at a point in a structure is known, Eq. (2.3.5) can be used to find three stress vectors, each of which is directed normal to the surface on which it acts. The magnitudes $\sigma^{(I)}$, $\sigma^{(II)}$, and $\sigma^{(III)}$ of these normal vectors are the *principal stresses* at the point in question, and the planes on which they act are called the *principal planes of stress*. Principal stresses are of great importance in structural analysis and design since, as we shall see shortly, the largest (in absolute value) of the principal stresses is the largest stress component for any orientation.

Once the principal stresses (eigenvalues) have been determined, the orientations $\mathbf{n}$ (eigenvectors) of the principal planes are determined using Eq. (2.3.3) and the condition

$$n_i n_i = 1 \qquad (2.3.6)$$

[1] It is possible to prove (see Ref. 2.3) that the eigenvalues of any symmetric second-order tensor are real numbers. In general the roots of Eq. (2.3.5) are distinct, although for certain special states of stress it is possible to have equal roots.

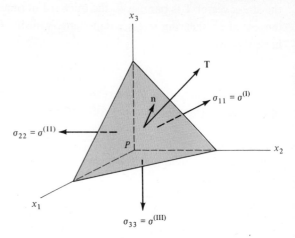

FIGURE 2.6
Principal stresses and principal axes.

For example, to find the direction cosines $n_i^{(I)}$ associated with the principal stress $\sigma^{(I)}$, the value $\sigma^{(I)}$ is substituted into (2.3.3). Because the determinant of the coefficients of $n_i^{(I)}$ is zero, (2.3.3) represents only two independent equations. Condition (2.3.6) supplies the needed third equation. Thus any two of Eqs. (2.3.3) and the relationship (2.3.6) may be solved simultaneously to obtain $n_1^{(I)}$, $n_2^{(I)}$, and $n_3^{(I)}$. The eigenvectors $n_i^{(II)}$ and $n_i^{(III)}$ associated with the eigenvalues $\sigma^{(II)}$ and $\sigma^{(III)}$, respectively, are obtained in a similar fashion. Furthermore, it can be shown (see Prob. 2.5) that the three eigenvectors $n_i^{(I)}$, $n_i^{(II)}$, and $n_i^{(III)}$ are mutually orthogonal.

After the principal stresses and principal planes have been established, it is generally convenient in any subsequent calculations to use a system of axes x_i which are directed perpendicular to the principal planes. With respect to these *principal axes* (Fig. 2.6), the shear stress components $\sigma_{12}, \sigma_{21}, \sigma_{23}, \sigma_{32}, \sigma_{31}$, and σ_{13} are zero, and the normal stress components are $\sigma_{11} = \sigma^{(I)}$, $\sigma_{22} = \sigma^{(II)}$, $\sigma_{33} = \sigma^{(III)}$. Then, using Eq. (2.2.5), the components of the stress vector $\mathbf{T}$ on an inclined plane having orientation $\mathbf{n}$ can be written simply as

$$T_1 = \sigma^{(I)}n_1$$
$$T_2 = \sigma^{(II)}n_2 \qquad (2.3.7)$$
$$T_3 = \sigma^{(III)}n_3$$

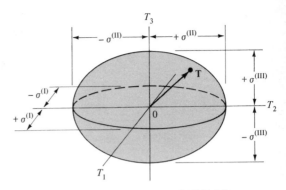

FIGURE 2.7
Stress ellipsoid of Lamé.

These relations may be used to construct a graphical representation of the state of stress at a point. Substituting Eqs. (2.3.7) into condition (2.3.6) gives

$$\left(\frac{T_1}{\sigma^{(I)}}\right)^2 + \left(\frac{T_2}{\sigma^{(II)}}\right)^2 + \left(\frac{T_3}{\sigma^{(III)}}\right)^2 = 1 \qquad (2.3.8)$$

which is the equation of an ellipsoid in the coordinates T_i. This ellipsoid, known as the *stress ellipsoid of Lamé*, is shown in Fig. 2.7. A position vector from the origin 0 to a point on the ellipsoid's surface has components T_i, and therefore represents the stress vector associated with a certain orientation. Since the semi-axes of the ellipsoid give the principal stresses, it is clear that the largest and smallest of the principal stresses are, respectively, the largest and smallest stresses for any orientation. Note, however, that the stress ellipsoid gives no information concerning a stress vector's orientation.

2.4   EQUATIONS OF EQUILIBRIUM

It was shown above that the state of stress at a point in a deformable body is specified completely by a knowledge of the stress tensor σ_{ij}. In general, the stress varies continuously from one point to another throughout the body. A set of equilibrium equations which govern the variation of stress will now be derived.

Consider a volume of material $\mathscr{V}$ bounded by a closed surface $\mathscr{S}$ within the body (Fig. 2.8). Let the body force per unit volume distributed throughout $\mathscr{V}$ be $\mathbf{f}$, and the stress vectors or *surface tractions* distributed over the boundary

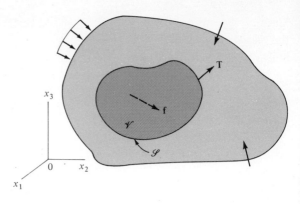

FIGURE 2.8
Region of a body in equilibrium.

$\mathscr{S}$ be **T**. If the medium is in equilibrium, then the summation of all forces acting on $\mathscr{V}$ must vanish; that is

$$\int_{\mathscr{V}} \mathbf{f} \, d\mathscr{V} + \int_{\mathscr{S}} \mathbf{T} \, d\mathscr{S} = 0 \qquad (2.4.1)$$

or in component form

$$\int_{\mathscr{V}} f_i \, d\mathscr{V} + \int_{\mathscr{S}} T_i \, d\mathscr{S} = 0 \qquad (2.4.2)$$

Using Eq. (2.2.5) we note that

$$\int_{\mathscr{S}} T_i \, d\mathscr{S} = \int_{\mathscr{S}} \sigma_{ji} n_j \, d\mathscr{S} \qquad (2.4.3)$$

Assuming that the stress components σ_{ji} and their first spatial derivatives are continuous, the divergence theorem[1] can be used to transform the surface integral in Eq. (2.4.3) to an integral over the volume $\mathscr{V}$, as

$$\int_{\mathscr{S}} \sigma_{ji} n_j \, d\mathscr{S} = \int_{\mathscr{V}} \sigma_{ji,j} \, d\mathscr{V} \qquad (2.4.4)$$

Eq. (2.4.2) then yields

$$\int_{\mathscr{V}} (f_i + \sigma_{ji,j}) \, d\mathscr{V} = 0 \qquad (2.4.5)$$

---

[1] See Prob. A.10 in Appendix A. Also note that the *comma convention* has been used in writing Eq. (2.4.4); that is $\sigma_{ji,j} = \partial \sigma_{ji}/\partial x_j$.

Since every element of volume in the medium is assumed to be in equilibrium, the region of integration $\mathscr{V}$ is arbitrary, and hence the integrand of Eq. (2.4.5) must vanish identically. Therefore, at every point we have

$$\sigma_{ji,j} + f_i = 0 \qquad (2.4.6)$$

These three equations of equilibrium, when expressed in expanded notation, are

$$\frac{\partial \sigma_{11}}{\partial x_1} + \frac{\partial \sigma_{21}}{\partial x_2} + \frac{\partial \sigma_{31}}{\partial x_3} + f_1 = 0$$

$$\frac{\partial \sigma_{12}}{\partial x_1} + \frac{\partial \sigma_{22}}{\partial x_2} + \frac{\partial \sigma_{32}}{\partial x_3} + f_2 = 0 \qquad (2.4.7)$$

$$\frac{\partial \sigma_{13}}{\partial x_1} + \frac{\partial \sigma_{23}}{\partial x_2} + \frac{\partial \sigma_{33}}{\partial x_3} + f_3 = 0$$

The student may recall that these same equations were derived in elementary strength of materials by considering the equilibrium of a differential element.

The resultant moment about the origin 0 of the forces acting on the volume $\mathscr{V}$ must also vanish for equilibrium. Letting $\mathbf{r}$ denote the position vector to an arbitrary point in the body, we have

$$\int_{\mathscr{V}} \mathbf{r} \times \mathbf{f} \, d\mathscr{V} + \int_{\mathscr{S}} \mathbf{r} \times \mathbf{T} \, d\mathscr{S} = 0 \qquad (2.4.8)$$

which may be written in index notation as

$$\int_{\mathscr{V}} \epsilon_{ijk} x_j f_k \, d\mathscr{V} + \int_{\mathscr{S}} \epsilon_{ijk} x_j T_k \, d\mathscr{S} = 0 \qquad (2.4.9)$$

where ϵ_{ijk} is the permutation symbol.[1] Using Eq. (2.2.5) and applying the divergence theorem, the surface integral becomes

$$\int_{\mathscr{S}} \epsilon_{ijk} x_j T_k \, d\mathscr{S} = \int_{\mathscr{S}} \epsilon_{ijk} x_j \sigma_{lk} n_l \, d\mathscr{S} = \int_{\mathscr{V}} (\epsilon_{ijk} x_j \sigma_{lk})_{,l} \, d\mathscr{V} \qquad (2.4.10)$$

Performing the differentiation and noting that $x_{j,l} = \delta_{jl}$ (the Kronecker delta), Eq. (2.4.9) can then be written as

$$\int_{\mathscr{V}} \epsilon_{ijk} (x_j f_k + \delta_{jl} \sigma_{lk} + x_j \sigma_{lk,l}) \, d\mathscr{V} = 0 \qquad (2.4.11)$$

Noting that $\delta_{jl} \sigma_{lk} = \sigma_{jk}$, and using the force-equilibrium equations $\sigma_{lk,l} + f_k = 0$, the foregoing expression becomes

$$\int_{\mathscr{V}} \epsilon_{ijk} \sigma_{jk} \, d\mathscr{V} = 0 \qquad (2.4.12)$$

---

[1] See Eq. (A.6.3) in Appendix A.

Since this equation must hold for every element of volume, the integrand itself must vanish. Thus

$$\epsilon_{ijk}\sigma_{jk} = 0 \qquad (2.4.13)$$

Expanding this expression gives

$$\sigma_{12} - \sigma_{21} = 0$$

$$\sigma_{23} - \sigma_{32} = 0 \qquad (2.4.14)$$

$$\sigma_{31} - \sigma_{13} = 0$$

or in general

$$\sigma_{ij} = \sigma_{ji} \qquad (2.4.15)$$

Thus the stress tensor is symmetric, and the state of stress at a point is completely characterized by six (rather than nine) independent components. Using this result, the relationship (2.2.5) between the stress vector and stress tensor may now be written in the more standard form

$$T_i = \sigma_{ij}n_j \qquad (2.4.16)$$

Similarly the equations of equilibrium (2.4.6) become

$$\sigma_{ij,j} + f_i = 0 \qquad (2.4.17)$$

Equations (2.4.16) and (2.4.17) will be referred to repeatedly throughout the text since they form the basis for the analysis of stress in a deformable body. At this point, however, we have more unknowns than we do equations. The additional relations which are needed for a unique determination of stress can only be obtained by considering the deformation and material behavior of the body. These topics will be discussed in the following two chapters.

PROBLEMS

2.1   The state of stress at a point P in a structure is given by

$\sigma_{11} = 10,000 \text{ lb}_f/\text{in.}^2$ $\qquad \sigma_{22} = -5,000 \text{ lb}_f/\text{in.}^2$ $\qquad \sigma_{33} = -5,000 \text{ lb}_f/\text{in.}^2$

$\sigma_{12} = 2,000 \text{ lb}_f/\text{in.}^2$ $\qquad \sigma_{23} = 2,000 \text{ lb}_f/\text{in.}^2$ $\qquad \sigma_{31} = 0$

Compute the scalar components T_1, T_2, and T_3 of the stress vector **T** on the plane passing through P whose outward normal vector **n** makes equal angles with the coordinate axes x_1, x_2, and x_3. Resolve this stress vector into normal and shear components.

2.2   At a certain point in a body $\sigma_{11} = \sigma_{22} = \sigma_{33} = -p$, and $\sigma_{12} = \sigma_{23} = \sigma_{31} = 0$. Using the stress-transformation law (2.2.6), show that the normal stresses are equal to $-p$ and the shear stresses vanish for any other cartesian coordinate system.

2.3 Write the stress-transformation law (2.2.6) in expanded notation for a two-dimensional state of stress. That is, assume that the stress components in the x_3 direction are zero ($\sigma_{13} = \sigma_{23} = \sigma_{33} = 0$), and consider a rotation about the x_3 axis. Compare the resulting expressions with the well-known Mohr's circle representation.

2.4 Find the principal stresses and their directions corresponding to the state of stress specified in Prob. 2.1.

2.5 Prove that the principal axes corresponding to distinct principal stresses are mutually orthogonal. Show, for example, that $\mathbf{n}^{(1)} \cdot \mathbf{n}^{(11)} = 0$.

2.6 Consider two elements of area passing through a point P. Let these elements have unit normals $\mathbf{n}^{(1)}$ and $\mathbf{n}^{(2)}$ and be subject to stress vectors $\mathbf{T}^{(1)}$ and $\mathbf{T}^{(2)}$, respectively. Prove that the projection of $\mathbf{T}^{(1)}$ on $\mathbf{n}^{(2)}$ is equal to the projection of $\mathbf{T}^{(2)}$ on $\mathbf{n}^{(1)}$; that is, prove that $T_i^{(1)}n_i^{(2)} = T_i^{(2)}n_i^{(1)}$.

REFERENCES

For an elementary discussion of stress:

2.1 SHAMES I. H.: "Mechanics of Deformable Solids," Prentice-Hall, Englewood Cliffs, N.J., 1964.

2.2 BISPLINGHOFF R. L., J. W. MAR, and T. H. H. PIAN: "Statics of Deformable Solids," Addison-Wesley, Reading, Mass., 1965.

For a more detailed treatment of the analysis of stress:

2.3 SOKOLNIKOFF, I. S.: "Mathematical Theory of Elasticity," 2d ed., McGraw-Hill, New York, 1956.

2.4 PEARSON, C. E.: "Theoretical Elasticity," Harvard, Cambridge, Mass., 1959.

3

DEFORMATION

3.1 INTRODUCTION

When a structure moves under the action of applied loads, its various elements will suffer translation, rotation, and deformation. In this chapter, we are concerned only with the deformation—or strain—rather than the rigid-body motions. The *strain tensor* is defined in terms of the change in distance between neighboring material particles. The relationship of the strain components to the displacement vector is examined. Next the assumptions of infinitesimal strain theory are discussed, and a physical interpretation of the *small strain tensor* is given. The chapter concludes with a discussion of *principal strains*.

3.2 STRAIN TENSOR

Let the position of a point P in an undeformed body be given by the coordinates X_i ($i = 1, 2, 3$) in the cartesian coordinate system X_1, X_2, X_3 (Fig. 3.1). When subjected to a system of applied loads, the body is deformed, and the material point P moves to a new location p having the coordinates x_i ($i = 1, 2, 3$) with respect to a second coordinate system x_1, x_2, x_3. The length of an infinitesimal

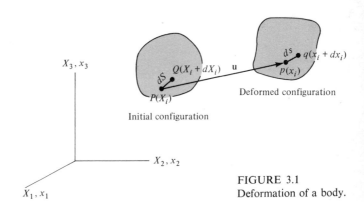

FIGURE 3.1
Deformation of a body.

line element connecting the point $P(X_i)$ and a neighboring point $Q(X_i + dX_i)$ in the initial configuration is denoted by dS. After deformation has occurred, the line connecting the same material points $p(x_i)$ and $q(x_i + dx_i)$ has a length ds. The squares of the lengths of these line elements are, by the Pythagorean theorem,

$$dS^2 = dX_1{}^2 + dX_2{}^2 + dX_3{}^2 = dX_i\, dX_i \qquad (3.2.1)$$

and
$$ds^2 = dx_1{}^2 + dx_2{}^2 + dx_3{}^2 = dx_i\, dx_i \qquad (3.2.2)$$

The difference between ds^2 and dS^2, which is a measure of the deformation of the body at point p, is

$$ds^2 - dS^2 = dx_i\, dx_i - dX_i\, dX_i \qquad (3.2.3)$$

Assuming that the final coordinates x_i of point p are differentiable functions of the initial coordinates X_i and vice versa, we have

$$X_i = X_i(x_1, x_2, x_3) \qquad (3.2.4)$$

and
$$dX_i = \frac{\partial X_i}{\partial x_1}\, dx_1 + \frac{\partial X_i}{\partial x_2}\, dx_2 + \frac{\partial X_i}{\partial x_3}\, dx_3 = X_{i,j}\, dx_j \qquad (3.2.5)$$

Therefore Eq. (3.2.3) can be written as

$$ds^2 - dS^2 = dx_i\, dx_i - X_{i,j}\, X_{i,k}\, dx_j\, dx_k$$
$$= (\delta_{jk} - X_{i,j}\, X_{i,k})\, dx_j\, dx_k \qquad (3.2.6)$$

We now define the *strain tensor* e_{jk} by

$$e_{jk} = \tfrac{1}{2}(\delta_{jk} - X_{i,j}\, X_{i,k}) \qquad (3.2.7)$$

so that
$$ds^2 - dS^2 = 2e_{jk}\, dx_j\, dx_k \qquad (3.2.8)$$

A more useful form of the strain tensor is obtained if we introduce the *displacement vector* **u** (Fig. 3.1) having components

$$u_i = x_i - X_i \qquad (3.2.9)$$

Therefore

$$X_{i,j} = x_{i,j} - u_{i,j} = \delta_{ij} - u_{i,j} \qquad (3.2.10)$$

and substitution of Eq. (3.2.10) into (3.2.7) gives

$$e_{jk} = \tfrac{1}{2}[\delta_{jk} - (\delta_{ij} - u_{i,j})(\delta_{ik} - u_{i,k})]$$
$$= \tfrac{1}{2}(u_{j,k} + u_{k,j} - u_{i,j} u_{i,k}) \qquad (3.2.11)$$

We note that the strain tensor is nonlinear in the derivatives of the displacements. If these displacement gradients are small compared to unity (that is, $u_{i,j} \ll 1$) so that their squares and products can be neglected in Eq. (3.2.11), then

$$e_{jk} \cong \tfrac{1}{2}(u_{j,k} + u_{k,j}) \qquad (3.2.12)$$

The set of quantities e_{jk} so defined is called the *small* or *infinitesimal strain tensor*. In unabridged notation, Eq. (3.2.12) becomes

$$e_{11} = \frac{\partial u_1}{\partial x_1} \qquad e_{22} = \frac{\partial u_2}{\partial x_2} \qquad e_{33} = \frac{\partial u_3}{\partial x_3}$$

$$e_{12} = e_{21} = \frac{1}{2}\left(\frac{\partial u_1}{\partial x_2} + \frac{\partial u_2}{\partial x_1}\right)$$

$$e_{23} = e_{32} = \frac{1}{2}\left(\frac{\partial u_2}{\partial x_3} + \frac{\partial u_3}{\partial x_2}\right) \qquad (3.2.13)$$

$$e_{31} = e_{13} = \frac{1}{2}\left(\frac{\partial u_3}{\partial x_1} + \frac{\partial u_1}{\partial x_3}\right)$$

Since the strain-displacement relations (3.2.13) are linear, they are much simpler to deal with than are the exact relations (3.2.11). Fortunately the linear approximation is justifiable for most engineering structures, at least for those constructed of relatively stiff materials such as metals. Hence, it will be used throughout this text. It should be mentioned, however, that situations do exist when the linear approximation is not valid. For example, in the case of a very long slender beam, the displacement gradients may be large, corresponding to large rotations.

The set of quantities e_{ij} defined by Eq. (3.2.12) constitutes a tensor of order 2. This follows from the fact that the displacement u_i is a tensor of order 1, and its partial derivatives $u_{i,j}$, therefore, represent a tensor of order 2. (See Prob. A.15 of Appendix A.) Hence we can determine the components of strain relative to a transformed coordinate system x_i' by means of the tensor transformation law

$$e_{ij}' = \alpha_{ik} \alpha_{jl} e_{kl} \qquad (3.2.14)$$

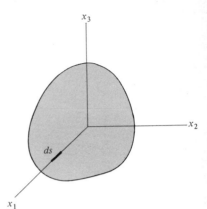

FIGURE 3.2
Line element in the deformed body.

3.3 PHYSICAL INTERPRETATION OF THE INFINITESIMAL STRAIN TENSOR

In order to understand the physical meaning of the components of the infinitesimal strain tensor, consider a line element, or material fiber, parallel to the x_1 coordinate axis (Fig. 3.2). In this case

$$dx_1 = ds \qquad dx_2 = dx_3 = 0 \qquad (3.3.1)$$

so that Eq. (3.2.8) gives

$$ds^2 - dS^2 = 2e_{11}\, dx_1\, dx_1 = 2e_{11}\, ds^2 \qquad (3.3.2)$$

Therefore

$$\frac{ds - dS}{dS} = 2e_{11}\frac{ds^2}{dS(ds + dS)} \qquad (3.3.3)$$

For the case of small deformations (although not necessarily small displacements) we assume that $ds \cong dS$, in which case

$$\frac{ds - dS}{dS} \cong e_{11} \qquad (3.3.4)$$

The *normal strain* e_{11} may therefore be interpreted as the extension, or fractional elongation, of a line element parallel to the x_1 coordinate axis. Similarly, components e_{22} and e_{33} represent the extensions of line elements parallel to the x_2 and x_3 axes, respectively.

The geometrical meaning of the infinitesimal *shear strain* e_{12} may be investigated by considering two line elements $ds^{(1)}$ and $ds^{(2)}$ which are parallel, respectively, to the x_1 and x_2 coordinate axes (Fig. 3.3). In this case

$$\begin{aligned} dx_1^{(1)} = ds^{(1)} \qquad dx_2^{(1)} = dx_3^{(1)} = 0 \\ dx_2^{(2)} = ds^{(2)} \qquad dx_1^{(2)} = dx_3^{(2)} = 0 \end{aligned} \qquad (3.3.5)$$

The angle between the corresponding lines $dS^{(1)}$ and $dS^{(2)}$ in the undeformed body is taken to be $\theta + \pi/2$, so that θ denotes the decrease in the right angle due to

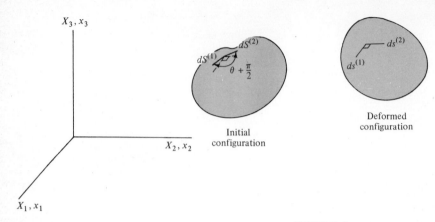

FIGURE 3.3
Shear deformation in a body.

the deformation. Making use of the definition of scalar multiplication of vectors $[A_i B_i = A B \cos (A,B)]$, we have

$$\cos \left(\theta + \frac{\pi}{2}\right) = \frac{dX_i^{(1)} \, dX_i^{(2)}}{dS^{(1)} \, dS^{(2)}}$$

$$= \frac{(\delta_{ij} - u_{i,j})(\delta_{ik} - u_{i,k}) \, dx_j^{(1)} \, dx_k^{(2)}}{dS^{(1)} \, dS^{(2)}}$$

$$= \frac{-2e_{12} \, dS^{(1)} \, dS^{(2)}}{dS^{(1)} \, dS^{(2)}} \tag{3.3.6}$$

Since $\cos(\theta + \pi/2) \cong -\theta$, $ds^{(1)} \cong dS^{(1)}$, and $ds^{(2)} \cong dS^{(2)}$ for infinitesimal deformation, it follows that

$$\theta \cong 2e_{12} \tag{3.3.7}$$

Therefore the strain component e_{12} is equal to one-half the decrease in the angle between two line elements which are parallel respectively to the x_1 and x_2 axes in the deformed configuration. A similar interpretation may be given to the other shear strains. It might be noted that in most elementary treatments of deformation the shear strains are defined as the *total* changes in the right angles, rather than *one-half* the changes. An advantage of the present definition is that the strain components e_{ij} represent a tensor, and the rules of tensor analysis are therefore applicable.

3.4   PRINCIPAL STRAINS

Based upon our understanding of the concept of principal stresses, we might expect an analogous situation for strain. That is, we might anticipate the existence of three orthogonal directions with respect to which all shearing strains vanish. Since the results derived in Sec. 2.3 followed directly from the fact that we were dealing with a symmetric second-order tensor, they are obviously applicable also to the strain tensor e_{ij}. We may conclude, therefore, that e_{ij} possesses three eigenvalues or *principal strains* which are the roots of the characteristic equation

$$|e_{ji} - e\delta_{ji}| = 0 \qquad (3.4.1)$$

The corresponding eigenvectors, or directions of line elements that suffer the principal strains, can be computed using the equation

$$(e_{ji} - e\delta_{ji})n_j = 0 \qquad (3.4.2)$$

and condition (2.3.6) in the same manner as before. These *principal directions* are mutually orthogonal, and moreover they coincide with the principal axes of stress for isotropic, elastic materials (to be discussed in Chap. 4). Since the shearing strains are absent in the case of principal strains, line elements oriented in the three principal directions remain perpendicular to each other under the deformation. Furthermore the largest extension experienced by any line element through a point is equal to the largest of the principal strains at that point.

PROBLEMS

3.1   Write Eq. (3.2.11) in unabridged form.

3.2   The strain components at a point p in a structure are experimentally found to be

$$e_{11} = e_{22} = -e_{33} = 0.002$$
$$e_{12} = 0.001 \quad e_{23} = e_{31} = 0$$

Compute the strain components at p associated with the x_i' coordinate system shown.

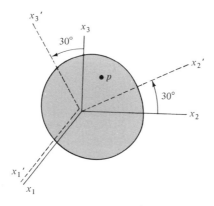

PROBLEM 3.2

3.3 Compute the infinitesimal strains, and sketch the deformed configuration of an element of volume (initially rectangular) which is subject to the following displacement fields:

    (a) Simple extension: $u_1 = Ax_1$, $u_2 = u_3 = 0$

    (b) Simple shear: $u_1 = Bx_2$, $u_2 = u_3 = 0$

    (c) Uniform dilatation: $u_i = Cx_i$

    (d) Homogeneous strain: $u_i = D_{ij}x_j$

where A, B, C, and D_{ij} are constants.

3.4 When subjected to a certain set of distributed forces, the rectangular bar shown below is in a state of *homogeneous strain* (i.e., the strain components e_{ij} do not vary from one point to another within the body). If the strains are

$$e_{11} = 0.001 \qquad e_{22} = 0.002 \qquad e_{12} = 0.0075$$

$$e_{13} = e_{23} = e_{33} = 0$$

determine the change in length of side AB, side BC, and diagonal BD of a cross section of the bar.

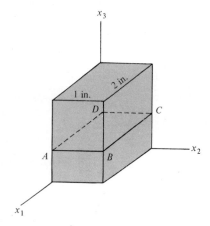

PROBLEM 3.4

3.5 Show that in the infinitesimal strain theory the quantity e_{ii} represents the expansion in volume per unit volume.

3.6 Derive an expression analogous to Eq. (3.2.8) for the strain measure $ds^2 - dS^2$ referred to the initial coordinates X_i rather than the final coordinates x_i. That is, let $ds^2 - dS^2 = 2E_{ij}\,dX_i\,dX_j$ and show that

$$E_{ij} = \frac{1}{2}\left(\frac{\partial u_i}{\partial X_j} + \frac{\partial u_j}{\partial X_i} + \frac{\partial u_k}{\partial X_i}\frac{\partial u_k}{\partial X_j}\right)$$

Then show that the *lagrangian strain tensor* E_{ij} is equal to the *eulerian strain tensor* e_{ij} in the case of small displacements and displacement gradients.

3.7 Decompose the strain gradients $u_{i,j}$ as follows

$$u_{i,j} = e_{ij} + \omega_{ij}$$

where e_{ij} denotes the small strain tensor. Verify that ω_{ij} is a tensor of order 2. Show that the nonzero components of ω_{ij} represent rigid-body rotations. For example, show that ω_{32} is the rigid-body rotation of an element of volume about the x_1 axis.

3.8 Investigate whether a set of six quantities e_{ij}, chosen arbitrarily, will automatically yield a physically realizable displacement field u_i. In particular, study the so-called "compatability relations" which the strain components must satisfy to insure a single-valued, continuous set of displacements. (See any one of Refs. 2.1 to 2.4.)

REFERENCES

Further discussions of the infinitesimal strain theory may be found in Refs. 2.1 to 2.4.

For a thorough treatment of finite deformations:

3.1 MURNAGHAN, F. D.: "Finite Deformation of an Elastic Solid," Wiley, New York, 1951.

3.2 NOVOZHILOV, V. V.: "Foundations of the Nonlinear Theory of Elasticity," Graylock, Rochester, N.Y., 1953.

3.3 GREEN, A. E., and W. ZERNA: "Theoretical Elasticity," Oxford University Press, London, 1954.

4

MATERIAL BEHAVIOR

4.1 INTRODUCTION

The concepts of stress and strain have been discussed independently in the preceding chapters. We shall now consider the relationship of one to the other. For the most part we will focus our attention upon linearly elastic materials (ones for which the stress components σ_{ij} are proportional to the strain components e_{ij}) since these are the simplest to deal with analytically. While certain types of inelastic behavior will be mentioned, many other important structural phenomena (creep, fatigue, etc.) have been omitted altogether from this introductory treatment.

4.2 EXPERIMENTAL OBSERVATIONS

Experimentally measured stress-strain data for most structural materials are available in the literature.[1] The majority of the data relates to one-dimensional

[1]Stress-strain data for various metals may be found in T. Lyman (ed.), Properties and Selection of Metals, in "Metals Handbook," 8th ed., vol. 1, American Society for Metals, Metals Park, Ohio, 1961.

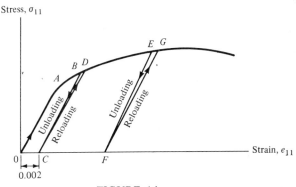

FIGURE 4.1
One-dimensional stress-versus-strain curve.

tensile tests (where σ_{11} is the only nonzero stress component) or shear tests (in which case σ_{12} is nonzero). Fig. 4.1 shows the general form of the stress-versus-strain curve obtained in a simple tensile test. The curve demonstrates several features which are common to nearly all materials, regardless of the particular type of test performed. For small values of strain the curve appears as a straight line, and in this region the well-known *Hooke's law* is valid. In many instances the point A where the curve deviates from a straight line, the so-called "proportional limit" of the material, is not clearly defined.

If the specimen is loaded beyond the material's proportional limit and is then unloaded, the sample may or may not suffer permanent deformation. If, upon complete removal of the load, there exists a measurable residual strain, then we say that the *elastic limit* of the material has been exceeded. The point B on the curve corresponding to a permanent or "plastic" strain equal to 0.002 is referred to as the *yield point*. For most materials the unloading portion of the curve (BC or EF) and the corresponding reloading portion (CD or FG) have slopes which are approximately equal to the slope of the initial loading curve ($0A$).

If the load is increased until the specimen ruptures, the stress at which rupture occurs is known as the *fracture stress*; the maximum stress measured during the entire test is referred to as the material's *ultimate stress*.

In order to analyze a structure fabricated from a material whose stress-versus-strain curve is known, it is necessary to express the stress-strain data in mathematical form. In so doing, it is naturally desirable to choose as simple a mathematical model as possible, yet one which will yield sufficiently accurate results. With this in mind let us first consider various idealizations to the rather complicated behavior exemplified by real materials. The simplest and most

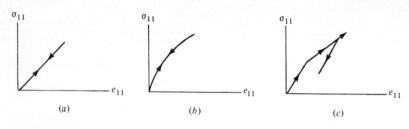

FIGURE 4.2
Idealized stress-strain diagrams: (a) linearly elastic behavior; (b) nonlinearly elastic behavior; (c) elastic-plastic response.

common idealization is the *linearly elastic* type of behavior shown in Fig. 4.2a. While no real material is truly linearly elastic, a linear relation is approximately valid for nearly all materials, providing the deformations encountered are relatively small.

The stress-versus-strain curves for some materials, including certain polymers and rubbers, have a low or nonexistent proportional limit but a relatively high yield point. If the unloading and reloading curves for such a material coincide, as illustrated in Fig. 4.2b, the response is called *nonlinearly elastic*. This behavior can often be described mathematically by means of a simple polynomial relation between stress and strain. The deformation of a structure whose behavior is nonlinearly elastic is analyzed in Chap. 5.

A material whose stress-strain curve exhibits a well-defined yield point can sometimes be represented adequately by a piecewise-linear model like the one shown in Fig. 4.2c. Such a material is said to be *elastic-plastic*; a rigorous analysis of elastic-plastic behavior is beyond the scope of this text. In fact, from this point on we shall generally restrict our attention to solids whose behavior is linearly elastic. The three-dimensional stress-strain laws for this class of materials are developed in the following section.

4.3 GENERALIZED HOOKE'S LAW

If each stress component σ_{ij} is assumed to be proportional to each strain component e_{ij}, the resulting stress-strain relation is known as the *generalized Hooke's law*. This law is expressed in tensor notation as

$$\sigma_{ij} = E_{ijkl}e_{kl} \qquad (4.3.1)$$

where summations on the repeated indices k and l are implied. The 81 constants of proportionality E_{ijkl}, called the *elastic moduli*, constitute a tensor of fourth order. Fortunately this tensor exhibits certain symmetry properties which reduce the number of the independent components. Owing to the symmetry of the stress tensor ($\sigma_{ij} = \sigma_{ji}$), it is evident from Eq. (4.3.1) that the tensor E_{ijkl} is unaltered if its first two indices are interchanged. Similarly, the symmetry of the strain tensor ($e_{kl} = e_{lk}$) requires that E_{ijkl} be symmetric with respect to its last two indices. Therefore

$$E_{ijkl} = E_{jikl}$$
$$E_{ijkl} = E_{ijlk}$$

(4.3.2)

Furthermore, from a consideration of thermodynamics it can be shown that if the deformation process is adiabatic or isothermal (see Ref. 4.3), then

$$E_{ijkl} = E_{klij} \qquad (4.3.3)$$

By taking into account the above symmetry properties, the equations for generalized Hooke's law (4.3.1) may be written in unabridged notation as

$$\sigma_{11} = E_{1111}e_{11} + E_{1122}e_{22} + E_{1133}e_{33} + 2E_{1112}e_{12} + 2E_{1123}e_{23}$$
$$+ 2E_{1131}e_{31}$$

$$\sigma_{22} = E_{1122}e_{11} + E_{2222}e_{22} + E_{2233}e_{33} + 2E_{2212}e_{12} + 2E_{2223}e_{23}$$
$$+ 2E_{2231}e_{31}$$

$$\sigma_{33} = E_{1133}e_{11} + E_{2233}e_{22} + E_{3333}e_{33} + 2E_{3312}e_{12} + 2E_{3323}e_{23}$$
$$+ 2E_{3331}e_{31}$$

$$\sigma_{12} = \sigma_{21} = E_{1112}e_{11} + E_{2212}e_{22} + E_{3312}e_{33} + 2E_{1212}e_{12}$$
$$+ 2E_{1223}e_{23} + 2E_{1231}e_{31}$$

$$\sigma_{23} = \sigma_{32} = E_{1123}e_{11} + E_{2223}e_{22} + E_{3323}e_{33} + 2E_{1223}e_{12}$$
$$+ 2E_{2323}e_{23} + 2E_{2331}e_{31}$$

$$\sigma_{31} = \sigma_{13} = E_{1131}e_{11} + E_{2231}e_{22} + E_{3331}e_{33} + 2E_{1231}e_{12}$$
$$+ 2E_{2331}e_{23} + 2E_{3131}e_{31}$$

(4.3.4)

These stress-strain relations involve 21 rather than 81 independent moduli and apply to the most general linearly elastic material; i.e., one having different elastic properties in different directions. Such a material is said to be *anisotropic*. The analysis of a structure made of an anisotropic material is extremely complicated. Fortunately, however, most of the important construction materials possess a basic internal structure which exhibits certain symmetry properties. For example, consider the fiber-reinforced composite beam shown in Fig. 4.3. Since the fibers are distributed in a rectangular array, the macroscopic elastic properties of the composite are symmetric with respect to the planes $x_1 = 0$,

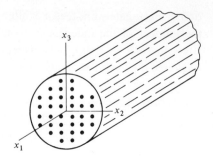

FIGURE 4.3
A fiber-reinforced composite material having orthotropic symmetry.

$x_2 = 0$, and $x_3 = 0$. A material of this type which possesses three orthogonal planes of symmetry is called *orthotropic*. Wood, sheet metal, and honeycomb are other examples of materials which demonstrate orthotropic behavior.

The stress-strain law for an orthotropic material may be obtained from the generalized Hooke's law by requiring that elastic moduli E_{ijkl} remain unchanged when any one of the coordinate axes is inverted (see Prob. 4.2). The following relations are so obtained:

$$\sigma_{11} = E_{1111}e_{11} + E_{1122}e_{22} + E_{1133}e_{33}$$
$$\sigma_{22} = E_{1122}e_{11} + E_{2222}e_{22} + E_{2233}e_{33}$$
$$\sigma_{33} = E_{1133}e_{11} + E_{2233}e_{22} + E_{3333}e_{33}$$
$$\sigma_{12} = 2E_{1212}e_{12}$$
$$\sigma_{23} = 2E_{2323}e_{23}$$
$$\sigma_{31} = 2E_{3131}e_{31}$$

(4.3.5)

Stress-strain equations for materials possessing additional symmetries may be obtained by requiring that E_{ijkl} remain invariant for additional coordinate transformations (see Refs. 4.2 to 4.4.).

The simplest linearly elastic material, one for which the elastic behavior is independent of the orientation of the axes, is called an *isotropic* material. In this case, there are just 2 independent moduli E_{ijkl} known as the *Lamé constants* μ and λ. The generalized Hooke's law reduces to the following form for an isotropic material

$$\sigma_{11} = 2\mu e_{11} + \lambda(e_{11} + e_{22} + e_{33})$$
$$\sigma_{22} = 2\mu e_{22} + \lambda(e_{11} + e_{22} + e_{33})$$
$$\sigma_{33} = 2\mu e_{33} + \lambda(e_{11} + e_{22} + e_{33})$$
$$\sigma_{12} = 2\mu e_{12}$$
$$\sigma_{23} = 2\mu e_{23}$$
$$\sigma_{31} = 2\mu e_{31}$$

(4.3.6)

These relations may be written in index notation as

$$\sigma_{ij} = 2\mu e_{ij} + \lambda \delta_{ij} e_{kk} \qquad (4.3.7)$$

Alternatively, by inverting Eq. (4.3.7), the strains may be expressed in terms of stresses as

$$e_{ij} = \frac{1}{2\mu} \sigma_{ij} - \frac{\lambda}{2\mu(2\mu + 3\lambda)} \delta_{ij} \sigma_{kk} \qquad (4.3.8)$$

Relations between the Lamé constants μ and λ and the more common "engineering" material constants may be found by considering special states of stress. For example, consider the case of uniaxial stress in which $\sigma_{11} = $ constant, all other $\sigma_{ij} = 0$. Defining *Young's modulus* E as the ratio of the normal stress σ_{11} to the strain e_{11}, Eq. (4.3.8) gives

$$E = \frac{\sigma_{11}}{e_{11}} = \frac{\mu(2\mu + 3\lambda)}{\mu + \lambda} \qquad (4.3.9)$$

Poisson's ratio v, which denotes the fractional lateral contraction during simple tension, may be computed using Eqs. (4.3.8) and (4.3.9), with the result

$$v = -\frac{e_{22}}{e_{11}} = -\frac{e_{33}}{e_{11}} = \frac{\lambda}{2(\mu + \lambda)} \qquad (4.3.10)$$

Secondly, consider a state of pure shear, where $\sigma_{12} = \sigma_{21} = $ constant, $\sigma_{ij} = 0$ otherwise. The *shear modulus* G, defined as the ratio of the shear stress to the change in right angle is, according to Eq. (4.3.8),

$$G = \frac{\sigma_{12}}{2e_{12}} = \mu \qquad (4.3.11)$$

Using Eqs. (4.3.9) to (4.3.11), the stress-strain relations (4.3.7) and the strain-stress relations (4.3.8) may be expressed in terms of the elastic constants E, v, G rather than the Lamé constants μ, λ. It should be remembered, however, that an isotropic linearly elastic material is characterized by just two independent moduli.

4.4 SUMMARY OF THE EQUATIONS OF ELASTICITY

The basic equations which govern the state of stress, strain, and displacement in an isotropic elastic solid have now been derived, and are summarized below. These 15 equations, together with appropriate boundary conditions, constitute the *classical theory of elasticity*. A solution to these equations for the 15 unknowns (6 independent stress components, 6 independent strain components, and 3 displacement components) is unique (see Ref. 4.4). A direct approach for finding such a solution is normally very difficult. In Chap. 5 we shall develop various energy theorems which will yield at least approximate solutions for relatively complicated structures. It should be kept in mind, however, that the energy principles are nothing more than reformulations of the equations of elasticity. The equations are now summarized in index as well as unabridged notation.

EQUATIONS OF LINEAR ELASTICITY

Equilibrium equations

$$\sigma_{ij,j} + f_i = 0 \begin{cases} \dfrac{\partial \sigma_{11}}{\partial x_1} + \dfrac{\partial \sigma_{12}}{\partial x_2} + \dfrac{\partial \sigma_{13}}{\partial x_3} + f_1 = 0 \\[2mm] \dfrac{\partial \sigma_{21}}{\partial x_1} + \dfrac{\partial \sigma_{22}}{\partial x_2} + \dfrac{\partial \sigma_{23}}{\partial x_3} + f_2 = 0 \\[2mm] \dfrac{\partial \sigma_{31}}{\partial x_1} + \dfrac{\partial \sigma_{32}}{\partial x_2} + \dfrac{\partial \sigma_{33}}{\partial x_3} + f_3 = 0 \end{cases} \tag{4.4.1}$$

Strain-displacement equations

$$e_{ij} = \tfrac{1}{2}(u_{i,j} + u_{j,i}) \begin{cases} e_{11} = \dfrac{\partial u_1}{\partial x_1} \qquad e_{22} = \dfrac{\partial u_2}{\partial x_2} \qquad e_{33} = \dfrac{\partial u_3}{\partial x_3} \\[2mm] e_{12} = e_{21} = \dfrac{1}{2}\left(\dfrac{\partial u_1}{\partial x_2} + \dfrac{\partial u_2}{\partial x_1}\right) \\[2mm] e_{23} = e_{32} = \dfrac{1}{2}\left(\dfrac{\partial u_2}{\partial x_3} + \dfrac{\partial u_3}{\partial x_2}\right) \\[2mm] e_{31} = e_{13} = \dfrac{1}{2}\left(\dfrac{\partial u_3}{\partial x_1} + \dfrac{\partial u_1}{\partial x_3}\right) \end{cases} \tag{4.4.2}$$

Generalized Hooke's law for an isotropic elastic solid

$$\sigma_{ij} = 2\mu e_{ij} + \lambda \delta_{ij} e_{kk}$$

or

$$e_{ij} = \dfrac{1}{E}\left[(1+\nu)\sigma_{ij} - \nu \delta_{ij}\sigma_{kk}\right]$$

$$\begin{cases} e_{11} = \dfrac{1}{E}\left[\sigma_{11} - \nu(\sigma_{22} + \sigma_{33})\right] \\[2mm] e_{22} = \dfrac{1}{E}\left[\sigma_{22} - \nu(\sigma_{33} + \sigma_{11})\right] \\[2mm] e_{33} = \dfrac{1}{E}\left[\sigma_{33} - \nu(\sigma_{11} + \sigma_{22})\right] \\[2mm] e_{12} = \dfrac{1}{E}(1+\nu)\sigma_{12} = \dfrac{1}{2G}\sigma_{12} \\[2mm] e_{23} = \dfrac{1}{E}(1+\nu)\sigma_{23} = \dfrac{1}{2G}\sigma_{23} \\[2mm] e_{31} = \dfrac{1}{E}(1+\nu)\sigma_{31} = \dfrac{1}{2G}\sigma_{31} \end{cases} \tag{4.4.3}$$

where
$$\mu = \dfrac{E}{2(1+\nu)} \qquad E = \dfrac{\mu(2\mu + 3\lambda)}{\mu + \lambda}$$

$$\lambda = \dfrac{E\nu}{(1+\nu)(1-2\nu)} \qquad \nu = \dfrac{\lambda}{2(\mu + \lambda)} \tag{4.4.4}$$

$$G = \mu$$

PROBLEMS

4.1   Show that the stress-strain relations (4.3.4) follow directly from Eqs. (4.3.1) to (4.3.3).

4.2   Determine the generalized Hooke's law for an orthotropic elastic solid (4.3.5), starting with the equations for an anisotropic material (4.3.4). *Hint:* Assume that the elastic properties of the solid are symmetric with respect to the three orthogonal planes $x_1 = 0$, $x_2 = 0$, $x_3 = 0$. Note that in this case the form of the stress-strain relations will not change if any one of the coordinate axes is inverted. The fourth-order tensor $E'_{ijkl} = \alpha_{iq}\alpha_{jr}\alpha_{ks}\alpha_{lt}E_{qrst}$ must therefore remain invariant under the two coordinate transformations shown in the figure below. Examine what restrictions this requirement places upon the elastic moduli E_{ijkl}.

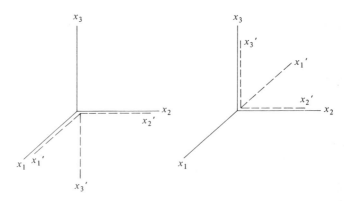

PROBLEM 4.2

4.3   Obtain Eq. (4.3.8) by inverting (4.3.7).

4.4   Verify Eqs. (4.3.9) to (4.3.11).

4.5   Show that the principal axes of stress coincide with the principal axes of strain for an isotropic, linearly elastic material.

REFERENCES

For an elementary account of the mechanical behavior of elastic and inelastic solids:

4.1   CRANDALL, S. H., et al.: "An Introduction to the Mechanics of Solids," 2d ed., McGraw-Hill, New York, 1972.

4.2   SHAMES, I. H.: "Mechanics of Deformable Solids," Prentice-Hall, Englewood Cliffs, N.J., 1964.

4.3   BISPLINGHOFF,. R. L., J. W. MAR, and T. H. H. PIAN: "Statics of Deformable Solids," Addison-Wesley, Reading, Mass., 1965.

For a more thorough treatment of isotropic elastic solids:

4.4   SOKOLNIKOFF, I. S.: "Mathematical Theory of Elasticity," 2nd ed., McGraw-Hill, New York, 1956.

For an account of anisotropic elastic behavior:

4.5   LEKHNITSKII, S. G.: "Theory of Elasticity of an Anisotropic Elastic Body," English translation by P. Fern, Holden-Day, San Francisco, 1963.

For a study of inelastic behaviors of structures:

4.6   LIN, T. H.: "Theory of Inelastic Structures," Wiley, New York, 1968.

ENERGY PRINCIPLES

5.1 INTRODUCTION

The equations which govern the distributions of stress and deformation in elastic structures have been developed in the preceding chapters. It is possible, in theory, to integrate these differential equations [(2.4.17), (3.2.12), and (4.3.1)] and satisfy the appropriate boundary conditions to obtain the displacement vector u_i and the stress vector T_i at any point in the structure. In practice, however, performing the integration analytically is nearly always an impossible task. An alternative to this direct "vectorial" approach can be developed from a consideration of the scalar quantities work and energy. In this chapter we shall consider two conjugate energy methods, one formulated in terms of displacements (*displacement method*), the other expressed in terms of forces (*force method*). The well-known principles of virtual work[1] and complementary virtual work, which

---

[1] The principle of virtual work was first formulated by J. J. Bernoulli (1667–1748). The mathematical tools required for its application were developed largely by J. L. Lagrange (1736–1813). An interesting review of the history of virtual work and of variational mechanics in general is given by: E. Mach, "The Science of Mechanics," 1883, translated by T. J. McCormack, Open Court, LaSalle, Ill., 1942; also R. Dugas, "A History of Mechanics," Central Book, New York, 1955.

follow directly from the equations of equilibrium (2.4.17) and the strain-displacement relations (3.2.12), form the basis for the displacement and force formulations, respectively. Several other important energy theorems, including the principles of minimum potential energy and minimum complementary energy are derived from virtual work concepts.

Besides providing convenient methods for computing unknown displacements and forces in structures, the energy principles are fundamental to the study of structural stability (Chap. 9) and structural dynamics (Chaps. 11 and 12). They are also the basis for various approximate methods of analysis, applicable to problems involving deformations, stability, and vibrations of elastic bodies. Hence a thorough familiarity with the principles presented in this chapter is a prerequisite to understanding the remainder of the text.

5.2 STRAIN ENERGY

Consider the deformation which occurs when an initially unstressed elastic structure is subjected to a system of applied loads (Fig. 5.1). The deformation process is governed by the first law of thermodynamics which requires that

$$W_E + \mathcal{Q} = \Delta E \qquad (5.2.1)$$

where W_E is the work done by the applied forces during the loading process, $\mathcal{Q}$ is the heat absorbed by the structure from its surroundings, and ΔE is the change of energy associated with the body as a result of the loading. We shall assume here that the deformation process is adiabatic, so that $\mathcal{Q} = 0$.

In general, the change in energy ΔE of an elastic structure consists of a change in kinetic energy T, plus a change in internal energy U. We will assume for the time being that the loads are applied very slowly, and that a state of equilibrium

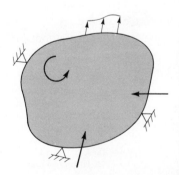

FIGURE 5.1
Elastic body subject to external forces.

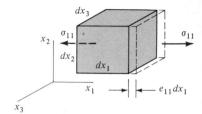

FIGURE 5.2
Elemental volume subject to normal
stress σ_{11}.

is maintained during the entire process.[1] In this case the kinetic energy is zero, and ΔE represents only a change in the internal energy U. Under the above assumptions, the conservation of energy expression (5.2.1) reduces to

$$W_E = U \qquad (5.2.2)$$

That is, the mechanical work done by the applied loads is equal to the change in the internal energy. From the assumption that the material is perfectly elastic, it follows that the mechanical work will be regained if the loads are removed very slowly. The work may be regarded, therefore, as being stored in the deformed structure in the form of energy. This stored energy is called *strain energy*. General expressions for the strain energy in terms of the stress tensor σ_{ij} and the strain tensor e_{ij} will now be derived.

Consider an element of volume $d\mathscr{V}$ of a structure which is subject to a single component of stress, say σ_{11} (Fig. 5.2). As the deformation proceeds, the normal stress and strain increase from zero to their final values, σ_{11} and e_{11}, respectively. At any stage of the loading process, the strain energy is equal to the work done by the stress, in accordance with Eq. (5.2.2). During an increment of strain de_{11}, the work is equal to the force $\sigma_{11}\, dx_2\, dx_3$ times the extension $de_{11}\, dx_1$. Therefore the energy dU stored in the element when the strain has reached its final value e_{11} is

$$dU = \int_0^{e_{11}} \sigma_{11}\, de_{11}\, dx_1\, dx_2\, dx_3 = \int_0^{e_{11}} \sigma_{11}\, de_{11}\, d\mathscr{V} \qquad (5.2.3)$$

Integrating over the entire volume $\mathscr{V}$ of the structure gives the total strain energy

$$U = \int_{\mathscr{V}} \left(\int_0^{e_{11}} \sigma_{11}\, de_{11} \right) d\mathscr{V} \qquad (5.2.4)$$

Now consider a structure whose volume elements are subject to a three-dimensional state of stress. The strain energy in this case may be computed

---

[1]In Part III of this text we will investigate situations in which the external loads are applied rapidly, thereby causing the structure to vibrate.

by summing expressions similar to Eq. (5.2.4) for each stress component σ_{ij}. In this fashion the strain energy is found to be

$$
\begin{aligned}
U = \int_{\mathscr{V}} \Bigg(& \int_0^{e_{11}} \sigma_{11}\,de_{11} + \int_0^{e_{22}} \sigma_{22}\,de_{22} + \int_0^{e_{33}} \sigma_{33}\,de_{33} \\
& + \int_0^{e_{12}} \sigma_{12}\,de_{12} + \int_0^{e_{21}} \sigma_{21}\,de_{21} + \int_0^{e_{23}} \sigma_{23}\,de_{23} \\
& + \int_0^{e_{32}} \sigma_{32}\,de_{32} + \int_0^{e_{31}} \sigma_{31}\,de_{31} + \int_0^{e_{13}} \sigma_{13}\,de_{13} \Bigg)\,d\mathscr{V} \\
= \int_{\mathscr{V}} & \left(\int_0^{e_{ij}} \sigma_{ij}\,de_{ij} \right) d\mathscr{V}
\end{aligned}
\tag{5.2.5}
$$

in which summations on the repeated indices i and j are implied.

The strain energy U in any elastic structure, linear or nonlinear, may be computed using Eq. (5.2.5). In the special case of an isotropic linearly elastic material the stress is related to the strain by Eq. (4.4.3)

$$
\sigma_{ij} = 2\mu e_{ij} + \lambda \delta_{ij} e_{kk} \tag{5.2.6}
$$

Substituting Eq. (5.2.6) into (5.2.5) and integrating gives

$$
U = \int_{\mathscr{V}} \left(\mu e_{ij} e_{ij} + \frac{\lambda}{2} e_{kk}{}^2 \right) d\mathscr{V} \tag{5.2.7}
$$

or, in terms of the stresses rather than the strains,

$$
U = \int_{\mathscr{V}} \left(\frac{1}{4\mu} \sigma_{ij} \sigma_{ij} - \frac{\lambda}{4\mu(2\mu + 3\lambda)} \sigma_{kk}{}^2 \right) d\mathscr{V} \tag{5.2.8}
$$

For convenience in later work, these relations are written in unabridged notation as follows:

$$
U = \int_{\mathscr{V}} \left[\frac{Ev}{2(1+v)(1-2v)} (e_{11} + e_{22} + e_{33})^2 + G(e_{11}{}^2 + e_{22}{}^2 \right.
$$
$$
\left. + e_{33}{}^2 + 2e_{12}{}^2 + 2e_{23}{}^2 + 2e_{31}{}^2) \right] d\mathscr{V} \tag{5.2.9}
$$

and
$$
U = \int_{\mathscr{V}} \left[\frac{1}{2E} (\sigma_{11}{}^2 + \sigma_{22}{}^2 + \sigma_{33}{}^2) - \frac{v}{E} (\sigma_{11}\sigma_{22} + \sigma_{22}\sigma_{33} \right.
$$
$$
\left. + \sigma_{33}\sigma_{11}) + \frac{1}{2G} (\sigma_{12}{}^2 + \sigma_{23}{}^2 + \sigma_{31}{}^2) \right] d\mathscr{V} \tag{5.2.10}
$$

where the engineering material constants E, v, G are related to the Lamé constants μ, λ by Eqs. (4.4.4).

The strain energy stored in a number of basic structural elements including prismatic bars, beams, and shear panels will now be computed. Extensive use will be made of the results in later chapters when we analyze relatively complicated

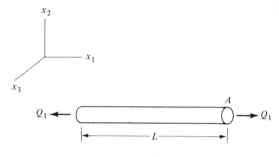

FIGURE 5.3
Axially loaded prismatic bar.

"built-up" structures, i.e., structures fabricated from a large number of simple structural components.

Example 5.1   Strain energy in an axially loaded prismatic bar   Consider a bar of uniform cross-sectional area A and length L subject to a tensile load Q_1 (Fig. 5.3). The state of stress at any point in the bar is assumed to be

$$\sigma_{11} = \frac{Q_1}{A} \qquad \sigma_{ij} = 0 \text{ otherwise} \qquad (5.2.11)$$

Substituting these stress components into Eq. (5.2.10) and performing the integration gives

$$U = \int_{\mathscr{V}} \frac{1}{2E} \sigma_{11}{}^2 \, d\mathscr{V} = \frac{L}{2EA} Q_1{}^2 \qquad (5.2.12)$$

////

Example 5.2   Strain energy in a symmetric beam element   As a second example, consider the bending of a symmetric beam about one of its principal axes, say x_3 (Fig. 5.4). The bending is produced by a system of transverse applied loads Q_i acting in the $x_1 x_2$ plane. The internal bending moment and shear force which act at an arbitrary cross section x_1 are denoted by $M = M(x_1)$ and $V = V(x_1)$,† respectively, as shown. From strength-of-materials we recall that the state of stress at a point having the coordinates x_i is given by

$$\sigma_{11} = -\frac{Mx_2}{I} \qquad \sigma_{12} = \frac{VQ}{Ib} \qquad \sigma_{ij} = 0 \text{ otherwise} \qquad (5.2.13)$$

† The sign convention used here for the internal forces is the same as that adopted for the components of stress. For example, the components of the force and moment vectors on a cross section whose outward normal is in the positive x_1 direction are positive when they are directed in the positive $x_1, x_2,$ and x_3 directions.

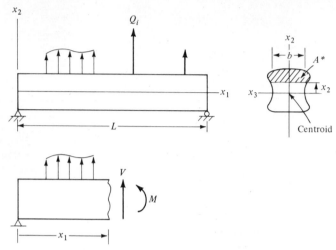

FIGURE 5.4
Symmetric beam element.

where I represents the moment of inertia of the cross-sectional area A with respect to the x_3 axis $\left(I = \int_A x_2^2 \, dA\right)$ and Q is the first moment of the area A^* shown in Fig. 5.4 about the x_3 axis $\left(Q = \int_{A^*} x_2 \, dA\right)$. Substituting the stress components (5.2.13) into (5.2.10) yields

$$U = \int_{\mathscr{V}} \left(\frac{1}{2E} \sigma_{11}^2 + \frac{1}{2G} \sigma_{12}^2\right) d\mathscr{V}$$

$$= \int_{\mathscr{V}} \left(\frac{1}{2E} \frac{M^2 x_2^2}{I^2} + \frac{1}{2G} \frac{V^2 Q^2}{I^2 b^2}\right) d\mathscr{V} \qquad (5.2.14)$$

Since M and V are functions of x_1 only, while Q and b are functions of x_2 alone, Eq. (5.2.14) may be simplified to

$$U = \int_0^L \frac{M^2 \, dx_1}{2EI} + \int_0^L \frac{V^2 \, dx_1}{2GI^2} \int_A \frac{Q^2}{b^2} \, dA \qquad (5.2.15)$$

By considering typical geometries and applied loads, it is found that, in general, the strain energy associated with shear [represented by the second term in Eq. (5.2.15)] is negligible compared to that associated with bending (the first integral).[1] Hence, in all the examples investigated hereafter we shall neglect the shear term.

---

[1] Problem 5.5 provides a comparison of the bending- and shear-strain energies for a beam.

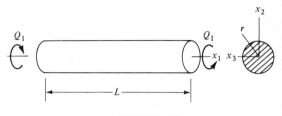

FIGURE 5.5
Circular shaft subject to a torque.

An alternate expression for the strain energy due to bending is obtained by recalling the moment-curvature expression

$$\frac{M}{EI} = \frac{d^2 u_2}{dx_1^2} \quad (5.2.16)$$

in which case

$$U = \int_0^L \frac{EI}{2} \left(\frac{d^2 u_2}{dx_1^2}\right)^2 dx_1 \quad (5.2.17)$$

It should be mentioned that the formulas (5.2.13) and (5.2.16) correspond to the so-called "elementary beam theory," and do not represent a solution to the equations of elasticity (4.4.1) to (4.4.3). Since the beam theory can be obtained from the theory of elasticity only by introducing several simplifying assumptions (see Ref. 5.1), we have no guarantee that the results will be accurate. However, experimental evidence indicates that the elementary beam theory is adequate for most engineering applications.

Example 5.3   Strain energy in a twisted circular bar   Next consider a shaft of circular cross section, subject to a constant twisting moment Q_1 (Fig. 5.5). The corresponding state of stress is given by

$$\sigma_{12} = -\frac{Q_1 x_3}{J} \qquad \sigma_{13} = \frac{Q_1 x_2}{J} \qquad \sigma_{ij} = 0 \text{ otherwise} \quad (5.2.18)$$

where J denotes the polar moment of inertia of the cross-sectional area A [$J = \int_A (x_2^2 + x_3^2)\, dA$]. This stress field represents an exact solution to the equations of elasticity (see Ref. 5.1). Substituting the relations (5.2.18) into (5.2.10) and using the definition of J, we obtain the following expression for the strain energy:

$$U = \int_{\mathcal{V}} \frac{1}{2G} (\sigma_{12}^2 + \sigma_{13}^2)\, d\mathcal{V} = \frac{L}{2GJ} Q_1^2 \quad (5.2.19)$$

////

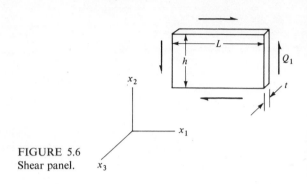

FIGURE 5.6
Shear panel. x_3

Example 5.4   Strain energy in a shear panel   As a last example we shall compute the strain energy in a thin plate, or panel, which is subject to a *shear flow* Q_1 (Fig. 5.6). (Shear flow is defined as the shear stress times the thickness of the panel; $Q_1 = \sigma_{12} t$ in this case.) For this state of stress the relation (5.2.10) gives

$$U = \int_{\mathscr{V}} \frac{1}{2G} \sigma_{12}^2 \, d\mathscr{V} = \frac{hL}{2Gt} Q_1^2 \qquad (5.2.20)$$

////

Let us define at this time another useful quantity known as the *complementary strain energy* U^*:

$$U^* = \int_{\mathscr{V}} \left(\int_0^{\sigma_{ij}} e_{ij} \, d\sigma_{ij} \right) d\mathscr{V} \qquad (5.2.21)$$

Whereas the strain energy U represents the energy stored in a deformed elastic body, the physical interpretation of U^* is less clear. However, let us examine these two quantities for the case of a nonlinearly elastic bar subject to a uniform tensile stress σ_{11}. From Eq. (5.2.4) the strain energy U is

$$U = \mathscr{V} \int_0^{e_{11}} \sigma_{11} \, de_{11} \qquad (5.2.22)$$

The strain energy per unit volume is therefore equal to the area under the material's stress-strain curve (Fig. 5.7). On the other hand, the complementary strain energy U^* is, from Eq. (5.2.21),

$$U^* = \mathscr{V} \int_0^{\sigma_{11}} e_{11} \, d\sigma_{11} \qquad (5.2.23)$$

Hence the complementary strain energy per unit volume corresponds to the area above the stress-strain curve. For a linearly elastic material the two areas are equal, and $U^* = U$.

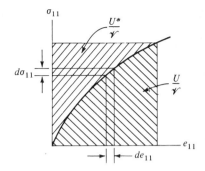

FIGURE 5.7
Stress-strain diagram for a nonlinearly
elastic bar.

5.3 WORK OF THE EXTERNAL FORCES

General expressions for the energy U stored in a deformed elastic structure
were derived in Sec. 5.2. We now wish to investigate the work W_E done by the
system of external loads which produce the deformation; in particular we wish
to express W_E in terms of the applied loads. From Eq. (5.2.2) we know that the
two quantities U and W_E will be equal if the deformation process is adiabatic and
quasi-static.

Again consider the equilibrium of a structure under the action of a system
of external loads (Fig. 5.8). It is assumed that the structure is supported in space
in a manner that precludes any rigid-body motion. The applied loads, denoted by
$Q_1, Q_2, \ldots, Q_n$ or Q_i ($i = 1, 2, \ldots, n$), are not restricted just to concentrated
forces, but may represent distributed forces, couples, shear flows, etc; for this
reason they will be referred to as *generalized forces*.

Corresponding to each generalized force Q_i, we may associate a *generalized
displacement* q_i, where q_i represents the deformation of the structure at the point
(or surface) of application of the generalized force Q_i in the direction of Q_i.

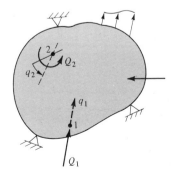

FIGURE 5.8
Body subject to generalized forces
Q_i ($i = 1, 2, \ldots, n$).

For example q_1 represents the linear deflection of point 1 in the direction of the concentrated force Q_1, and q_2 denotes the rotation of a line element at point 2 in the direction of the couple Q_2. We see, therefore, that the product of a generalized force Q_i and the corresponding generalized displacement q_i represents work. A more precise definition of generalized displacements will be given shortly.

In general, the deformation at a point in the structure is produced by the combined action of all the applied loads. That is, the displacement q_1 is due not only to Q_1, but to $Q_2, Q_3, \ldots, Q_n$ as well. For a linearly elastic structure, one for which the strains are proportional to the stresses and hence the generalized displacements are proportional to the generalized forces, we may write

$$
\begin{aligned}
q_1 &= c_{11} Q_1 + c_{12} Q_2 + \cdots + c_{1n} Q_n \\
q_2 &= c_{21} Q_1 + c_{22} Q_2 + \cdots + c_{2n} Q_n \\
&\vdots \\
q_n &= c_{n1} Q_1 + c_{n2} Q_2 + \cdots + c_{nn} Q_n
\end{aligned}
\tag{5.3.1}
$$

or
$$
q_i = \sum_{j=1}^{n} c_{ij} Q_j \qquad i = 1, 2, \ldots, n
\tag{5.3.2}
$$

where the constants of proportionality c_{ij} are called *flexibility influence coefficients*. Equation (5.3.2) may be conveniently expressed in matrix notation (see Appendix B) as

$$
\{q\} = [c]\{Q\}
\tag{5.3.3}
$$

In order to see the physical meaning of the flexibility influence coefficients c_{ij}, consider the deformation produced when $Q_1 = 1$, but all other $Q_i = 0$. Then, according to Eq. (5.3.2),

$$
q_i = c_{i1}
\tag{5.3.4}
$$

Thus c_{i1} is equal to the generalized displacement at point i in the direction of Q_i due to a unit generalized force Q_1. In general then:

The *flexibility influence coefficient* c_{ij} represents the generalized displacement at i in the direction of Q_i due to a unit generalized force Q_j.

The flexibility coefficients for an idealized airplane wing subject to a set of generalized applied forces are illustrated in Fig. 5.9.

From Eq. (5.3.2) it follows that

$$
c_{ij} = \frac{\partial q_i}{\partial Q_j}
\tag{5.3.5}
$$

which is sometimes taken as the definition of the flexibility influence coefficient.

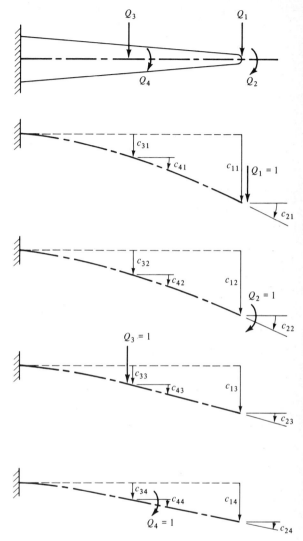

FIGURE 5.9
Flexibility influence coefficients for an airplane wing.

We now assume that the system of Eqs. (5.3.1) can be inverted so that the generalized forces may be expressed in terms of the generalized displacements as

$$Q_1 = k_{11}q_1 + k_{12}q_2 + \cdots + k_{1n}q_n$$
$$Q_2 = k_{21}q_1 + k_{22}q_2 + \cdots + k_{2n}q_n \qquad (5.3.6)$$
$$\vdots$$
$$Q_n = k_{n1}q_1 + k_{n2}q_2 + \cdots + k_{nn}q_n$$

or
$$Q_i = \sum_{j=1}^{n} k_{ij}q_j \qquad i = 1, 2, \ldots, n \qquad (5.3.7)$$

where the constants k_{ij} are called the *stiffness influence coefficients*. Equation (5.3.7) may be written in matrix notation as

$$\{Q\} = [k]\{q\} \qquad (5.3.8)$$

Comparing Eqs. (5.3.3) and (5.3.8), it is clear that $[k]$ is the inverse of $[c]$, or

$$[k] = [c]^{-1} \qquad (5.3.9)$$

Later we shall prove that $[c]$ and $[k]$ are symmetric matrices.

A physical interpretation of the stiffness coefficients may be obtained by considering the deformation state corresponding to the case $q_1 = 1$, all other $q_i = 0$. Then, according to Eq. (5.3.7),

$$Q_i = k_{i1} \qquad (5.3.10)$$

Hence k_{i1} is equal to the generalized force required at point i when the generalized displacement q_1 is unity and all other generalized displacements are zero.

The *stiffness influence coefficient* k_{ij} is the generalized force required at i for a unit displacement at j, all other generalized displacements being zero.

The stiffness coefficients for an idealized wing are shown in Fig. 5.10.

It is obvious from Eq. (5.3.7) that

$$k_{ij} = \frac{\partial Q_i}{\partial q_j} \qquad (5.3.11)$$

This relation is sometimes taken as the definition of the stiffness coefficient.

Let us now examine the work W_E done by the generalized forces Q_i during the loading process. For simplicity we first consider the special case of a single concentrated force Q_1. The corresponding generalized displacement q_1 is the translation at point 1 in the direction of Q_1. Since the structure is assumed to be linearly elastic, we have from Eq. (5.3.6)

$$Q_1 = k_{11}q_1 \qquad (5.3.12)$$

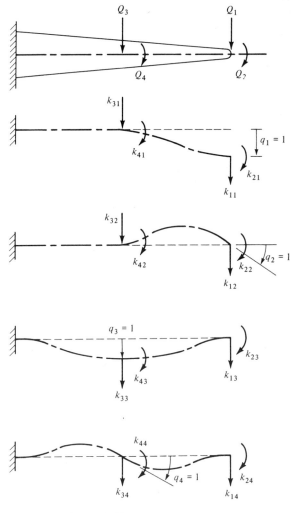

FIGURE 5.10
Stiffness influence coefficients for an airplane wing.

The work in this case is

$$W_E = \int_0^{q_1} Q_1 \, dq_1 = \int_0^{q_1} k_{11}q_1 \, dq_1 = \tfrac{1}{2}k_{11}q_1{}^2$$
$$= \tfrac{1}{2}q_1 Q_1 \qquad\qquad (5.3.13)$$

The presence of the factor $\tfrac{1}{2}$ in this relation may be attributed to the fact that the average value of the force during the loading process is $\tfrac{1}{2}Q_1$, while the total displacement is q_1. Generalizing to the case where all forces Q_i act simultaneously, the total work done is

$$W_E = \tfrac{1}{2} \sum_{i=1}^{n} q_i Q_i \qquad (5.3.14)$$

By virtue of Eqs. (5.3.2) and (5.3.7), the work may also be written as

$$W_E = \tfrac{1}{2} \sum_{i=1}^{n} \sum_{j=1}^{n} c_{ij} Q_i Q_j \qquad (5.3.15)$$

or

$$W_E = \tfrac{1}{2} \sum_{i=1}^{n} \sum_{j=1}^{n} k_{ij} q_i q_j \qquad (5.3.16)$$

Equation (5.3.14) serves as a criterion for determining what type of generalized displacement corresponds to a particular type of generalized force. In other words, the generalized displacement q_i is defined as that quantity which must multiply the force Q_i in order that the work W_E will be given by Eq. (5.3.14). From this definition it is clear that if Q_i is a concentrated force, then q_i is indeed a translation at the point of application and in the direction of Q_i; likewise if Q_i denotes a concentrated couple, then q_i represents a rotation about the axis of the couple. A physical interpretation of the generalized displacement corresponding to a distributed force is less obvious. However, if we remember that the displacement q_i must be such that the work done during the loading process is $\tfrac{1}{2}q_i Q_i$, then if Q_i is a surface force, q_i may be interpreted as an "average" displacement of the area acted upon by Q_i.

We now have at our disposal expressions for the strain energy U and the work W_E done by the external loads during the deformation of an elastic structure. These expressions, in conjunction with the principle of conservation of energy (5.2.2), $U = W_E$, provide one means of calculating generalized displacements in structures. An example of this approach follows.

Example 5.5   Deflection of a beam   Consider a simply supported beam of uniform cross section subject to a concentrated force Q_1 at its midpoint (Fig. 5.11). The strain energy in the beam may be calculated using Eq. (5.2.15); neglecting the energy associated with shear, this gives

$$U = \int_0^L \frac{M^2 \, dx_1}{2EI} \qquad (5.3.17)$$

From symmetry considerations it is obvious that the same amount of energy will be stored in the right and left halves of the beam, and since $M = Q_1 x_1/2$ for $0 < x_1 < L/2$, we have

$$U = 2 \int_0^{L/2} \frac{1}{2EI} \left(\frac{Q_1 x_1}{2} \right)^2 dx_1 = \frac{Q_1^2 L^3}{96EI} \qquad (5.3.18)$$

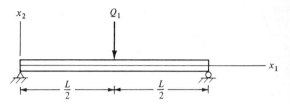

FIGURE 5.11
Simply supported beam.

The work done by Q_1 is

$$W_E = \tfrac{1}{2} q_1 Q_1 \qquad (5.3.19)$$

Equating U and W_E and solving for q_1 gives

$$q_1 = \frac{Q_1 L^3}{48EI} \qquad (5.3.20)$$

The student may verify that this result agrees with the strength-of-materials solution; i.e., the solution to the flexure equation $EI(d^2 u_2/dx_1^{\,2}) = M$ and the prescribed boundary conditions $u_2(0) = u_2(L) = M(0) = M(L) = 0$. ////

The direct application of the principle of conservation of energy is limited to the calculation of a displacement produced by a single force at its point of application. Quite clearly the one scalar equation $W_E = U$ is insufficient for a determination of more than one unknown. Hence, a more general method is required to handle structures which are subject to several loads simultaneously. In the following sections we shall develop several such methods, all of which are based on the concepts of work and energy.

5.4   VIRTUAL WORK

An extremely useful approach to many problems in structural mechanics involves a concept known as *virtual work*. Virtual work methods are applicable to problems involving inelastic as well as elastic behaviors, thermal as well as mechanical loadings (Chap. 8), and problems involving structural stability (Chap. 9) and structural dynamics (Chaps. 11 and 12). In this section we shall see how the concept of virtual work can be used to calculate forces and displacements in structures subject to purely mechanical loads.

Let us once again consider a structure supported against rigid-body motion and subjected to a system of generalized forces Q_i (Fig. 5.12a). The state of stress

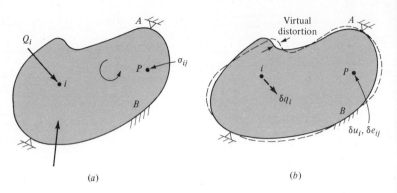

FIGURE 5.12
(a) Body subject to generalized forces Q_i ($i = 1, 2, \ldots, n$); (b) virtual distortion of the body.

at an arbitrary point P caused by the loads Q_i is specified by the stress tensor σ_{ij}. Since the body is assumed to be in equilibrium, the stress components satisfy the equilibrium equations (4.4.1).

Let us now imagine that the deformed structure is subjected to a second system of forces which produces *kinematically admissible* displacements (Fig. 5.12b). By kinematically admissible we mean that the imaginary deformation does not violate any of the prescribed boundary conditions. For example, if a hinge support prevents translation of the structure at point A, then during our hypothetical distortion, the displacement at A must remain zero. Similarly if the structure is fully constrained at point B, the imaginary distortion must be such that the rotation and translation of B remain zero. An imaginary, infinitesimal deformation of this nature is called a *virtual distortion*. We define the *virtual displacement* at point i in the direction of Q_i to be δq_i; similarly, the virtual displacement at the arbitrary point P is taken to be δu_i, and the corresponding virtual strains are $\delta e_{ij} = \frac{1}{2}(\delta u_{i,j} + \delta u_{j,i})$. It is important to keep in mind that the virtual deformation is assumed to occur after the deformation produced by the real loads Q_i has taken place. We now examine the so-called "external virtual work" δW_E, or the work done by the real loads Q_i during the virtual distortion. We shall see that this virtual work is equal to the change in the strain energy δU resulting from the real stresses σ_{ij} undergoing the virtual strains δe_{ij}.

In deriving the virtual work expression, let us consider the case of a structure having arbitrary material behavior and subject to arbitrary surface and body forces. No additional complexity is introduced by considering this rather general situation, and reduction to the simpler case of a linearly elastic structure subject only to surface loads follows naturally.

The virtual work done by a system of forces distributed over the surface of the structure is found by multiplying the surface tractions T_i by the corresponding virtual displacements δu_i and integrating over the total surface $\mathscr{S}$; the virtual work of forces distributed throughout the volume is obtained by an integration over the volume $\mathscr{V}$ of the product of the body forces f_i and the virtual displacements δu_i. Hence the work done during the virtual distortion δu_i is

$$\delta W_E = \int_{\mathscr{S}} T_i \, \delta u_i \, d\mathscr{S} + \int_{\mathscr{V}} f_i \, \delta u_i \, d\mathscr{V} \qquad (5.4.1)$$

Since $T_i = \sigma_{ij} n_j$ from Eq. (2.4.16), the surface integral may be written as

$$\int_{\mathscr{S}} T_i \, \delta u_i \, d\mathscr{S} = \int_{\mathscr{S}} \sigma_{ij} n_j \, \delta u_i \, d\mathscr{S} = \int_{\mathscr{V}} (\sigma_{ij} \, \delta u_i)_{,j} \, d\mathscr{V} \qquad (5.4.2)$$

where use has been made of the *divergence theorem* in transforming the surface integral to a volume integral. Substituting Eq. (5.4.2) into (5.4.1) gives

$$\delta W_E = \int_{\mathscr{V}} [(\sigma_{ij,j} + f_i)\delta u_i + \sigma_{ij} \, \delta u_{i,j}] \, d\mathscr{V} \qquad (5.4.3)$$

Since the structure is in equilibrium, $\sigma_{ij,j} + f_i = 0$, and Eq. (5.4.3) may therefore be expressed as

$$\delta W_E = \int_{\mathscr{V}} \sigma_{ij} \, \delta u_{i,j} \, d\mathscr{V} \qquad (5.4.4)$$

It can easily be shown, by using the fact that $\sigma_{ij} = \sigma_{ji}$ and introducing the strain-displacement relations for the virtual distortion $\delta e_{ij} = \frac{1}{2}(\delta u_{i,j} + \delta u_{j,i})$, that

$$\sigma_{ij} \, \delta u_{i,j} = \sigma_{ij} \, \delta e_{ij} \qquad (5.4.5)$$

A comparison of Eqs. (5.4.1) and (5.4.4) then yields

$$\int_{\mathscr{S}} T_i \, \delta u_i \, d\mathscr{S} + \int_{\mathscr{V}} f_i \, \delta u_i \, d\mathscr{V} = \int_{\mathscr{V}} \sigma_{ij} \, \delta e_{ij} \, d\mathscr{V} \qquad (5.4.6)$$

We therefore have succeeded in proving that

$$\delta W_E = \delta U \qquad (5.4.7)$$

where

$$\delta W_E = \int_{\mathscr{S}} T_i \, \delta u_i \, d\mathscr{S} + \int_{\mathscr{V}} f_i \, \delta u_i \, d\mathscr{V}$$
$$\qquad (5.4.8)$$
$$\delta U = \int_{\mathscr{V}} \sigma_{ij} \, \delta e_{ij} \, d\mathscr{V}$$

The quantity δU, defined by Eq. (5.4.8), represents the work done by the actual stresses during the virtual distortion; it is referred to as the *internal virtual work*.

Equation (5.4.7) is a mathematical statement of the *principle of virtual work*, also known as the *principle of virtual displacements*. This principle may be expressed as follows:

> *Principle of virtual work* If a structure is in equilibrium and remains in equilibrium while it is subject to a virtual distortion, the external virtual work δW_E done by the external forces acting on the structure is equal to the internal virtual work δU done by the internal stresses.

The converse of this principle is also true. That is, if $\delta W_E = \delta U$ for an arbitrary virtual distortion, then the body is in equilibrium.

Note that the principle of virtual work has been derived without introducing the requirement that the virtual distortion be consistent with the structure's boundary conditions. Equation (5.4.7) is in fact valid for an arbitrary infinitesimal deformation. In practice, however, it is generally convenient to restrict attention to displacement fields which are kinematically admissible.

In the case of a structure subject to a system of n discrete generalized forces Q_i and zero body forces (Fig. 5.12a), the expression for the external virtual work δW_E reduces from an integral of continuous functions $T_i \, \delta u_i$ to a summation of the discrete quantities $Q_i \, \delta q_i$; that is,

$$\delta W_E = \sum_{i=1}^{n} Q_i \, \delta q_i \qquad (5.4.9)$$

It should be noted that the factor of $\frac{1}{2}$ which is present in the expression for the work W_E (5.3.14) does not appear in the virtual work expression (5.4.9). This is as it should be, since it was assumed that the forces Q_i had reached their final value before the virtual distortion occurred. The virtual work done by one of the applied forces, say Q_i, is therefore equal to the magnitude of this force times the corresponding total virtual displacement δq_i.

The fact that δW_E must be equal to δU in order for a structure to be in equilibrium may be used to advantage in calculating unknown forces. In particular, consider a structure whose deformation is known. Let us assume, for instance, that the displacements q_i in the structure shown in Fig. 5.13a have been measured, and we wish to compute one of the force, say Q_1, required for equilibrium in this deformed configuration. We imagine a virtual distortion of the structure, such that the point of application of Q_1 suffers a virtual displacement δq_1, while the positions of all other loads remain fixed (Fig. 5.13b). In this case the external virtual work δW_E is simply $Q_1 \, \delta q_1$, and the principle of virtual work (5.4.7) gives

$$Q_1 \, \delta q_1 = \int_{\mathscr{V}} \sigma_{ij} \, \delta e_{ij} \, d\mathscr{V} \qquad (5.4.10)$$

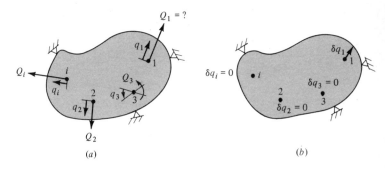

FIGURE 5.13.
(a) Body with an unknown force Q_i; (b) virtual displacement δq_1 in the body.

Since the virtual displacement δq_1 is completely arbitrary, we assume for simplicity that $\delta q_1 = 1$. Then

$$Q_1 = \int_{\mathcal{V}} \sigma_{ij} \, \delta e_{ij} \, d\mathcal{V} \qquad (5.4.11)$$

Thus, by a clever choice of the virtual distortion, we have obtained an expression for the unknown force Q_1. Providing the real stresses σ_{ij} and the virtual strains δe_{ij} (corresponding to $\delta q_1 = 1$) are available, we can easily compute the value of Q_1 using Eq. (5.4.11). Any other force Q_i can be obtained in a similar fashion by introducing a unit virtual displacement at point i in the direction of Q_i. This technique of calculating forces is known as the *dummy-displacement method*. Application of the method to a particular problem is illustrated in the following example.

Example 5.6 Dummy-displacement method for a three-bar truss The dummy-displacement method provides a useful means for computing unknown forces in statically indeterminate as well as determinate structures. (Indeterminate structures will be treated more fully in Chap. 6.) Suppose we wish to compute the forces Q_1 and Q_2 required to produce a vertical displacement q_1 in the indeterminate three-bar truss shown in Fig. 5.14a. All bars are assumed to have a cross-sectional area A and are made of a linearly elastic material having Young's modulus E.

From a consideration of the geometry (Fig. 5.14b), it is clear that during the real deformation, bar AD elongates a distance $\frac{3}{5}q_1$ (assuming small deformations); similarly bar CD contracts a distance $\frac{3}{5}q_1$; and the length of bar BD remains unchanged. Based upon the assumption of frictionless joints, each bar is

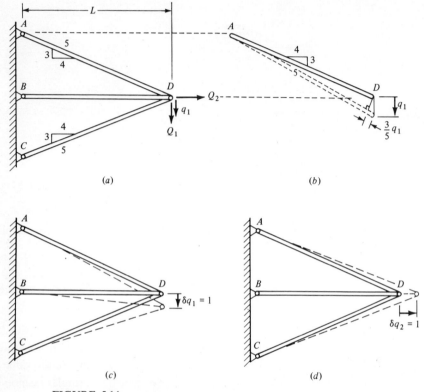

FIGURE 5.14
(a) Three-bar truss; (b) deformation of bar AD; (c) dummy displacement $\delta q_1 = 1$; (d) dummy displacement $\delta q_2 = 1$.

subject to an axial stress only. Denoting the axial stress in bars AD, BD, and CD by $\sigma_{11}{}^{(AD)}$, $\sigma_{11}{}^{(BD)}$, and $\sigma_{11}{}^{(CD)}$, respectively, we have

$$\sigma_{11}{}^{(AD)} = E\,\frac{\frac{3}{5}q_1}{\frac{5}{4}L} = \frac{12Eq_1}{25L}$$

$$\sigma_{11}{}^{(CD)} = E\,\frac{-\frac{3}{5}q_1}{\frac{5}{4}L} = -\frac{12Eq_1}{25L} \qquad (5.4.12)$$

All other
$$\sigma_{ij}{}^{(AD)} = \sigma_{ij}{}^{(BD)} = \sigma_{ij}{}^{(CD)} = 0$$

In order to find the force Q_1, we introduce the unit virtual displacement $\delta q_1 = 1$ shown in Fig. 5.14c. From a consideration of the geometry associated with this virtual distortion, the following strain components are obtained:

$$\delta e_{11}^{(AD)} = \frac{12}{25}\frac{1}{L} \qquad \delta e_{22}^{(AD)} = \delta e_{33}^{(AD)} = -\frac{12}{25}\frac{v}{L}$$

$$\delta e_{11}^{(CD)} = -\frac{12}{25}\frac{1}{L} \qquad \delta e_{22}^{(CD)} = \delta e_{33}^{(CD)} = \frac{12}{25}\frac{v}{L} \qquad (5.4.13)$$

All other $\quad \delta e_{ij}^{(AD)} = \delta e_{ij}^{(BD)} = \delta e_{ij}^{(CD)} = 0$

Applying the dummy-displacement relation (5.4.11), where the volume integration may be performed for each bar separately, gives

$$Q_1 = \int_{\mathcal{V}(AD)} \frac{12Eq_1}{25L}\frac{12}{25L}\,d\mathcal{V} + \int_{\mathcal{V}(CD)} \frac{-12Eq_1}{25L}\frac{-12}{25L}\,d\mathcal{V}$$

$$= \frac{72}{125}\frac{EA}{L}q_1 \qquad (5.4.14)$$

The force Q_2 may be found in a similar fashion by considering the virtual displacement $\delta q_2 = 1$ shown in Fig. 5.14d. The corresponding virtual strain components are

$$\delta e_{11}^{(AD)} = \frac{16}{25}\frac{1}{L} \qquad \delta e_{22}^{(AD)} = \delta e_{33}^{(AD)} = -\frac{16}{25}\frac{v}{L}$$

$$\delta e_{11}^{(BD)} = \frac{1}{L} \qquad \delta e_{22}^{(BD)} = \delta e_{33}^{(BD)} = -\frac{v}{L} \qquad (5.4.15)$$

$$\delta e_{11}^{(CD)} = \frac{16}{25}\frac{1}{L} \qquad \delta e_{22}^{(CD)} = \delta e_{33}^{(CD)} = -\frac{16}{25}\frac{v}{L}$$

Application of the principle of virtual work in this case yields

$$Q_2 = \int_{\mathcal{V}(AD)} \frac{12Eq_1}{25L}\frac{16}{25L}\,d\mathcal{V} + \int_{\mathcal{V}(CD)} \frac{-12Eq_1}{25L}\frac{16}{25L}\,d\mathcal{V} = 0 \qquad (5.4.16)$$

Hence the vertical displacement q_1 is produced solely by the force Q_1. The result that $Q_2 = 0$ for this deformation could have been anticipated from a consideration of the structure's symmetry. ////

5.5   COMPLEMENTARY VIRTUAL WORK

In the previous section we considered the work done by a system of forces during an imaginary or virtual distortion. This led to a method for calculating unknown forces, known as the dummy-displacement method. We shall now develop a conjugate or complementary concept, known as complementary virtual work.

Rather than considering real forces and fictitious displacements, we will investigate the work done by a system of *imaginary forces* during the *actual deformation*. The structure under consideration is again assumed to have arbitrary material properties. We now introduce an imaginary system of surface forces δT_i and body forces δf_i which produce a state of stress $\delta\sigma_{ij}$ within the structure. These imaginary forces and stresses are assumed to satisfy the equations of equilibrium (4.4.1), so that

$$(\delta\sigma_{ij})_{,j} + \delta f_i = 0 \qquad (5.5.1)$$

but otherwise they are completely arbitrary and are independent of the actual forces and stresses existing in the structure. The work done by these *virtual forces* during the actual deformation u_i will be referred to as the *complementary virtual work* δW_E^*. It is expressed mathematically as

$$\delta W_E^* = \int_{\mathscr{S}} \delta T_i u_i \, d\mathscr{S} + \int_{\mathscr{V}} \delta f_i u_i \, d\mathscr{V} \qquad (5.5.2)$$

Proceeding in a manner similar to that used in the development of the principle of virtual work in Sec. 5.4 (with the roles of the actual and virtual quantities interchanged), one obtains the result

$$\int_{\mathscr{S}} \delta T_i u_i \, d\mathscr{S} + \int_{\mathscr{V}} \delta f_i u_i \, d\mathscr{V} = \int_{\mathscr{V}} \delta\sigma_{ij} e_{ij} \, d\mathscr{V} \qquad (5.5.3)$$

The integral on the right-hand side of Eq. (5.5.3) represents the work done by the virtual stresses $\delta\sigma_{ij}$ under the actual strains e_{ij}. We shall denote this quantity by δU^*, that is,

$$\delta U^* = \int_{\mathscr{V}} \delta\sigma_{ij} e_{ij} \, d\mathscr{V} \qquad (5.5.4)$$

Using Eqs. (5.5.2) and (5.5.4), Eq. (5.5.3) may be written as

$$\delta W_E^* = \delta U^* \qquad (5.5.5)$$

This equation is known as the *principle of complementary virtual work* or the *principle of virtual forces*; it may be stated as follows:

> *Principle of complementary virtual work* The complementary virtual work δW_E^* done by an external virtual force system under the actual deformation of a structure is equal to the complementary work δU^* done by the virtual stresses under the actual strains.

The duality between the principle of virtual work (pvw) and the principle of complementary virtual work (pcvw) is now clear. We recall that both apply in situations involving small deformations of structures having arbitrary material

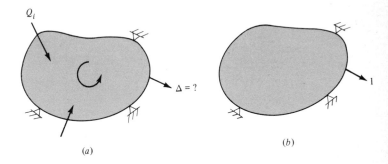

FIGURE 5.15
(a) Body with an unknown displacement Δ; (b) virtual-force system.

behavior (elastic or inelastic). It might also be noted that since the pvw involves real forces and stresses, it may be interpreted as a requirement for equilibrium. On the other hand, in the pcvw we deal with real strains and displacements, and hence we may regard this principle as a requirement for a consistent deformation configuration. This point will be made clearer through the illustrative examples presented below.

The pcvw provides an extremely useful means for calculating unknown displacements. Consider, for example, a structure subject to a system of n discrete generalized forces Q_i and zero body forces (Fig. 5.15a). The expression (5.5.2) for the external complementary virtual work in this case becomes

$$\delta W_E^* = \sum_{i=1}^{n} \delta Q_i q_i \qquad (5.5.6)$$

In order to compute a generalized displacement Δ in the structure, we introduce a unit virtual force applied at the point where the displacement is desired, and acting in the direction of this displacement (Fig. 5.15b). The complementary virtual work δW_E^* is then simply $1 \cdot \Delta$, and application of the pcvw (5.5.5) yields

$$\Delta = \int_V \delta\sigma_{ij} e_{ij} \, dV \qquad (5.5.7)$$

Here $\delta\sigma_{ij}$ are the virtual stresses in the structure produced by the unit virtual force, and e_{ij} are the strains resulting from the actual applied loads Q_i. Hence the unknown displacement Δ can be computed providing the virtual stresses and real strains can be found. Because we have introduced a fictitious force at the point where the deflection is desired, this technique is sometimes called the *dummy-force method*. As the following illustrative problems will indicate, this method is particularly useful in the analysis of beams and frames. Since it is usually easier

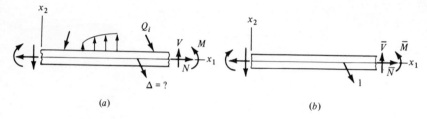

FIGURE 5.16
(a) Beam element with an unknown displacement Δ; (b) dummy-force system.

to express the bending stresses and strains in terms of given applied loads rather than in terms of prescribed displacements, application of the dummy-force method is often simpler than use of the dummy-displacement method.

Example 5.7   Dummy-force method for a beam element   As a first example, we will obtain an expression for the generalized displacement Δ of a linearly elastic beam element which bends about one of its principal axes (Fig. 5.16a). Flexure is produced by an arbitrary set of generalized forces Q_i. No a priori assumption is made with regard to boundary conditions. We denote the shear force, axial force, and bending moment at a distance x_1 along the centroidal axis by V, N, and M respectively. Neglecting the effect of shear,[1] the nonzero strain components at this cross section are

$$e_{11} = \frac{N}{EA} - \frac{Mx_2}{EI}$$

$$e_{22} = e_{33} = -\frac{vN}{EA} + \frac{vMx_2}{EI}$$

(5.5.8)

Next consider the internal forces associated with the unit dummy-force shown in Fig. 5.16b. The shear force, axial force, and bending moment produced by this virtual loading are taken to be $\bar{V}$, $\bar{N}$, and $\bar{M}$ respectively. Again neglecting the effect of shear, the only nonzero virtual stress component is

$$\delta\sigma_{11} = \frac{\bar{N}}{A} - \frac{\bar{M}x_2}{I}$$

(5.5.9)

[1]This assumption was discussed in Example 5.2.

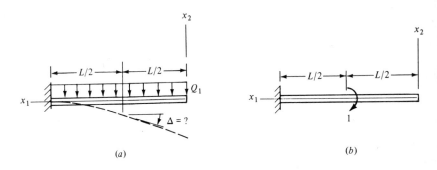

FIGURE 5.17
(a) Cantilever beam subject to a uniformly distributed load; (b) dummy-force system for computing the rotation Δ.

Substituting Eqs. (5.5.8) and (5.5.9) into (5.5.7) gives

$$\Delta = \int_{\mathcal{V}} \delta\sigma_{11} e_{11} \, d\mathcal{V}$$

$$= \int_{\mathcal{V}} \left(\frac{\bar{N}}{A} - \frac{\bar{M}x_2}{I} \right) \left(\frac{N}{EA} - \frac{Mx_2}{EI} \right) d\mathcal{V}$$

$$= \int_0^L \int_A \left(\frac{N\bar{N}}{EA^2} - \frac{N\bar{M}x_2}{EAI} - \frac{\bar{N}Mx_2}{EAI} + \frac{M\bar{M}x_2{}^2}{EI^2} \right) dA \, dx_1 \qquad (5.5.10)$$

Noting that $\int_A x_2 \, dA = 0$ and $\int_A x_2{}^2 \, dA = I$ since x_3 is a centroidal axis, Eq. (5.5.10) may be simplified to

$$\Delta = \int_0^L \left(\frac{N\bar{N}}{EA} + \frac{M\bar{M}}{EI} \right) dx_1 \qquad (5.5.11)$$

This equation may be used to compute a generalized displacement in any linearly elastic beam element, providing the axial forces and bending moments produced by the actual loads and the dummy load are available. We will make use of this simple expression rather than the more general relation (5.5.7) whenever we deal with beams.   ////

Example 5.8   Dummy-force method for a cantilever beam   Let us now apply the results of Example 5.7 in order to compute the rotation Δ at the midpoint of the cantilever beam shown in Fig. 5.17a. The beam is assumed to have a constant moment of inertia I, and is subject to a uniformly distributed load Q_1. For this applied loading the axial force is $N = 0$, and the bending moment at a distance x_1 from the free end is

$$M = \frac{-Q_1 x_1^2}{2} \qquad 0 < x_1 < L \qquad (5.5.12)$$

In order to compute the *rotation* of the beam's midpoint, we must introduce a unit *couple* at this point (Fig. 5.17b). The normal force is $\overline{N} = 0$ for this assumed virtual-force system, and the bending moment distribution is

$$\overline{M} = \begin{cases} 0 & 0 < x_1 < \dfrac{L}{2} \\[2mm] -1 & \dfrac{L}{2} < x_1 < L \end{cases} \qquad (5.5.13)$$

Equation (5.5.11) then gives

$$\Delta = \int_0^{L/2} 0 + \frac{1}{EI} \int_{L/2}^{L} \frac{-Q_1 x_1^2}{2}(-1)\,dx_1 = \frac{7Q_1 L^3}{48EI} \qquad (5.5.14)$$

////

Example 5.9 Dummy-force method for a nonlinearly elastic beam element

Since the pvw and the pcvw were developed without introducing a particular stress-strain law, these principles are valid for structures having arbitrary material behavior. For the most part we shall restrict our attention to linearly elastic materials; however, let us now examine a structure which possesses a nonlinearly elastic response.

Consider a beam of rectangular cross section, supported and loaded as shown in Fig. 5.18a. Assume that the material's stress-strain law has the following form (Fig. 5.18b):

$$\sigma_{11} = \pm E|e_{11}|^{1/n} \qquad (5.5.15)$$

where the constant E and the integer n represent material constants. The plus sign in Eq. (5.5.15) applies when e_{11} is positive, and the minus sign applies when e_{11} is negative.

In order to determine the transverse displacement Δ at an arbitrary point in the beam, a unit dummy force is introduced at that point (Fig. 5.18c). The displacement Δ is then found by substituting the actual strains e_{ij} (caused by Q_1) and the virtual stresses $\delta\sigma_{ij}$ (produced by the dummy force) into Eq. (5.5.7). The relationships of stress and strain to the bending moment in a nonlinear beam element are not the same as those for a linear beam (5.5.8) and (5.5.9); the

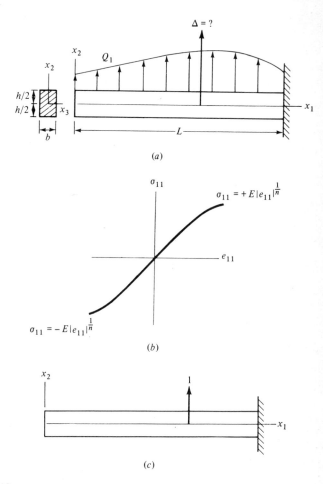

FIGURE 5.18
(a) Nonlinearly elastic cantilever beam; (b) stress-strain curve; (c) dummy-force system for computing the displacement Δ.

former will now be derived. We recall from elementary beam theory that the normal strain e_{11} is related to the radius of curvature ρ of the deformed neutral axis of the beam by

$$e_{11} = \frac{-x_2}{\rho} \qquad (5.5.16)$$

From statics, the bending moment at an arbitrary cross section of the beam is

$$M = -\int_A \sigma_{11} x_2 \, dA = -2b \int_0^{h/2} \sigma_{11} x_2 \, dx_2 \qquad (5.5.17)$$

Consequently
$$M = 2b \int_0^{h/2} E\left(\frac{x_2}{\rho}\right)^{1/n} x_2 \, dx_2 = \frac{EI_n}{\rho^{1/n}} \qquad (5.5.18)$$

where I_n is defined by
$$I_n = 2b \int_0^{h/2} x_2^{1+1/n} \, dx_2 \qquad (5.5.19)$$

Combining Eqs. (5.5.16) and (5.5.18) yields the normal strain component

$$e_{11} = -\left(\frac{M}{EI_n}\right)^n x_2 \qquad (5.5.20)$$

The corresponding stress, computed using Eq. (5.5.15), is

$$\sigma_{11} = -\frac{M}{I_n} x_2^{1/n} \qquad (5.5.21)$$

Letting $\overline{M}$ denote the bending moment produced by the dummy force shown in Fig. 5.18c, the virtual stress $\delta\sigma_{11}$ is

$$\delta\sigma_{11} = -\frac{\overline{M}}{I_n} x_2^{1/n} \qquad (5.5.22)$$

Substituting Eqs. (5.5.20) and (5.5.22) into (5.5.7) gives the displacement expression

$$\Delta = \int_{\mathscr{V}} \delta\sigma_{11} e_{11} \, d\mathscr{V}$$
$$= 2 \int_0^L \int_0^{h/2} \left(-\frac{\overline{M}}{I_n} x_2^{1/n}\right)\left[-\left(\frac{M}{EI_n}\right)^n x_2\right] b \, dx_2 \, dx_1 \qquad (5.5.23)$$

Finally, noting that M, $\overline{M}$ and I_n are functions of x_1 alone, and using the definition of I_n, we obtain

$$\Delta = \int_0^L \overline{M}\left(\frac{M}{EI_n}\right)^n dx_1 \qquad (5.5.24)$$

Equation (5.5.24) reduces to the expression derived earlier for a linearly elastic beam when we let $n = 1$.   ////

Example 5.10   Dummy-force method for a curved bar   Curved bars are often used as stiffening members in airplane fuselages, missiles, submarines, and many other types of structures. We shall now investigate the deformation of a curved beam; in particular we wish to compute the displacement Δ of the quarter-circle bar shown in Fig. 5.19a. Let us assume that the cross-sectional dimensions of the bar are very small[1] in comparison with the radius of curvature R. In

---

[1] If the cross-sectional dimensions of a curved bar are not small in comparison with the radius of curvature, the stress distribution is rather complicated (Ref. 5.1) and the strength-of-materials equations for a straight beam do not apply.

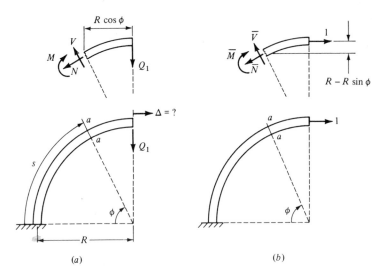

FIGURE 5.19
(a) Curved beam; (b) dummy-force system for computing the displacement Δ.

this case, every differential element ds along the centroidal axis of the beam is approximately a straight line segment, and Eq. (5.5.11) for bending of straight bars may be used; that is,

$$\Delta = \int \left(\frac{N\bar{N}}{EA} + \frac{M\bar{M}}{EI} \right) ds \qquad (5.5.25)$$

where the integration now extends along the total arc length of the bar. The normal force N and the bending moment M at the cross section aa, found by writing equations of equilibrium for the free body shown in Fig. 5.19a, are as follows:

$$N = -Q_1 \cos \phi$$
$$M = -Q_1 R \cos \phi \qquad (5.5.26)$$

In order to calculate the horizontal displacement Δ, we consider the virtual-force system shown in Fig. 5.19b. The virtual normal force $\bar{N}$ and the virtual bending moment $\bar{M}$ at section aa, are given by

$$\bar{N} = \sin \phi$$
$$\bar{M} = -R(1 - \sin \phi) \qquad (5.5.27)$$

Substituting Eqs. (5.5.26) and (5.5.27) into (5.5.25), and letting $ds = R\,d\phi$, gives

$$\Delta = \int_0^{\pi/2} \left[-\frac{Q_1}{EA} \cos \phi \sin \phi + \frac{Q_1 R^2}{EI} \cos \phi (1 - \sin \phi) \right] R\,d\phi$$

$$= -\frac{Q_1 R}{2EA} + \frac{Q_1 R^3}{2EI} \tag{5.5.28}$$

Under our assumption that the radius of curvature is much larger than the cross-sectional dimensions of the bar, it follows that the first term in expression (5.5.28) for Δ is negligibly small compared to the second term. Hence, in the analysis of the deformation of thin rings or beams, the effect of the normal force N, as well as the shear force V, may generally be disregarded.

5.6 RECIPROCAL THEOREMS

The pvw (or alternatively the pcvw) can be used to formulate two reciprocal theorems which often prove to be useful in the analysis of linearly elastic structures. Let us begin by considering a structure which is supported against rigid-body motion, and is subjected to a system of generalized forces Q_i ($i = 1, 2, \ldots, n$) (Fig. 5.20a). The stresses and strains induced in the structure by the forces Q_i will be denoted by σ_{ij} and e_{ij}, respectively. Next we consider a second set of generalized forces $\bar{Q}_j$ ($j = 1, 2, \ldots, m$) which induce stresses $\bar{\sigma}_{ij}$ and strains $\bar{e}_{ij}$ in the structure (Fig. 5.20b). If we imagine that the Q_i forces are applied first, and then the $\bar{Q}_j$ forces produce a virtual distortion, the pvw states that

$$\sum_{i=1}^{n} Q_i \bar{q}_i = \int_{\mathcal{V}} \sigma_{ij} \bar{e}_{ij}\,d\mathcal{V} \tag{5.6.1}$$

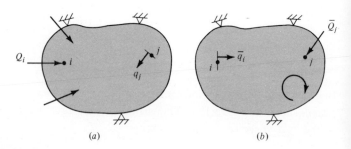

(a)   (b)

FIGURE 5.20
(a) System of generalized forces Q_i ($i = 1, 2, \ldots, n$); (b) second system of generalized forces $\bar{Q}_j$ ($j = 1, 2, \ldots, m$).

where $\bar{q}_i$ represents the displacement of the point of application of Q_i in the direction of Q_i produced by the system of forces $\bar{Q}_j$. On the other hand, if the Q_i forces are assumed to produce a virtual distortion of the structure after the $\bar{Q}_j$ forces have been applied, the pvw gives

$$\sum_{j=1}^{m} \bar{Q}_j q_j = \int_{\mathcal{V}} \bar{\sigma}_{ij} e_{ij} \, d\mathcal{V} \qquad (5.6.2)$$

Here q_j is the generalized displacement of the point of application of $\bar{Q}_j$ in the direction of $\bar{Q}_j$ caused by the Q_i-force system.

Since the body is linearly elastic, the stresses σ_{ij} are related to the corresponding strains e_{ij} by the relation (4.3.1)

$$\sigma_{ij} = E_{ijkl} e_{kl} \qquad (5.6.3)$$

and similarly

$$\bar{\sigma}_{ij} = E_{ijkl} \bar{e}_{kl} \qquad (5.6.4)$$

Equations (5.6.1) and (5.6.2) can therefore be rewritten as

$$\sum_{i=1}^{n} Q_i \bar{q}_i = \int_{\mathcal{V}} E_{ijkl} e_{kl} \bar{e}_{ij} \, d\mathcal{V} \qquad (5.6.5)$$

and

$$\sum_{j=1}^{m} \bar{Q}_j q_j = \int_{\mathcal{V}} E_{ijkl} \bar{e}_{kl} e_{ij} \, d\mathcal{V} \qquad (5.6.6)$$

By interchanging the dummy indices within the volume integral in Eq. (5.6.6) and recalling that $E_{ijkl} = E_{klij}$, we find that the right-hand sides of Eqs. (5.6.5) and (5.6.6) are identical; thus

$$\sum_{i=1}^{n} Q_i \bar{q}_i = \sum_{j=1}^{m} \bar{Q}_j q_j \qquad (5.6.7)$$

This is the reciprocal law of Betti,[1] which may be stated as the following theorem:

> *Betti's reciprocal theorem* If a linearly elastic structure is subjected to two separate generalized force systems, the work done by the first system during the distortion produced by the second system is equal to the work performed by the second system during the deformation caused by the first system.

As a special case of Betti's theorem let us assume that each force system consists of a single generalized force of unit magnitude. For example, we consider the forces $Q_i = 1$ and $\bar{Q}_j = 1$ shown in Fig. 5.21. Equation (5.6.7) then gives

$$\bar{q}_i = q_j \qquad (5.6.8)$$

The quantity $\bar{q}_i$ now represents the generalized displacement at i in the direction of Q_i due to a unit force $\bar{Q}_j$. This definition is equivalent to that given to the

---

[1]The reciprocal theorem was first stated by E. Betti in 1872.

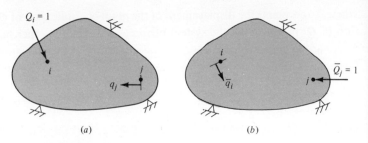

FIGURE 5.21
(a) Unit generalized force $Q_i = 1$; (b) unit generalized force $\bar{Q}_j = 1$.

flexibility coefficient c_{ij} in Sec. 5.3; that is, $\bar{q}_i = c_{ij}$. Likewise q_j is the generalized displacement at j in the direction of $\bar{Q}_j$ due to Q_i, so that $q_j = c_{ji}$. Therefore Eq. (5.6.8) expresses the fact that

$$c_{ij} = c_{ji} \qquad (5.6.9)$$

This result is known as *Maxwell's law*[1] and may be stated as follows:

> *Maxwell's reciprocal theorem* In a linearly elastic structure the generalized displacement at point i in the direction of Q_i produced by a unit generalized force Q_j is equal in magnitude to the generalized displacement at point j in the direction of Q_j due to the unit generalized force Q_i.

5.7   PRINCIPLE OF MINIMUM POTENTIAL ENERGY

In this section we shall restate the pvw as it applies to elastic structures. The symbol δ used to represent an imaginary or virtual quantity in Sec. 5.4 will now be regarded as the *variational operator* used in the calculus of variations.[2]

Let us once again consider a structure in equilibrium whose deformed configuration is characterized by the displacement field u_i. We then consider a class of arbitrary displacements $\bar{u}_i$ which are consistent with all constraints imposed on the body. These arbitrary displacements will, in general, differ from the actual displacements by some amount, say δu_i. That is,

$$\bar{u}_i = u_i + \delta u_i \qquad (5.7.1)$$

---

[1]The law of reciprocal displacements, which is a special case of Betti's law, was stated by J. C. Maxwell in 1864.
[2]If the student is not yet familiar with the calculus of variations, he should refer to Appendix C for a brief introduction to the subject and for a list of suggested references.

where the variation δu_i must vanish at those points in the structure where constraints (supports) exist. Quite clearly, the *variation* δu_i is equivalent to the virtual displacement δu_i introduced earlier.

Likewise, the internal virtual work $\delta U = \int_{\mathcal{V}} \sigma_{ij} \delta e_{ij} \, d\mathcal{V}$ defined by Eq. (5.4.8) may be interpreted as the variation in the strain energy resulting from the variation in the displacement field. To see this, recall that the strain energy U stored in a deformed, isotropic linearly elastic material is, according to Eq. (5.2.7),

$$U = \int_{\mathcal{V}} \left(\mu e_{ij} e_{ij} + \frac{\lambda}{2} e_{kk}^2 \right) d\mathcal{V} \qquad (5.7.2)$$

We now calculate the first variation of U for a variation in the deformation, that is, for a variation in the strains e_{ij}

$$\delta U = \delta \int_{\mathcal{V}} \left(\mu e_{ij} e_{ij} + \frac{\lambda}{2} e_{kk}^2 \right) d\mathcal{V}$$

$$= \int_{\mathcal{V}} \left(\mu 2 e_{ij} \, \delta e_{ij} + \frac{\lambda}{2} 2 e_{kk} \, \delta e_{ii} \right) d\mathcal{V}$$

$$= \int_{\mathcal{V}} (2\mu e_{ij} + \lambda \delta_{ij} e_{kk}) \delta e_{ij} \, d\mathcal{V}$$

$$= \int_{\mathcal{V}} \sigma_{ij} \, \delta e_{ij} \, d\mathcal{V} \qquad (5.7.3)$$

This expression is identical to that used to define the internal virtual work in Sec. 5.4. Thus we see that the internal virtual work may be regarded as the first variation of the strain energy U due to variations in the strain components e_{ij}. Although we have considered an isotropic, linearly elastic material, Eq. (5.7.3) is valid also for anisotropic and nonlinearly elastic materials (see Probs. 5.15 and 5.16).

Similarly, the external virtual work $\delta W_E = \int_{\mathcal{S}} T_i \, \delta u_i \, d\mathcal{S} + \int_{\mathcal{V}} f_i \, \delta u_i \, d\mathcal{V}$ may be regarded as the work done by the surface and body forces during a variation δu_i in the displacement field. For most types of external forces it is possible to define potential functions which, when differentiated with respect to the displacement components, yield the corresponding force components. For example, if $G(u_i)$ and $g(u_i)$ represent the potentials of the surface forces T_i and body forces f_i, respectively, then

$$T_i = -\frac{\partial G}{\partial u_i} \qquad \text{and} \qquad f_i = -\frac{\partial g}{\partial u_i} \qquad (5.7.4)$$

When such potential functions exist, the external virtual work becomes

$$\delta W_E = \int_{\mathscr{S}} -\frac{\partial G}{\partial u_i} \delta u_i \, d\mathscr{S} + \int_{\mathscr{V}} -\frac{\partial g}{\partial u_i} \delta u_i \, d\mathscr{V}$$

$$= -\delta \int_{\mathscr{S}} G \, d\mathscr{S} - \delta \int_{\mathscr{V}} g \, d\mathscr{V} \qquad (5.7.5)$$

or
$$\delta W_E = -\delta V_E \qquad (5.7.6)$$

where the *potential of the external forces* V_E is given by

$$V_E = \int_{\mathscr{S}} G \, d\mathscr{S} + \int_{\mathscr{V}} g \, d\mathscr{V} \qquad (5.7.7)$$

Suppose for instance that the surface and body forces are functions of position only; i.e., they are independent of the deformation of the structure. Such external forces are said to be *conservative*, and from Eq. (5.7.4) it follows that

$$G = -T_i u_i \qquad \text{and} \qquad g = -f_i u_i \qquad (5.7.8)$$

In this case the potential of the external force becomes

$$V_E = -\int_{\mathscr{S}} T_i u_i \, d\mathscr{S} - \int_{\mathscr{V}} f_i u_i \, d\mathscr{V} \qquad (5.7.9)$$

Thus, for a structure which possesses a strain energy U and an external potential V_E, the pvw (5.4.7) may be written as

$$\delta U - \delta W_E = \delta(U + V_E) = 0 \qquad (5.7.10)$$

or
$$\delta \Pi = 0 \qquad (5.7.11)$$

in which
$$\Pi = U + V_E \qquad (5.7.12)$$

is called the *total potential energy* of the structure. Equation (5.7.11) is a mathematical statement of the *principle of minimum potential energy* (pmpe); it may be stated as follows:

> *Principle of minimum potential energy* Of all displacement fields which satisfy the prescribed constraint conditions, the correct state is that which makes the total potential energy of the structure a minimum.

It should be mentioned that Eq. (5.7.11) only implies that Π is an extremum (see Appendix C) for the true displacement field. A proof that Π actually assumes a *minimum* value in the case of stable equilibrium may be found in Ref. 5.8.

It is important to understand that while the pvw is valid for both elastic and inelastic structures subject to arbitrary loads, the pmpe is applicable only to elastic structures (linear or nonlinear) acted upon by forces which are derivable from potential functions.

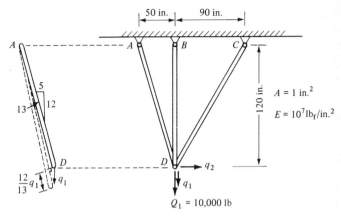

FIGURE 5.22
Three-bar truss.

Let us now restrict our attention to an isotropic, linearly elastic structure subject to n discrete conservative forces Q_i. The strain energy U is then given by Eq. (5.7.2), and according to Eq. (5.7.9) the potential V_E of the discrete force system is

$$V_E = -\sum_{i=1}^{n} Q_i q_i \qquad (5.7.13)$$

Whenever we compute the variation of expression (5.7.13), we must remember that it is the displacements q_i which are to be varied; the generalized forces Q_i are to be treated as constants.

The following example illustrates how the pmpe can be used in order to compute displacements in elastic structures.

Example 5.11 Application of the pmpe for the determination of displacements in a truss We wish to compute the displacements q_1 and q_2 in the three-bar truss shown in Fig. 5.22. In order to apply the pmpe, the total potential $\Pi = U + V_E$ must be expressed in terms of q_1 and q_2. To find U, we note that the strain energy stored in a prismatic bar of length L and cross-sectional area A, subject to an elongation Δ, is (Prob. 5.3)

$$U = \frac{EA}{2L} \Delta^2 \qquad (5.7.14)$$

Furthermore, from a consideration of the geometry, it is clear that bar AD elongates a distance $\frac{12}{13}q_1$ when point D experiences a vertical displacement q_1

(assuming small deformations); the bar elongates an amount $\frac{5}{13}q_2$ when D moves horizontally a distance q_2. The total extension of bar AD is therefore

$$\Delta^{(AD)} = \tfrac{12}{13}q_1 + \tfrac{5}{13}q_2 \qquad (5.7.15)$$

Similarly for bars BD and CD we have

$$\Delta^{(BD)} = q_1$$

$$\Delta^{(CD)} = \tfrac{4}{5}q_1 - \tfrac{3}{5}q_2 \qquad (5.7.16)$$

The strain energy for each member of the truss is found by substituting the appropriate values of E, A, L, and Δ into Eq. (5.7.14). If the units of q_1 and q_2 are inches, then for bar AD we obtain

$$U^{(AD)} = \frac{10^7}{2(130)}\left(\frac{12}{13}q_1 + \frac{5}{13}q_2\right)^2 \qquad \text{in.-lb} \qquad (5.7.17)$$

and likewise for bars BD and CD

$$U^{(BD)} = \frac{10^7}{2(120)}q_1{}^2 \qquad \text{in.-lb}$$

$$U^{(CD)} = \frac{10^7}{2(150)}\left(\frac{4}{5}q_1 - \frac{3}{5}q_2\right)^2 \qquad \text{in.-lb} \qquad (5.7.18)$$

The strain energy for the truss is equal to the sum of the energies stored in all its members

$$U = U^{(AD)} + U^{(BD)} + U^{(CD)} \qquad (5.7.19)$$

which gives

$$U = 95{,}800\,q_1{}^2 - 4{,}700\,q_1 q_2 + 17{,}700\,q_2{}^2 \qquad \text{in.-lb} \qquad (5.7.20)$$

Since Q_1 is the only external force acting on the truss, the external potential V_E is

$$V_E = -Q_1 q_1 = -10{,}000\,q_1 \qquad \text{in.-lb} \qquad (5.7.21)$$

and the total potential energy is therefore

$$\Pi = U + V_E$$
$$= 95{,}800\,q_1{}^2 - 4{,}700\,q_1 q_2 + 17{,}700\,q_2{}^2 - 10{,}000\,q_1 \qquad \text{in.-lb} \qquad (5.7.22)$$

The pmpe states that the potential energy is a minimum for arbitrary variations in the displacements q_1 and q_2. We therefore compute the variation of Π

$$\delta\Pi = 95{,}800\,(2q_1\,\delta q_1) - 4{,}700\,(\delta q_1 q_2 + q_1\,\delta q_2)$$
$$+ 17{,}700\,(2q_2\,\delta q_2) - 10{,}000\,\delta q_1$$
$$= (192{,}000\,q_1 - 4{,}700\,q_2 - 10{,}000)\,\delta q_1$$
$$+ (-4{,}700\,q_1 + 35{,}400\,q_2)\delta q_2 \qquad \text{in.-lb} \qquad (5.7.23)$$

Since $\delta\Pi = 0$ for arbitrary variations δq_1 and δq_2, the quantities within the parentheses in Eq. (5.7.23) must necessarily vanish. Thus

$$192,000\, q_1 - 4,700\, q_2 = 10,000$$
$$-4,700\, q_1 + 35,400\, q_2 = 0 \qquad (5.7.24)$$

Solving this pair of equations for q_1 and q_2 yields the desired results

$$q_1 = 0.0523 \text{ in.} \qquad q_2 = 0.00695 \text{ in.} \qquad (5.7.25)$$

Note that in using this approach, we have not written a single equation of statics. Hence a worthwhile check on our results involves examining whether or not the truss is in equilibrium. The axial force in each bar may be computed from a knowledge of the bar's elongation. Then if our results are correct, the resultant of these internal axial forces and the external force Q_1 will be zero (i.e., we consider equilibrium of the concurrent forces acting on joint D). The details of this check are left to the student as an exercise. ////

5.8 PRINCIPLE OF MINIMUM COMPLEMENTARY ENERGY

In Sec. 5.5 we observed that a duality exists between the concept of virtual work (work done by real forces during a virtual deformation) and the concept of complementary virtual work (work done by virtual forces during the real deformation). We then saw in Sec. 5.7 that pvw could be interpreted in terms of variations in the work and energy associated with a varied displacement field δu_i. It should now be quite obvious that the pcvw can likewise be interpreted in terms of mathematical variations rather than in terms of imaginary or virtual quantities. To show this, let us consider a structure which is in equilibrium under a set of externally applied loads. The state of stress at a point within the structure is described by the stress components σ_{ij} which satisfy the equations of equilibrium (4.4.1); that is, $\sigma_{ij,j} + f_i = 0$. We next introduce a set of stresses $\bar{\sigma}_{ij}$ which also satisfy the equilibrium equations, but are otherwise arbitrary. The difference between these arbitrary stresses $\bar{\sigma}_{ij}$ and the real stresses σ_{ij} represents the variation in the stress tensor σ_{ij}; that is,

$$\bar{\sigma}_{ij} = \sigma_{ij} + \delta\sigma_{ij} \qquad (5.8.1)$$

Note that the present definition of $\delta\sigma_{ij}$ as a variation in the stress field is equivalent to the definition given to virtual stresses in Sec. 5.5.

The internal complementary virtual work $\delta U^* = \int_V \delta\sigma_{ij}\, e_{ij}\, d\mathscr{V}$ may now be interpreted as the variation in the complementary strain energy resulting from the varied stresses $\delta\sigma_{ij}$. To show this, recall that the complementary strain energy U^* for an isotropic, linearly elastic material is

$$U^* = U = \int_{\mathscr{V}} \left(\frac{1+v}{2E} \sigma_{ij}\sigma_{ij} - \frac{v}{2E} \sigma_{kk}{}^2 \right) d\mathscr{V} \qquad (5.8.2)$$

The variation of U^* for a variation in the stresses σ_{ij} becomes

$$\delta U^* = \int_{\mathscr{V}} \left(\frac{1+v}{2E} 2\sigma_{ij}\,\delta\sigma_{ij} - \frac{v}{2E} 2\sigma_{kk}\,\delta\sigma_{ii} \right) d\mathscr{V}$$

$$= \int_{\mathscr{V}} \left(\frac{1+v}{E} \sigma_{ij} - \frac{v}{E} \delta_{ij}\sigma_{kk} \right) \delta\sigma_{ij}\,d\mathscr{V}$$

$$= \int_{\mathscr{V}} e_{ij}\,\delta\sigma_{ij}\,d\mathscr{V} \qquad (5.8.3)$$

This formula is identical to that used to define the internal complementary virtual work (5.5.4), and it is valid for nonlinear as well as linearly elastic materials.

Similarly, the external complementary virtual work δW_E^* may be interpreted as the work done by a variation in the external forces. For a conservative force system we may introduce a complementary potential function V_E^*, defined by

$$-\delta V_E^* = \delta W_E^* = \int_{\mathscr{S}} \delta T_i u_i\, d\mathscr{S} + \int_{\mathscr{V}} \delta f_i u_i\, d\mathscr{V} \qquad (5.8.4)$$

in which case

$$V_E^* = -\int_{\mathscr{S}} T_i u_i\, d\mathscr{S} - \int_{\mathscr{V}} f_i u_i\, d\mathscr{V} \qquad (5.8.5)$$

Application of the pcvw (5.5.5) then gives

$$\delta U^* - \delta W_E^* = \delta(U^* + V_E^*) = 0 \qquad (5.8.6)$$

or

$$\delta\Pi^* = 0 \qquad (5.8.7)$$

where the total complementary energy Π^* is defined as

$$\Pi^* = U^* + V_E^* \qquad (5.8.8)$$

This is the *principle of minimum complementary energy* (pmce).

> *Principle of minimum complementary energy* Of all states of stress which satisfy the equations of equilibrium, the correct state is that which makes the total complementary energy of the structure a minimum.

In the case of an isotropic, linearly elastic structure subject to a system of discrete generalized forces Q_i $(i = 1, 2, \ldots, n)$, the complementary strain energy U^* is equal to U and is given by Eq. (5.8.2). Furthermore it follows from Eq. (5.8.5) that the external complementary potential energy is

$$V_E^* = -\sum_{i=1}^{n} Q_i q_i = V_E \qquad (5.8.9)$$

Although V_E^* is equal to V_E in this case, it must be remembered that in applying the pmce it is the *forces* which are to be varied, while in the pmpe it is the *displacements* which are varied.

5.9 CASTIGLIANO'S THEOREMS

The principles of minimum potential energy and minimum complementary energy will now be used to derive two additional theorems which are extremely useful in the analysis of elastic structures.

For a structure in equilibrium under a set of discrete generalized forces Q_i ($i = 1, 2, \ldots, n$), the pmpe (5.7.11) states that

$$\delta \Pi = \delta(U + V_E) = 0 \qquad (5.9.1)$$

Substitution of the expression (5.7.13) for the external potential V_E into Eq. (5.9.1) gives

$$\delta\left[U(q_i) - \sum_{i=1}^{n} Q_i q_i \right] = 0 \qquad (5.9.2)$$

where the strain energy U is expressed in terms of the n generalized displacements q_i. Performing the variation indicated in Eq. (5.9.2) yields

$$\frac{\partial U}{\partial q_1} \delta q_1 + \frac{\partial U}{\partial q_2} \delta q_2 + \cdots + \frac{\partial U}{\partial q_n} \delta q_n - Q_1 \delta q_1 - Q_2 \delta q_2 - \cdots - Q_n \delta q_n = 0$$
$$(5.9.3)$$

Rearranging terms gives

$$\left(\frac{\partial U}{\partial q_1} - Q_1 \right) \delta q_1 + \left(\frac{\partial U}{\partial q_2} - Q_2 \right) \delta q_2 + \cdots + \left(\frac{\partial U}{\partial q_n} - Q_n \right) \delta q_n = 0 \qquad (5.9.4)$$

Since the variations δq_i are arbitrary, the terms within the parentheses must vanish; hence

$$\frac{\partial U}{\partial q_1} = Q_1, \frac{\partial U}{\partial q_2} = Q_2, \ldots, \frac{\partial U}{\partial q_n} = Q_n \qquad (5.9.5)$$

or in general

$$\frac{\partial U}{\partial q_i} = Q_i \qquad i = 1, 2, \ldots, n \qquad (5.9.6)$$

Equation (5.9.6) is a mathematical statement of *Castigliano's first theorem*.[1]

Castigliano's first theorem If the strain energy U stored in an elastic structure is expressed as a function of the generalized displacements q_i, then the first partial derivative of U with respect to any one of the generalized displacements q_i is equal to the corresponding generalized force Q_i.

We can make use of Eq. (5.9.6) in order to express the stiffness influence coefficients k_{ij} in terms of the strain energy of the structure. According to Eq. (5.3.11)

$$k_{ij} = \frac{\partial Q_i}{\partial q_j} \qquad (5.9.7)$$

Using Eq. (5.9.6) we may now write

$$k_{ij} = \frac{\partial^2 U}{\partial q_i\, \partial q_j} \qquad (5.9.8)$$

The example presented below illustrates the use of Castigliano's first theorem in the analysis of a simple structure. Applications of the method to more complicated, three-dimensional structures will be seen in Chap. 7.

Example 5.12 Castigliano's first theorem for a truss Let us now use Castigliano's theorem to compute the displacements q_1 and q_2 in the three-bar truss shown in Fig. 5.22. In Example 5.11 we found that the strain energy for this truss could be expressed in terms of the displacements as

$$U = 95{,}800\, q_1{}^2 - 4{,}700\, q_1 q_2 + 17{,}700\, q_2{}^2 \qquad \text{in.-lb} \qquad (5.9.9)$$

Castigliano's first theorem states that the first partial derivative of U with respect to a displacement q_i is equal to the corresponding force Q_i; therefore

$$\frac{\partial U}{\partial q_1} = 192{,}000\, q_1 - 4{,}700\, q_2 = Q_1$$

$$\frac{\partial U}{\partial q_2} = -4{,}700\, q_1 + 35{,}400\, q_2 = Q_2$$

$$(5.9.10)$$

[1]A. Castigliano, an Italian engineer, published a thesis in 1873 in which he presented several important energy principles. Among these were his first (5.9.6) and second theorems (5.9.18), and the theorem of least work which will be discussed in Chap. 6.

For this particular problem $Q_1 = 10,000$ lb and $Q_2 = 0$ (there is no horizontal force applied to the structure at point D). Thus

$$192,000 \, q_1 - 4,700 \, q_2 = 10,000$$
$$-4,700 \, q_1 + 35,400 \, q_2 = 0 \qquad (5.9.11)$$

Solving these equations gives

$$q_1 = 0.0523 \text{ in.} \qquad q_2 = 0.00695 \text{ in.} \qquad (5.9.12)$$

Note that Eqs. (5.9.11) are identical to Eqs. (5.7.24), obtained using the pmpe thus demonstrating the equivalence of the two methods.   ////

The pmce leads to another useful theorem, known as *Castigliano's second theorem*. For an elastic structure which is in equilibrium under a system of applied forces Q_i ($i = 1, 2, \ldots, n$), the pmce (5.8.7) states that

$$\delta \Pi^* = \delta(U^* + V_E^*) = 0 \qquad (5.9.13)$$

Let us assume that the complementary strain energy U^* in a structure has been calculated and can be expressed in terms of the applied forces Q_i. Then, using the definition of the complementary potential function V_E^* (5.8.9), we obtain

$$\delta \left[U^*(Q_i) - \sum_{i=1}^{n} q_i Q_i \right] = 0 \qquad (5.9.14)$$

which may be expanded to read

$$\frac{\partial U^*}{\partial Q_1} \delta Q_1 + \frac{\partial U^*}{\partial Q_2} \delta Q_2 + \cdots + \frac{\partial U^*}{\partial Q_n} \delta Q_n - q_1 \delta Q_1 - q_2 \delta Q_2 - \cdots - q_n \delta Q_n = 0$$
$$(5.9.15)$$

or
$$\left(\frac{\partial U^*}{\partial Q_1} - q_1 \right) \delta Q_1 + \left(\frac{\partial U^*}{\partial Q_2} - q_2 \right) \delta Q_2 + \cdots + \left(\frac{\partial U^*}{\partial Q_n} - q_n \right) \delta Q_n = 0$$
$$(5.9.16)$$

Since the variations δQ_i are arbitrary, Eq. (5.9.16) requires that

$$\frac{\partial U^*}{\partial Q_i} = q_i \qquad i = 1, 2, \ldots, n \qquad (5.9.17)$$

This result, known as *Engesser's theorem*, is valid for any elastic structure.[1] If the

---

[1] Equation (5.9.17), which is valid for nonlinear as well as linearly elastic structures and is therefore more general than Castigliano's second theorem (5.9.18), was derived by F. Engesser in 1889.

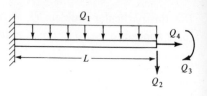

FIGURE 5.23
Cantilever beam subject to generalized forces.

structure is linearly elastic, the strain energy U and the complementary strain energy U^* are equal, and we obtain Castigliano's second theorem

$$\frac{\partial U}{\partial Q_i} = q_i \qquad i = 1, 2, \ldots, n \qquad (5.9.18)$$

which may be stated as follows

> *Castigliano's second theorem* If the strain energy U in a linearly elastic structure is expressed as a function of the generalized forces Q_i, then the first partial derivative of U with respect to any one of the generalized forces Q_i is equal to the corresponding displacement q_i.

Substituting Eq. (5.9.18) into (5.3.5), we see that the influence coefficients c_{ij}, which characterize the flexibility of a linearly elastic structure, may now be written as

$$c_{ij} = \frac{\partial^2 U}{\partial Q_i \, \partial Q_j} \qquad (5.9.19)$$

Equations (5.9.18) and (5.9.19) provide a straightforward and simple way of computing displacements and influence coefficients in linearly elastic structures. The amount of labor involved in the computation is comparable to that which is required when the dummy-force method is used. The choice of one of these methods over the other is generally a matter of personal taste. Several examples which illustrate the application of Castigliano's second theorem are given below.

Example 5.13 Castigliano's second theorem for a cantilevered beam element The strain energy U in a symmetric, linearly elastic cantilever beam subjected to a uniformly distributed load Q_1, a transverse force Q_2, a couple Q_3, and an axial force Q_4 (Fig. 5.23) is found to be (Prob. 5.6)

$$U = \frac{1}{2EI}\left(\frac{Q_1{}^2 L^5}{20} + \frac{Q_2{}^2 L^3}{3} + Q_3{}^2 L + \frac{Q_1 Q_2 L^4}{4} + \frac{Q_1 Q_3 L^3}{3} + Q_2 Q_3 L^2\right) + \frac{Q_4{}^2 L}{2EA}$$

$$(5.9.20)$$

Let us now compute the vertical displacement q_2 at the free end of the beam resulting from the applied forces. Castigliano's second theorem (5.9.18) may be used to obtain

$$q_2 = \frac{\partial U}{\partial Q_2} = \frac{Q_1 L^4}{8EI} + \frac{Q_2 L^3}{3EI} + \frac{Q_3 L^2}{2EI} \qquad (5.9.21)$$

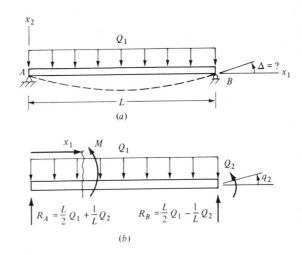

FIGURE 5.24
(a) Simply supported beam; (b) force system for computing the unknown rotation Δ.

The flexibility coefficient c_{21}, which represents the vertical displacement of the beam's end point due to a unit distributed force $Q_1 = 1$, is, according to Eq. (5.9.19)

$$c_{21} = \frac{\partial q_2}{\partial Q_1} = \frac{\partial^2 U}{\partial Q_2\, \partial Q_1} = \frac{L^4}{8EI} \qquad (5.9.22)$$

Likewise
$$c_{22} = \frac{L^3}{3EI} \qquad c_{23} = \frac{L^2}{2EI} \qquad c_{24} = 0 \qquad (5.9.23)$$

The remaining displacements and flexibility coefficients may be obtained in a similar fashion. While there is no confusion in interpreting the displacements corresponding to the discrete generalized forces Q_2, Q_3, and Q_4 (q_2 and q_4 are translations and q_3 is a rotation at the free end of the beam), the physical meaning of the displacement associated with the force Q_1 is less clear. We recall that, by definition, q_1 is that quantity which must multiply $\frac{1}{2}Q_1$ in order to obtain the work done by this force during the loading process. In this sense we may interpret q_1 as an "average" displacement along the length of the beam. Such a displacement is generally of no particular interest from an engineering standpoint. ////

Example 5.14 Castigliano's second theorem for a simply supported beam We wish to compute the rotation Δ at one end of a linearly elastic, symmetric beam which is subject to a uniformly distributed loading Q_1 (Fig. 5.24a). In order to compute the rotation using Castigliano's second theorem, we must differentiate the strain energy with respect to the corresponding generalized force (in this case a couple applied at the point B where the rotation is desired). Since no such generalized force is applied to the beam, a direct application of Castigliano's theorem will not be fruitful. We shall modify our approach as follows: (1) Assume that a couple Q_2 acts at B (Fig. 5.24b); (2) compute the rotation q_2 which results from Q_1 and Q_2 by using Castigliano's theorem; and (3) set $Q_2 = 0$ to obtain the rotation resulting from Q_1 alone. Following this outline, we first compute the reactive forces R_A and R_B required for equilibrium; these values are shown in Fig. 5.24b. The bending moment M at a distance x_1 along the beam is

$$M = R_A x_1 - Q_1 \frac{x_1{}^2}{2}$$

$$= Q_1 \frac{Lx_1 - x_1{}^2}{2} + Q_2 \frac{x_1}{L} \qquad (5.9.24)$$

We recall that the strain energy U stored in a linearly elastic beam element is, from Eq. (5.3.17),

$$U = \int_0^L \frac{M^2 \, dx_1}{2EI} \qquad (5.9.25)$$

Since Castigliano's second theorem states that

$$q_2 = \frac{\partial U(Q_1, Q_2)}{\partial Q_2} \qquad (5.9.26)$$

we have

$$q_2 = \frac{\partial}{\partial Q_2} \int_0^L \frac{M^2}{2EI} \, dx_1 = \int_0^L \frac{M}{EI} \frac{\partial M}{\partial Q_2} \, dx_1 \qquad (5.9.27)$$

Substituting Eq. (5.9.24) into (5.9.27) and integrating yields

$$q_2 = \frac{Q_1 L^3}{24EI} + \frac{Q_2 L}{3EI} \qquad (5.9.28)$$

Finally

$$\Delta = q_2 \Big|_{Q_2 = 0} = \frac{Q_1 L^3}{24EI} \qquad (5.9.29)$$

////

Example 5.15   Castigliano's second theorem for a curved beam   We now shall use Castigliano's second theorem to compute the displacement Δ in the curved beam shown in Fig. 5.25a. The student should note a similarity between the present analysis and the dummy-force solution to this problem which was given in Example 5.10. For simplicity we shall assume that the cross-sectional dimensions of the beam are small compared with the radius of curvature R, in

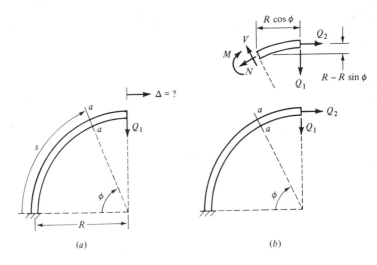

FIGURE 5.25
(a) Curved beam; (b) force system for computing the unknown displacement Δ.

which case we may neglect the strain energy associated with the axial force N and shear force V in the beam. The strain energy U is then given by

$$U = \int \frac{M^2}{2EI} \, ds = \int_0^{\pi/2} \frac{M^2}{2EI} R \, d\phi \qquad (5.9.30)$$

As in the previous example, we shall first assume that a force Q_2 is applied at the point where the displacement is desired, and later set $Q_2 = 0$. Castigliano's theorem then says

$$q_2 = \frac{\partial U(Q_1, Q_2)}{\partial Q_2} = \int_0^{\pi/2} \frac{M}{EI} \frac{\partial M}{\partial Q_2} R \, d\phi \qquad (5.9.31)$$

The bending moment M at an arbitrary cross section aa, found by writing an equation of moment equilibrium for the portion of the beam shown in Fig. 5.25b, is

$$M = - Q_1 R \cos \phi - Q_2 R(1 - \sin \phi) \qquad (5.9.32)$$

Substituting this expression into Eq. (5.9.31), solving for q_2, and then setting $Q_2 = 0$ gives

$$\Delta = q_2 \bigg|_{Q_2 = 0} = \frac{Q_1 R^3}{2EI} \qquad (5.9.33)$$

This, of course, agrees with the result (5.5.28) obtained using the dummy-force method. ////

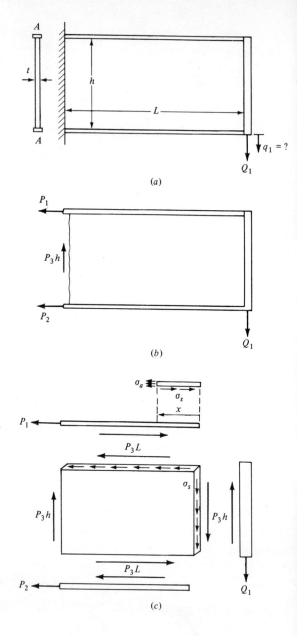

FIGURE 5.26
(a) Stiffened beam; (b) free-body diagram of the beam; (c) free-body diagrams of the individual members.

Example 5.16 Castigliano's second theorem for a stiffened beam Let us now compute the displacement q_1 in the stiffened or built-up structure shown in Fig. 5.26a. Finding an exact solution for such a structure is rarely possible, and we must therefore introduce certain simplifying assumptions. The nature of the assumptions naturally depends upon the desired degree of accuracy of the solution, and should be based upon good engineering judgment and experience. Below we introduce a set of assumptions which reduce the beam from a statically indeterminate to a statically determinate structure (i.e., one in which the forces in every member may be computed solely from a consideration of equilibrium). The assumptions are typical of the type which are often made in the analysis of a modern spacecraft structure. They are as follows:

1 The two horizontal flanges transmit axial loads only (no bending), and the axial stress in each flange is distributed uniformly over its cross section.
2 The web transmits pure shear (no tension or compression in the horizontal or vertical directions), and the shear stress is constant throughout the web.
3 The vertical flange is rigid.

Free-body diagrams for the entire structure and for each separate structural element are shown in Fig. 5.26b and c. Here P_1 and P_2 represent the axial forces at the built-in ends of the horizontal flanges, and P_3 denotes the shear flow in the web. (Recall from Example 5.4 that the shear flow is equal to the product of the shear stress σ_s and the thickness of the web t.) From statics we see that these internal forces are related to the applied force Q_1 as follows:

$$P_1 = \frac{L}{h} Q_1 \qquad P_2 = -\frac{L}{h} Q_1 \qquad P_3 = \frac{1}{h} Q_1 \qquad (5.9.34)$$

The above expression for P_3 was obtained by requiring that the sum of the vertical forces shown in Fig. 5.26b be equal to zero; note that the resultant vertical force along the built-in end of the structure is $P_3 h = \sigma_s th$.

The total strain energy U is computed by summing the strain energy of each of the individual components. First let us consider the upper horizontal flange. The axial stress σ_a at a distance x from the right end (Fig. 5.26c) is given by

$$\sigma_a = \frac{\sigma_s tx}{A} = \frac{P_3 x}{A} = \frac{Q_1 x}{Ah} \qquad (5.9.35)$$

Using Eq. (5.2.10), the strain energy in this member is found to be

$$U_{\text{top flange}} = \int_{\mathscr{V}} \frac{1}{2E} \sigma_a{}^2 \, d\mathscr{V} = \int_0^L \int_A \frac{1}{2E} \left(\frac{Q_1 x}{Ah}\right)^2 dA \, dx = \frac{Q_1{}^2 L^3}{6EAh^2} \qquad (5.9.36)$$

Similarly, the strain energy in the bottom flange is

$$U_{\text{bottom flange}} = \frac{Q_1{}^2 L^3}{6EAh^2} \qquad (5.9.37)$$

The internal energy in the web is, from Eq. (5.2.20),

$$U_{\text{web}} = \frac{hL}{2Gt} P_3{}^2 = \frac{L}{2Gth} Q_1{}^2 \qquad (5.9.38)$$

No energy is stored in the vertical flange since it is assumed to be rigid (i.e., the strain components are zero). Finally, adding the strain energies of all the elements we arrive at the total value

$$U = \left(\frac{L^3}{3EAh^2} + \frac{L}{2Gth} \right) Q_1{}^2 \qquad (5.9.39)$$

The application of Castigliano's second theorem yields the desired deflection

$$q_1 = \frac{\partial U}{\partial Q_1} = \left(\frac{2L^3}{3EAh^2} + \frac{L}{Gth} \right) Q_1 \qquad (5.9.40)$$

////

5.10   RAYLEIGH-RITZ METHOD

We have now seen that energy methods provide a convenient means for computing unknown forces and displacements in elastic structures. The solutions which are obtained using this approach (the displacements in a beam, for example) are exact within the framework of the theory (classical beam theory). Energy methods can also be used to derive approximate solutions in situations where exact solutions are difficult or impossible to obtain. The most notable approximate procedure is the Rayleigh-Ritz method,[1] in which the structure's displacement field is approximated by functions which contain a finite number of independent coefficients. The assumed functions are chosen to satisfy the kinematic boundary conditions (those involving translations and rotations), but they need not satisfy the static boundary conditions (ones involving forces and moments). The unknown coefficients in the assumed solution are determined by invoking the pmpe. Suppose, for example, that we choose a solution which has n independent coefficients a_i ($i = 1, 2, \ldots, n$). Since the approximate state of

[1]This method was proposed by Lord Rayleigh in 1877 and was refined and generalized by W. Ritz in 1908.

deformation of the structure is characterized by these n constants, the degree of freedom[1] of the structure has in effect been reduced from ∞ to n. In order to determine the coefficients a_i, we require that the total potential Π of the structure be a minimum; consequently

$$\delta\Pi = \frac{\partial\Pi}{\partial a_1}\,\delta a_1 + \frac{\partial\Pi}{\partial a_2}\,\delta a_2 + \cdots + \frac{\partial\Pi}{\partial a_n}\,\delta a_n = 0 \qquad (5.10.1)$$

Since the variations δa_i are arbitrary, Eq. (5.10.1) implies that

$$\frac{\partial\Pi}{\partial a_i} = 0 \qquad i = 1, 2, \ldots, n \qquad (5.10.2)$$

The conditions (5.10.2) represent a system of n simultaneous algebraic equations which can be solved to obtain the coefficients a_i.

Before considering detailed applications of the Rayleigh-Ritz method, a few comments with regard to the selection of a suitable form of solution are in order. As was stated earlier, the assumed functions must satisfy all the kinematic constraint conditions, but need not satisfy the static boundary conditions. The latter will be satisfied in an approximate way when the total potential of the structure is minimized. It is desirable, however, to satisfy as many boundary conditions as possible since then, in general, fewer terms will be required to achieve a given level of accuracy. Although the accuracy is usually improved by increasing the number of independent functions, the computations become correspondingly more complicated.

The decision as to what functions to select for a particular problem should be based upon an intuitive idea of what the true deformation of the structure looks like. From an analytical standpoint, trigonometric or polynomial functions are normally the most convenient functions to use.

As noted above, the basic idea of the Rayleigh-Ritz method is to approximate the continuous structure by a system having a finite number of degrees of freedom. Reducing the degrees of freedom of a structure is equivalent to introducing additional geometric constraints. Due to these imaginary constraints, the idealized structure is stiffer than the actual one. Thus the displacements predicted by the Rayleigh-Ritz method are generally smaller than the ones actually encountered.

Finally, if the approximate deformation state is to be used in order to calculate internal forces or stresses, the latter results should be viewed with caution.

---

[1] The degree of freedom of a structure is defined as the number of independent coordinates needed to describe the structure's displacement field. A continuous structure then has, by definition, an infinite number of degrees of freedom.

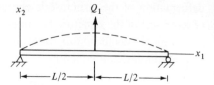

FIGURE 5.27
Uniform, simply supported beam.

This is because the stress components depend upon the derivatives of displacements, and although the displacements themselves may be relatively accurate, their derivatives may be considerably less so.

Two examples are presented below to illustrate the Rayleigh-Ritz method. While rather simple problems have been chosen in order to clarify the details of the procedure, the value of the method is fully appreciated only after it has been used in more complicated situations. This and other approximate techniques will be used later in our studies of elastic stability and structural dynamics.

Example 5.17   Rayleigh-Ritz method for a simply supported beam   As a first example we shall compute an approximate solution for the deflection $u_2(x_1)$ of a simply supported beam subject to a force Q_1 at its midpoint (Fig. 5.27). The boundary conditions for a beam on simple supports require that the displacement u_2 and the bending moment $M = EI \, d^2u_2/dx_1^2$ vanish at each end of the beam; that is,

$$u_2(0) = 0 \qquad u_2(L) = 0 \qquad (5.10.3)$$

and

$$\frac{d^2u_2(0)}{dx_1^2} = 0 \qquad \frac{d^2u_2(L)}{dx_1^2} = 0 \qquad (5.10.4)$$

The relations (5.10.3) are kinematic boundary conditions and therefore must be satisfied by our assumed solution; (5.10.4) are static boundary conditions and therefore need not be satisfied. Based upon an intuitive idea of what the actual deformation of the beam looks like, we shall choose a sine curve to represent the deformation. In particular we assume

$$u_2 = a_1 \sin \frac{\pi x_1}{L} \qquad (5.10.5)$$

where a_1 is an undetermined coefficient. The function (5.10.5) does indeed satisfy the kinematic boundary conditions (5.10.3), and moreover it also satisfies the static conditions (5.10.4). The total potential $\Pi = U + V_E$ of the beam is now computed on the basis of the assumed function. The strain energy of bending is, according to Eq. (5.2.17),

$$U = \int_0^L \frac{EI}{2} \left(\frac{d^2 u_2}{dx_1^2}\right)^2 dx_1 \quad (5.10.6)$$

Assuming that the beam has a constant flexural rigidity EI, the strain energy becomes

$$U = \frac{EI}{2} \int_0^L \left(-\frac{a_1 \pi^2}{L^2} \sin \frac{\pi x_1}{L}\right)^2 dx_1 = \frac{\pi^4 EI}{4L^3} a_1^2 \quad (5.10.7)$$

To determine the potential of the external force Q_1, we note that the displacement of the beam at the point of application and in the direction of Q_1 is $u_2(L/2) = a_1$. Therefore

$$V_E = -Q_1 a_1 \quad (5.10.8)$$

and the total potential energy $\Pi = U + V_E$ is

$$\Pi = \frac{\pi^4 EI}{4L^3} a_1^2 - Q_1 a_1 \quad (5.10.9)$$

The undetermined coefficient a_1 is found by minimizing Π. Thus we set $\delta\Pi = 0$, or what is equivalent [see Eq. (5.10.2)], we require that

$$\frac{\partial \Pi}{\partial a_1} = \frac{\pi^4 EI}{2L^3} a_1 - Q_1 = 0 \quad (5.10.10)$$

Hence

$$a_1 = \frac{2Q_1 L^3}{\pi^4 EI} = 0.0205 \frac{Q_1 L^3}{EI} \quad (5.10.11)$$

Substituting Eq. (5.10.11) into (5.10.5) yields the approximate solution

$$u_2 = \frac{0.0205 \, Q_1 L^3}{EI} \sin \frac{\pi x_1}{L} \quad (5.10.12)$$

From Eq. (5.10.12) it is clear that the maximum displacement occurs at $x_1 = L/2$ and has the value $0.0205 Q_1 L^3/EI$; this differs from the exact displacement at the beam's midpoint by less than 2 percent. An approximate value of the bending moment may also be computed by substituting Eq. (5.10.12) into $M =$

$EI\,d^2u_2/dx_1{}^2$; the moment at $x_1 = L/2$ is found to be 20 percent smaller than the corresponding exact value.

Let us now approximate the deformation by a somewhat more complicated function. One which has three undetermined coefficients and still satisfies the four boundary conditions is

$$u_2 = a_1 \sin \frac{\pi x_1}{L} + a_2 \sin \frac{2\pi x_1}{L} + a_3 \sin \frac{3\pi x_1}{L} \quad (5.10.13)$$

The strain energy of flexure, found by substituting Eq. (5.10.13) into (5.10.6), is

$$U = \frac{EI}{2} \int_0^L \left(-\frac{a_1 \pi^2}{L^2} \sin \frac{\pi x_1}{L} - \frac{a_2 4\pi^2}{L^2} \sin \frac{2\pi x_1}{L} - \frac{a_3 9\pi^2}{L^2} \sin \frac{3\pi x_1}{L} \right)^2 dx_1$$

$$= \frac{\pi^4 EI}{4L^3}(a_1{}^2 + 16a_2{}^2 + 81a_3{}^2) \quad (5.10.14)$$

In evaluating the integral in Eq. (5.10.14), use was made of the orthogonality relation for sine functions; namely,

$$\int_0^L \sin \frac{i\pi x_1}{L} \sin \frac{j\pi x_1}{L} dx_1 = \begin{cases} \dfrac{L}{2} & i = j \\ 0 & i \neq j \end{cases} \quad (5.10.15)$$

The potential of the force Q_1 is

$$V_E = -Q_1 u_2\left(\frac{L}{2}\right) = -Q_1(a_1 + 0 - a_3) \quad (5.10.16)$$

and the total potential energy $\Pi = U + V_E$ is

$$\Pi = \frac{\pi^4 EI}{4L^3}(a_1{}^2 + 16a_2{}^2 + 81a_3{}^2) - Q_1(a_1 - a_3) \quad (5.10.17)$$

For Π to be a minimum, we require that

$$\frac{\partial \Pi}{\partial a_1} = \frac{\pi^4 EI}{2L^3} a_1 - Q_1 = 0$$

$$\frac{\partial \Pi}{\partial a_2} = \frac{8\pi^4 EI}{L^3} a_2 = 0 \quad (5.10.18)$$

$$\frac{\partial \Pi}{\partial a_3} = \frac{81\pi^4 EI}{2L^3} a_3 + Q_1 = 0$$

Equations (5.10.18) can be solved for the unknown coefficients

$$a_1 = \frac{2Q_1 L^3}{\pi^4 EI} \qquad a_2 = 0 \qquad a_3 = \frac{-2Q_1 L^3}{81\pi^4 EI} \quad (5.10.19)$$

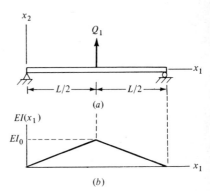

FIGURE 5.28
(a) Simply supported beam of variable cross section; (b) variation of the flexural rigidity.

Substituting Eq. (5.10.19) into (5.10.13) yields the approximate deflection curve

$$u_2 = \frac{2Q_1L^3}{\pi^4 EI}\left(\sin\frac{\pi x_1}{L} - \frac{1}{81}\sin\frac{3\pi x_1}{L}\right) \quad (5.10.20)$$

Note that the term proportional to $\sin(2\pi x_1/L)$ does not appear in the final expression (5.10.20). This is because $\sin(2\pi x_1/L)$ is antisymmetric with respect to the center of the beam, whereas the actual deformation is symmetrical. We could have anticipated this result and omitted the term $a_2 \sin(2\pi x_1/L)$ in our original approximation (5.10.13).

It can be verified that Eq. (5.10.20) yields more accurate values of displacements and moments than does the single-term approximation (5.10.12). It is also worth noting that if we had chosen an infinite trigonometric series to represent the deflected shape of the beam, the Rayleigh-Ritz method would have yielded the exact solution (see Ref. 5.5). ////

Example 5.18 Rayleigh-Ritz method for a beam of variable cross section We shall again consider a simply supported beam subject to a concentrated force Q_1, but it will now be assumed that the flexural rigidity of the beam has the linear variation shown in Fig. 5.28. Hence, in the region $0 < x_1 < L/2$ we have

$$EI(x_1) = EI_0\frac{2x_1}{L} \qquad 0 < x_1 < \frac{L}{2} \quad (5.10.21)$$

The deflection curve will again be represented by the sine function

$$u_2 = a_1 \sin\frac{\pi x_1}{L} \quad (5.10.22)$$

Substituting Eq. (5.10.22) into the expression for the bending strain energy (5.10.6) and recognizing that the same amount of energy is stored in the right and left halves of the beam, we obtain

$$U = 2 \int_0^{L/2} \frac{EI}{2} \left(\frac{d^2 u_2}{dx_1^2} \right)^2 dx_1$$

$$= \int_0^{L/2} \left(EI_0 \frac{2x_1}{L} \right) \left(-\frac{a_1 \pi^2}{L^2} \sin \frac{\pi x_1}{L} \right)^2 dx_1$$

$$= \frac{\pi^2 (1 + \pi^2/4) EI_0}{2L^3} a_1^2 \qquad (5.10.23)$$

The potential of the external force Q_1 is $-Q_1 a_1$, so that the total potential energy is

$$\Pi = \frac{\pi^2 (1 + \pi^2/4) EI_0}{2L^3} a_1^2 - Q_1 a_1 \qquad (5.10.24)$$

For Π to be a minimum, we require that

$$\frac{\partial \Pi}{\partial a_1} = \frac{\pi^2 (1 + \pi^2/4) EI_0}{L^3} a_1 - Q_1 = 0 \qquad (5.10.25)$$

Solving Eq. (5.10.25) for a_1 and substituting this value into (5.10.22) yields the approximate deflection curve

$$u_2 = 0.0293 \frac{Q_1 L^3}{EI_0} \sin \frac{\pi x_1}{L} \qquad (5.10.26)$$

The maximum displacement predicted by Eq. (5.10.26) is 6 percent smaller than the corresponding exact value. ////

5.11 SUMMARY OF THE ENERGY THEOREMS

A summary of the various energy principles derived in this chapter follows. It is important to note the duality which exists between those principles and theorems involving displacements as the varied quantities (displacement methods) and those involving variations in the forces (force methods). The equations apply to nonlinear as well as linearly elastic materials, except where noted to the contrary.

Displacement methods	Force methods

Principle of virtual work (pvw)

$$\delta W_E = \delta U$$

$$\delta W_E = \sum_{i=1}^{n} Q_i \delta q_i$$

$$\delta U = \int_{\mathscr{V}} \sigma_{ij} \delta e_{ij} \, d\mathscr{V}$$

Principle of complementary virtual work (pcvw)

$$\delta W_E^* = \delta U^*$$

$$\delta W_E^* = \sum_{i=1}^{n} q_i \delta Q_i$$

$$\delta U^* = \int_{\mathscr{V}} e_{ij} \delta \sigma_{ij} \, d\mathscr{V}$$

Principle of minimum potential energy (pmpe)

$$\delta \Pi = \delta(U + V_E) = 0$$

$$U = \int_{\mathscr{V}} \left(\int_0^{e_{ij}} \sigma_{ij} \, de_{ij} \right) d\mathscr{V}$$

$$\boxed{ = \int_{\mathscr{V}} \left(\mu e_{ij} e_{ij} + \frac{\lambda}{2} e_{kk}^2 \right) d\mathscr{V} }$$

$$V_E = -\sum_{i=1}^{n} Q_i q_i$$

Principle of minimum complementary energy (pmce)

$$\delta \Pi^* = \delta(U^* + V_E^*) = 0$$

$$U^* = \int_{\mathscr{V}} \left(\int_0^{\sigma_{ij}} e_{ij} \, d\sigma_{ij} \right) d\mathscr{V}$$

$$\boxed{ = \int_{\mathscr{V}} \left(\frac{1 + \nu}{2E} \sigma_{ij} \sigma_{ij} - \frac{\nu}{2E} \sigma_{kk}^2 \right) d\mathscr{V} = U }$$

$$V_E^* = -\sum_{i=1}^{n} q_i Q_i = V_E$$

Castigliano's first theorem

$$Q_i = \frac{\partial U}{\partial q_i}$$

$$\boxed{ k_{ij} = \frac{\partial^2 U}{\partial q_i \, \partial q_j} }$$

Castigliano's second theorem

$$q_i = \frac{\partial U^*}{\partial Q_i} = \boxed{\frac{\partial U}{\partial Q_i}}$$

$$\boxed{ c_{ij} = \frac{\partial^2 U}{\partial Q_i \, \partial Q_j} }$$

Note: Boxed-in terms are valid for linearly elastic materials only.

PROBLEMS

5.1 Verify that by substituting Eq. (5.2.6) into (5.2.5) one obtains the strain energy expression (5.2.7).

5.2 Show that the strain energy in an anisotropic, linearly elastic material ($\sigma_{ij} = E_{ijkl} e_{kl}$) is given by

$$U = \int_{\mathscr{V}} \tfrac{1}{2} E_{ijkl} e_{kl} e_{ij} \, d\mathscr{V} = \int_{\mathscr{V}} \tfrac{1}{2} \sigma_{ij} e_{ij} \, d\mathscr{V}$$

Hint: For convenience assume that all strain components increase from zero to their final values simultaneously; that is, let $e_{ij} = A_{ij} e$ where A_{ij} denote the final values of e_{ij} and e is a function varying from 0 to 1.

5.3 Using Eq. (5.2.9) show that the strain energy in a circular bar subject to an elongation q_1 is $U = (EA/2L)q_1{}^2$.

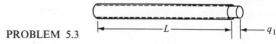

PROBLEM 5.3

5.4 Using Eq. (5.2.9) show that the strain energy in a circular bar twisted through an angle q_1 is $U = (GJ/2L)q_1{}^2$, where J denotes the polar moment of inertia of the bar's cross section. Begin by noting that the displacement field in the bar is (Ref. 5.1): $u_1 = 0$, $u_2 = -\alpha x_1 x_3$, and $u_3 = \alpha x_1 x_2$ in which $\alpha = q_1/L$ is the angle of twist per unit length along the x_1 axis.

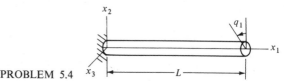

PROBLEM 5.4

5.5 Compute the strain energy U in the aluminum beam shown, and compare the strain energy associated with bending with that resulting from shear. Let $E = 10 \times 10^6$ lb$_f$/in.^{2} and $G = 4 \times 10^6$ lb$_f$/in.^{2}.

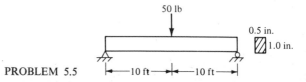

PROBLEM 5.5

5.6 Compute the strain energy U in a symmetric cantilever beam subjected to the uniformly distributed load Q_1, the transverse force Q_2, the couple Q_3, and the axial force (applied along the centroidal axis) Q_4 shown. Neglect the strain energy associated with shear.

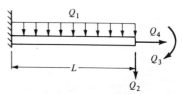

PROBLEM 5.6

5.7 A couple Q_1 is applied to one end of a simply supported beam. Use the principle of conservation of energy to compute the corresponding rotation q_1.

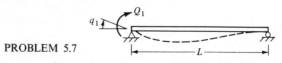

PROBLEM 5.7

5.8 Use the dummy-displacement method to compute the forces Q_1 and Q_2 required to produce a horizontal deflection q_2 of the structure in Example 5.6.

5.9   Use the dummy-displacement method to find the forces Q_1 and Q_2 required to produce the deformation of the structure shown. Assume small deformations.

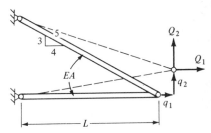

PROBLEM 5.9

5.10   Use the dummy-force method to obtain the displacements q_1 and q_2 caused by the forces Q_1 and Q_2 in Prob. 5.9.

5.11   The beam shown is subjected to a uniformly distributed force Q_1 and a concentrated force Q_2. Compute the vertical deflection at point C using the dummy-force method.

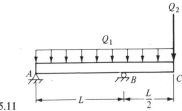

PROBLEM 5.11

5.12   Use the dummy-force method to find the rotation Δ at the right end of the simply supported beam shown.

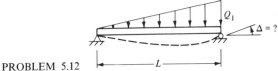

PROBLEM 5.12

5.13   Find the vertical displacement Δ of point B of the semicircular beam shown. Neglect the deformation due to shear and normal forces in the ring.

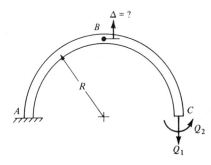

PROBLEM 5.13

5.14 Compute the vertical displacement at the free end of the cantilever beam shown, assuming that the stress-strain law for the material has the form $\sigma_{11} = \pm E|e_{11}|^{1/2}$.

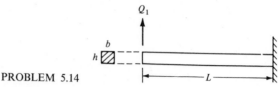

Q_1

PROBLEM 5.14

5.15 Show that Eq. (5.7.3) is valid for an anisotropic, linearly elastic material.

5.16 Show that Eq. (5.7.3) is valid for a nonlinear elastic material. Consider only a one-dimensional state of stress, for which $\sigma_{11} = f(e_{11})$.

5.17 Solve Prob. 5.9 using Castigliano's first theorem.

5.18–5.22 Solve Probs. 5.7, 5.10 to 5.13 using Castigliano's second theorem.

5.23 Compute the force in each bar of the truss shown, and find the vertical displacement of the hinge C. Assume that all bars have the same cross-sectional area A.

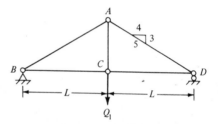

PROBLEM 5.23

5.24 Compute the vertical displacement Δ of point B of the frame shown if all three members have the same bending stiffness EI.

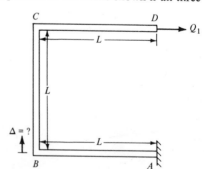

PROBLEM 5.24

5.25 Determine the flexibility matrix $[c]$ for the beam shown.

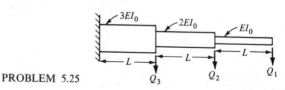

PROBLEM 5.25

5.26   A stepped circular shaft is subjected to the concentrated torques Q_1 and Q_2 shown. Compute the flexibility matrix $[c]$ and the stiffness matrix $[k]$ for this structure.

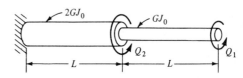

PROBLEM 5.26

5.27   Use the Rayleigh-Ritz method to find the maximum displacement and the maximum bending moment in the uniform cantilever beam shown. Represent the beam's deflected shape by the expression

$$u_2 = a_1\left(1 - \cos\frac{\pi x_1}{2L}\right)$$

where a_1 is an unknown coefficient. Compare your results with the exact solutions.

PROBLEM 5.27

5.28   Repeat Prob. 5.27 assuming the deflection is of the form

$$u_2 = a_1 + a_2\frac{x_1}{L} + a_3\left(\frac{x_1}{L}\right)^2 + a_4\left(\frac{x_1}{L}\right)^3$$

Hint: Before proceeding with the Rayleigh-Ritz method, examine the restrictions which the kinematic boundary conditions impose upon the constants a_i.

5.29   A simply supported beam of uniform cross section is loaded by a couple Q_1 at its midpoint. Find an approximate solution for the deflection curve $u_2(x_1)$ using the Rayleigh-Ritz method. Also compute the maximum slope and the maximum bending moment in the beam. Compare your results with the exact solutions.

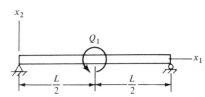

PROBLEM 5.29

5.30   Use the Rayleigh-Ritz method to obtain an approximate value for the maximum displacement of a simply supported beam subject to a uniformly distributed load Q_1.

Assume that the flexural rigidity EI of the beam is constant. Compare your results with the exact solution.

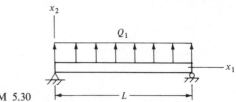

PROBLEM 5.30

5.31 Find an approximate value for the maximum displacement of the beam in Prob. 5.30, assuming that the beam's bending rigidity EI has the spanwise distribution shown.

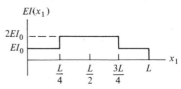

PROBLEM 5.31

REFERENCES

For an introductory account of energy principles:

5.1 SHAMES, I. H.: " Mechanics of Deformable Solids," Prentice-Hall, Englewood Cliffs, N.J., 1964.

5.2 BISPLINGHOFF, R. L., J. W. MAR, and T. H. H. PIAN: " Statics of Deformable Solids," Addison-Wesley, Reading, Mass., 1965.

5.3 ODEN, J. T.: " Mechanics of Elastic Structures," McGraw-Hill, New York, 1967.

5.4 RIVELLO, R. M.: " Theory and Analysis of Flight Structures," McGraw-Hill, New York, 1969.

For a more detailed treatment of energy methods:

5.5 HOFF, N. J.: " The Analysis of Structures," Wiley, New York, 1956.

5.6 LANGHAAR, H. L.: " Energy Methods in Applied Mechanics," Wiley, New York, 1962.

5.7 FUNG, Y. C.: " Foundations of Solid Mechanics," Prentice-Hall, Englewood Cliffs, N.J., 1965.

For a discussion on the application of variational methods in elasticity (5.7 and 5.8) and plasticity (5.9):

5.8 SOKOLNIKOFF, I. S.: " Mathematical Theory of Elasticity," 2d ed., McGraw-Hill, New York, 1956.

5.9 WASHIZU, K.: " Variational Methods in Elasticity and Plasticity," Pergamon, Oxford, 1968.

PART TWO

Static Behavior of Structures

STATICALLY INDETERMINATE
STRUCTURES

6.1 INTRODUCTION

In Chap. 5 we developed several methods for computing generalized forces and displacements in elastic structures. We saw that unknown forces can be obtained directly by applying either the pvw (the dummy-displacement method) or the pmpe (in the form of Castigliano's first theorem); unknown displacements can be computed directly using either the pcvw (the dummy-force method) or the pmce (Castigliano's second theorem). Aside from a few examples involving plane trusses, the structures considered so far have been statically determinate. That is, the stress distributions in these structures could be found by applying the principles of statics. We now shall consider *statically indeterminate* structures or ones for which the stresses and forces cannot be obtained from a consideration of statics alone; in addition it will be necessary to consider the structures' deformations. Either of the two dual approaches, the displacement or the force method, can be used to analyze statically indeterminate systems. In this chapter we shall consider relatively simple indeterminate structures; more complicated ones will be dealt with using matrix methods in Chap. 7.

6.2   DISPLACEMENT METHOD

In the displacement method we consider a system of arbitrary generalized displacements which satisfy prescribed constraint conditions. The pmpe (or the pvw) is then applied in order to determine the actual deformation of the structure, i.e., the deformation which corresponds to a state of equilibrium. The general procedure is the same whether the structure is statically determinate or indeterminate. First, the strain energy U is expressed in terms of the generalized displacements q_i ($i = 1, 2, \ldots, n$). Application of the pmpe in the form of Castigliano's first theorem leads to a set of n equations relating the displacements q_i and the corresponding generalized forces Q_i. This approach was used to analyze a planar truss in Chap. 5 (Example 5.12). A second example, involving a space truss, is presented below.

Example 6.1   Displacement method for a statically indeterminate space truss   Consider a four-bar space truss which is subject to three applied forces Q_1, Q_2, and Q_3 and experiences the displacements q_1, q_2, and q_3 (Fig. 6.1a). It will be assumed that the members of the truss are connected at their ends by frictionless hinges, in which case the bars transmit purely axial forces, say P_i ($i = 1, 2, 3, 4$). We shall now use the displacement method to compute the force-displacement relationships Q_i versus q_i and the unknown axial forces P_i. In Sec. 6.3 we shall solve this same problem using the force method.

First, the strain energy U in the structure is expressed in terms of the unknown displacements q_i. Recall (Prob. 5.3) that the strain energy in a bar of length L subject to an elongation Δ is

$$U = \frac{EA}{2L} \Delta^2 \qquad (6.2.1)$$

As a preliminary step we shall derive an expression for the total elongation Δ of a pinned bar, when one end of the bar is displaced through distances q_1, q_2, q_3 in the directions x_1, x_2, x_3, respectively. From Fig. 6.2 it can be concluded that

$$(L + \Delta)^2 = (L_1 + q_1)^2 + (L_2 + q_2)^2 + (L_3 + q_3)^2 \qquad (6.2.2)$$

where L_1, L_2, and L_3 represent the projections of line L on the coordinate axes. Squaring each term in Eq. (6.2.2) gives

$$L^2 + 2L\,\Delta + \Delta^2 = L_1{}^2 + 2L_1 q_1 + q_1{}^2 + L_2{}^2 + 2L_2 q_2 + q_2{}^2$$
$$+ L_3{}^2 + 2L_3 q_3 + q_3{}^2 \qquad (6.2.3)$$

Dividing by L^2 and noting that $L^2 = L_1{}^2 + L_2{}^2 + L_3{}^2$, we obtain

$$2\frac{\Delta}{L} + \left(\frac{\Delta}{L}\right)^2 = 2\frac{L_1}{L}\frac{q_1}{L} + \left(\frac{q_1}{L}\right)^2 + 2\frac{L_2}{L}\frac{q_2}{L} + \left(\frac{q_2}{L}\right)^2 + 2\frac{L_3}{L}\frac{q_3}{L} + \left(\frac{q_3}{L}\right)^2 \qquad (6.2.4)$$

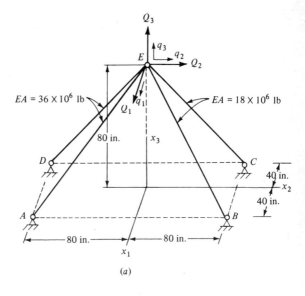

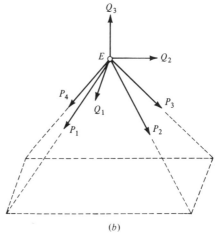

(b)

FIGURE 6.1
(a) Space truss; (b) free-body diagram of joint E.

By assuming that the deformation is so small that the squares of the quantities Δ/L, q_1/L, q_2/L, and q_3/L can be neglected in Eq. (6.2.4), the following approximation is obtained

$$\frac{\Delta}{L} = \frac{L_1}{L}\frac{q_1}{L} + \frac{L_2}{L}\frac{q_2}{L} + \frac{L_3}{L}\frac{q_3}{L} \qquad (6.2.5)$$

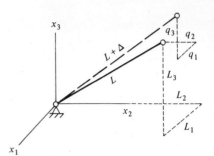

FIGURE 6.2
Elongation of a bar.

Therefore the elongation of the bar is

$$\Delta = \frac{L_1}{L} q_1 + \frac{L_2}{L} q_2 + \frac{L_3}{L} q_3 \qquad (6.2.6)$$

Table 6.1 shows the values of L_1, L_2, L_3, and L (in inches) for each bar of the space truss. The signs of the projections are given by the signs of the scalar components of the position vector from the fixed end of the bar to the displaced end; note, for example, that the components of the position vector from A to E are $L_1 = -40$ in., $L_2 = 80$ in., and $L_3 = 80$ in.

Substituting these values into Eq. (6.2.6), the extensions of the individual members are found to be

$$\Delta^{(AE)} = -\tfrac{40}{120} q_1 + \tfrac{80}{120} q_2 + \tfrac{80}{120} q_3$$

$$\Delta^{(BE)} = -\tfrac{40}{120} q_1 - \tfrac{80}{120} q_2 + \tfrac{80}{120} q_3$$

$$\Delta^{(CE)} = \tfrac{40}{120} q_1 - \tfrac{80}{120} q_2 + \tfrac{80}{120} q_3 \qquad (6.2.7)$$

$$\Delta^{(DE)} = \tfrac{40}{120} q_1 + \tfrac{80}{120} q_2 + \tfrac{80}{120} q_3$$

The strain energy of each bar is found by substituting the relations (6.2.7) into Eq. (6.2.1). If the units of q_1, q_2, and q_3 are inches, then for bar AE we obtain

$$U^{(AE)} = \frac{36 \times 10^6}{2 \times 120} \left(-\frac{40}{120} q_1 + \frac{80}{120} q_2 + \frac{80}{120} q_3 \right)^2$$

$$= \frac{10^6}{60} (q_1{}^2 + 4q_2{}^2 + 4q_3{}^2 - 4q_1 q_2 - 4q_1 q_3 + 8q_2 q_3) \qquad \text{in.-lb} \qquad (6.2.8)$$

Table 6.1

Bar	L_1	L_2	L_3	L
AE	-40	80	80	120
BE	-40	-80	80	120
CE	40	-80	80	120
DE	40	80	80	120

and similarly for bars BE, CE, and DE

$$U^{(BE)} = \frac{10^6}{120}(q_1{}^2 + 4q_2{}^2 + 4q_3{}^2 + 4q_1q_2 - 4q_1q_3 - 8q_2q_3) \qquad \text{in.-lb}$$

$$U^{(CE)} = \frac{10^6}{120}(q_1{}^2 + 4q_2{}^2 + 4q_3{}^2 - 4q_1q_2 + 4q_1q_3 - 8q_2q_3) \qquad \text{in.-lb} \qquad (6.2.9)$$

$$U^{(DE)} = \frac{10^6}{60}(q_1{}^2 + 4q_2{}^2 + 4q_3{}^2 + 4q_1q_2 + 4q_1q_3 + 8q_2q_3) \qquad \text{in.-lb}$$

The total strain energy is equal to the sum of the energies stored in the four bars, or

$$U = \frac{10^6}{60}(3q_1{}^2 + 12q_2{}^2 + 12q_3{}^2 + 8q_2q_3) \qquad \text{in.-lb} \qquad (6.2.10)$$

Application of Castigliano's first theorem (5.9.6) gives the desired force-displacement relations

$$Q_1 = \frac{\partial U}{\partial q_1} = 10^5 q_1 \qquad \text{lb}$$

$$Q_2 = \frac{\partial U}{\partial q_2} = 10^5(4q_2 + \tfrac{4}{3}q_3) \qquad \text{lb} \qquad (6.2.11)$$

$$Q_3 = \frac{\partial U}{\partial q_3} = 10^5(\tfrac{4}{3}q_2 + 4q_3) \qquad \text{lb}$$

In the event that the forces have been specified and the displacements are to be calculated, we may invert Eqs. (6.2.11) to obtain the displacement-force expressions

$$q_1 = \frac{1}{10^5}Q_1 \qquad \text{in.}$$

$$q_2 = \frac{3}{32 \times 10^5}(3Q_2 - Q_3) \qquad \text{in.} \qquad (6.2.12)$$

$$q_3 = \frac{3}{32 \times 10^5}(-Q_2 + 3Q_3) \qquad \text{in.}$$

in which the units of Q_1, Q_2, and Q_3 are pounds.

The extensions of the bars may be obtained by substituting Eqs. (6.2.12) into (6.2.7). This gives

$$\Delta^{(AE)} = \frac{1}{24 \times 10^5}(-8Q_1 + 3Q_2 + 3Q_3) \quad \text{in.}$$

$$\Delta^{(BE)} = \frac{1}{12 \times 10^5}(-4Q_1 - 3Q_2 + 3Q_3) \quad \text{in.}$$

$$\Delta^{(CE)} = \frac{1}{12 \times 10^5}(4Q_1 - 3Q_2 + 3Q_3) \quad \text{in.}$$

$$\Delta^{(DE)} = \frac{1}{24 \times 10^5}(8Q_1 + 3Q_2 + 3Q_3) \quad \text{in.}$$

(6.2.13)

The axial force in each bar is then found by computing the strain Δ/L, the stress $E\,\Delta/L$, and finally the force $EA\,\Delta/L$. In this way we obtain

$$P_1 = -Q_1 + \tfrac{3}{8}Q_2 + \tfrac{3}{8}Q_3$$
$$P_2 = -\tfrac{1}{2}Q_1 - \tfrac{3}{8}Q_2 + \tfrac{3}{8}Q_3$$
$$P_3 = \tfrac{1}{2}Q_1 - \tfrac{3}{8}Q_2 + \tfrac{3}{8}Q_3$$
$$P_4 = Q_1 + \tfrac{3}{8}Q_2 + \tfrac{3}{8}Q_3$$

(6.2.14)

The final results can and should be checked by verifying that the forces P_i satisfy the equations of equilibrium. The student may show that the joint E (Fig. 6.1b) is indeed in equilibrium when the axial forces are given by Eqs. (6.2.14).

////

Obtaining an expression for the strain energy $U(q_i)$ in a structure is not always as simple as it was in the preceding problem. Consider for example the plane truss shown in Fig. 6.3. We know that the total strain energy is equal to the sum of the energies stored in the individual bars of the truss. Furthermore we have a formula (6.2.1) for computing the energy in a bar which is subject to an elongation Δ, where Δ represents the displacement of one end of the bar *relative* to the other end. Note for instance that the extension of bar CD depends upon the displacement of joint C relative to D. Although the vertical and horizontal translations of D have been specified (q_1 and q_2) and the vertical translation of C is known (zero), the horizontal translation of C is unknown. Thus, in order to compute U for bar CD, we must first determine the unknown horizontal displacement of C. The general procedure which we shall use for finding unknown joint displacements is outlined below.

Consider a structure in which there are n prescribed generalized displacements q_i ($i = 1, 2, \ldots, n$) and an additional r unknown joint displacements which

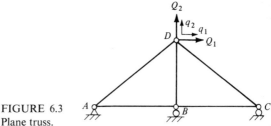

FIGURE 6.3
Plane truss.

we shall call y_i $(i = 1, 2, \ldots, r)$. Since the quantities y_i cannot be expressed in terms of the prescribed displacements q_i from a consideration of kinematics, they are sometimes referred to as *kinematically indeterminate displacements*; r denotes the *degree of kinematic indeterminacy*.

Once each joint displacement or degree of freedom of the structure has been designated by either a symbol q_i or y_i, the strain energy may be written in the form

$$U = U(q_1, \ldots, q_n, y_1, \ldots, y_r) \qquad (6.2.15)$$

Assuming that there is a generalized force Q_i corresponding to each generalized displacement q_i, but that no other forces are applied to the structure (in particular there are no forces acting at those joints designated by a displacement y_i), then the external potential (5.7.13) is

$$V_E = -\sum_{i=1}^{n} Q_i q_i \qquad (6.2.16)$$

The pmpe (5.7.11) requires that the total potential energy $\Pi = U + V_E$ be a minimum in which case

$$\delta\Pi = \frac{\partial U}{\partial q_1}\delta q_1 + \cdots + \frac{\partial U}{\partial q_n}\delta q_n + \frac{\partial U}{\partial y_1}\delta y_1 + \cdots + \frac{\partial U}{\partial y_r}\delta y_r$$
$$- Q_1 \delta q_1 - \cdots - Q_n \delta q_n = 0 \qquad (6.2.17)$$

or $\left(\dfrac{\partial U}{\partial q_1} - Q_1\right)\delta q_1 + \cdots + \left(\dfrac{\partial U}{\partial q_n} - Q_n\right)\delta q_n + \dfrac{\partial U}{\partial y_1}\delta y_1 + \cdots$

$$+ \frac{\partial U}{\partial y_r}\delta y_r = 0 \qquad (6.2.18)$$

Since the variations are arbitrary, Eq. (6.2.18) yields

$$\frac{\partial U}{\partial q_i} = Q_i \qquad i = 1, 2, \ldots, n \qquad (6.2.19)$$

and

$$\frac{\partial U}{\partial y_i} = 0 \qquad i = 1, 2, \ldots, r \qquad (6.2.20)$$

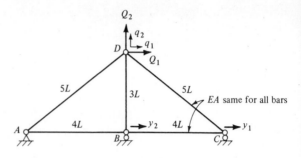

FIGURE 6.4
Kinematically indeterminate truss.

We recognize Eq. (6.2.19) as being Castigliano's first theorem. Equation (6.2.20) represents a set of r simultaneous equations which can be solved to obtain the kinematically indeterminate displacements y_i. Once these displacements have been determined, they can be eliminated from the strain energy expression (6.2.15). The unknown forces are then obtained using Castigliano's theorem (6.2.19).

Example 6.2   Displacement method for a plane truss   We shall now use the procedure outlined above to compute the forces Q_1 and Q_2 which produce the displacements q_1 and q_2 in the truss shown in Fig. 6.4. Note that there are two unspecified joint displacements in the structure, namely, the horizontal translations of joints C and B.[1] We denote these unknowns by y_1 and y_2, as shown. From kinematic considerations we can express the elongation Δ of any member of the truss in terms of the four degrees of freedom q_1, q_2, y_1, and y_2; that is,

$$\Delta^{(AD)} = \tfrac{4}{5}q_1 + \tfrac{3}{5}q_2$$

$$\Delta^{(CD)} = -\tfrac{4}{5}q_1 + \tfrac{3}{5}q_2 + \tfrac{4}{5}y_1$$

$$\Delta^{(BD)} = q_2 \qquad\qquad (6.2.21)$$

$$\Delta^{(AB)} = y_2$$

$$\Delta^{(BC)} = y_1 - y_2$$

[1]The joint displacements in a structure are treated as generalized displacements, and as such they include rotations as well as translations. However, if the joints are pin-connected, the rotation of a joint will not produce any deformation in the structure. Hence, only translations need to be considered in the case of an ideal truss.

Substituting the appropriate values of Δ, E, A, and L into the strain energy expression (6.2.1) gives

$$U^{(AD)} = \frac{1}{2} \frac{EA}{5L} \left(\frac{4}{5} q_1 + \frac{3}{5} q_2 \right)^2$$

$$U^{(CD)} = \frac{1}{2} \frac{EA}{5L} \left(-\frac{4}{5} q_1 + \frac{3}{5} q_2 + \frac{4}{5} y_1 \right)^2$$

$$U^{(BD)} = \frac{1}{2} \frac{EA}{3L} q_2^2 \tag{6.2.22}$$

$$U^{(AB)} = \frac{1}{2} \frac{EA}{4L} y_2^2$$

$$U^{(BC)} = \frac{1}{2} \frac{EA}{4L} (y_1 - y_2)^2$$

Summing the above energies yields the total strain energy

$$U = \frac{EA}{3000L} (384q_1^2 + 716q_2^2 - 384q_1 y_1 + 288q_2 y_1$$

$$+ 567y_1^2 + 750y_2^2 - 750y_1 y_2) \tag{6.2.23}$$

Utilizing Eq. (6.2.20), we obtain

$$\frac{\partial U}{\partial y_1} = \frac{EA}{3000L} (-384q_1 + 288q_2 + 1134y_1 - 750y_2) = 0$$

$$\tag{6.2.24}$$

$$\frac{\partial U}{\partial y_2} = \frac{EA}{3000L} (-750y_1 + 1500y_2) = 0$$

Solving Eq. (6.2.24) for the kinematically indeterminate displacements gives

$$y_1 = 0.506q_1 - 0.379q_2$$
$$\tag{6.2.25}$$
$$y_2 = 0.253q_1 - 0.190q_2$$

Equations (6.2.25) may be used to eliminate y_1 and y_2 from the strain energy expression (6.2.23). Application of Castigliano's theorem (6.2.19) then yields the desired forces

$$Q_1 = (0.191q_1 + 0.049q_2) \frac{EA}{L}$$

$$\tag{6.2.26}$$

$$Q_2 = (0.049q_1 + 0.441q_2) \frac{EA}{L}$$

////

6.3   FORCE METHOD

In the force method we choose generalized forces rather than displacements as the unknown quantities. Let us begin by considering a structure for which there is a total of m unknown forces but only s independent equations of statics. We say that the structure is *indeterminate to the Rth degree*, where R is the difference between m and s. That is,

$$R = m - s \qquad (6.3.1)$$

R	m	s
Degree of static indeterminacy	Number of unknown generalized forces, including both internal forces and external reactions	Number of independent equations of equilibrium

The number of independent equations s is generally rather easy to determine if the student remembers the basic principles he learned in statics. For example, he should recall that there are six scalar equations of statics for any three-dimensional structure or for any portion of a structure; these include three force-equilibrium equations and three moment-equilibrium equations. For a structure subject to a coplanar system of forces, there are only three nontrivial equations, two of which represent force equilibrium in the plane, and the third represents moment equilibrium about an axis perpendicular to the plane. For a concurrent force system, the number of equations is reduced still further since the moment equations are not independent of the force equations.

We may think of a structure which is indeterminate to the Rth degree as being one for which R of the unknown forces are not required for equilibrium. Let us designate the extra or *redundant* forces by Y_i $(i = 1, 2, \ldots, R)$.† The structure which would remain if the forces Y_i were removed is called the *base structure*. The decision as to which of the unknown forces shall be considered redundant is an arbitrary one. We may designate any R forces as the redundants,

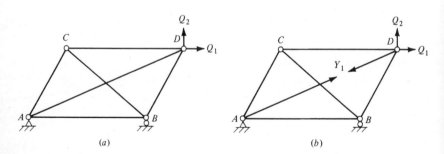

FIGURE 6.5
(*a*) Statically indeterminate plane truss; (*b*) a base structure.

†The number of redundants Y_i is, in general, *not* equal to the number of kinematically indeterminate displacements y_i used in the displacement method.

the only restriction being that the base structure must remain in equilibrium. (For example, we may not choose as redundant a force which is required to prevent rigid body motion of the base structure.)

Let us now investigate the degree of indeterminacy of several different types of structures. First consider the plane truss shown in Fig. 6.5a. The unknown forces in this structure include the six axial bar forces, two external reactive forces at hinge A, plus a vertical reaction at the roller B, or a total of $m = 9$ unknowns. We may write two equations of force equilibrium at each of the four joints so we have $s = 8$ independent equations of statics. Thus the degree of indeterminacy is $R = 9 - 8 = 1$. Any one of the internal axial forces may be chosen as the redundant. We may not, however, designate one of the external reactions as the redundant since all three reactions are needed to prevent rigid body motions of the structure. A suitable base structure for the truss is given in Fig. 6.5b.

Now let us consider the space truss shown in Fig. 6.6a. Since each of the six external reactions (at $A, B, \ldots, F$) is equal to the axial force in the corresponding

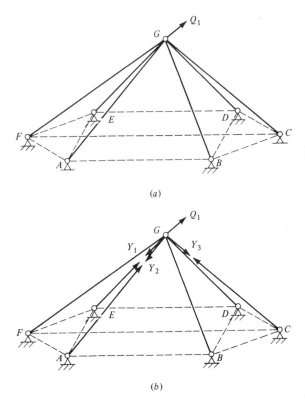

(a)

(b)

FIGURE 6.6
(a) Statically indeterminate space truss; (b) a base structure.

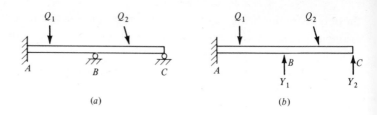

FIGURE 6.7
(a) Statically indeterminate beam subject to a coplanar force system; (b) a base structure.

bar $(AG, BG, \ldots, FG)$, there are just $m = 6$ unknown forces. Because the force system is concurrent at joint G, we have $s = 3$ equations of statics, so that $R = 6 - 3 = 3$. Arbitrarily designating three of the internal forces as redundant, we arrive at the base structure given in Fig. 6.6b.

Now consider the flexure of a beam subject to a coplanar system of forces and supported as shown in Fig. 6.7a. The unknowns include a couple and a horizontal and vertical force at A, and vertical reactions at B and C. Since there are three independent equations of equilibrium for a coplanar force system, the structure is statically indeterminate to the $R = 5 - 3 = 2$d degree. With the exception of the horizontal reaction at A (which is the only force capable of preventing a rigid body motion in the horizontal direction), any two of the unknown reactive forces may be selected as redundant. One possibility is shown in Fig. 6.7b.

For a beam which is clamped at both ends and is subject to a general three-dimensional system of applied loads (Fig. 6.8a), there are twelve unknowns: the three unknown force components and three moment components at each end of the beam. Since there are six equations of equilibrium, the degree of indeterminacy is $R = 12 - 6 = 6$. Figure 6.8b shows a suitable base structure.

The circular ring shown in Fig. 6.9a is in equilibrium under a system of self-equilibrating applied loads. If we consider the entire ring as a free body, the

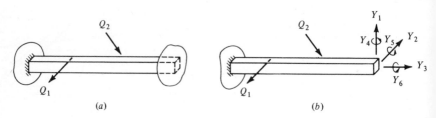

FIGURE 6.8
(a) Statically indeterminate beam subject to a three-dimensional force system;
(b) a base structure.

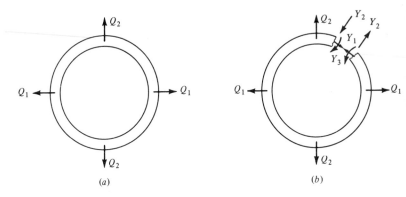

FIGURE 6.9
(a) Statically indeterminate ring; (b) a base structure.

equations of statics are trivial. Consider instead the equilibrium of a segment of the ring of arbitrary arc length. Each end of the segment is subject to an unknown bending moment, an axial force, and a shear force. The total number of internal forces applied to a segment of the ring is therefore six. Since there are only three equations of equilibrium for a coplanar force system, the degree of indeterminacy is $R = 6 - 3 = 3$. Figure 6.9b shows an appropriate base structure in which the unknown forces at an arbitrary cross section have been chosen as redundant. Note that the internal forces at any other cross section of the base structure can be obtained from statics if the redundant forces are known.

Next consider the frame shown in Fig. 6.10a. To find the degree of indeterminacy of this type of structure, a somewhat different approach is convenient.

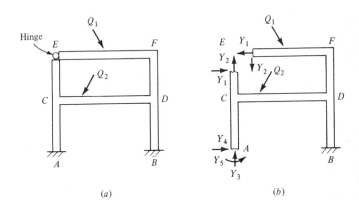

FIGURE 6.10
(b) Statically indeterminate frame; (b) a base structure.

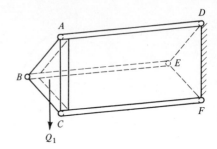

FIGURE 6.11
Stiffened-panel structure.

Let us try to determine the number of generalized forces which must be removed in order to reduce the indeterminate structure to a statically determinate one. The removed forces then constitute a suitable system of redundants, and the structure which remains represents the corresponding base structure. For example, if we remove the two internal forces at E and the three external reactions at A, the resulting structure (Fig. 6.10b) is statically determinate. (It is noted that the bending moment at a hinge is necessarily zero, and hence there are only two rather than three internal forces at E.) We conclude therefore that the degree of indeterminacy is $R = 5$.

A stiffened structure consisting of a rigid vertical bulkhead (ABC), three horizontal flanges, and three shear webs is shown in Fig. 6.11. If we make the same assumptions as those introduced in Example 5.16, then there are a total of six unknown generalized forces: the three forces at the built-in ends of the flanges and the three shear flows in the webs. By considering the entire structure as a free body, we can write six equations of equilibrium, and hence the system is statically determinate.

Similarly, the unknown forces in a structure stiffened by four flanges (Fig. 6.12a) consist of four axial forces and four shear flows. Let us now imagine that one of the stiffeners (AE) were removed. The shear flows in the two adjacent webs ($AEHD$ and $AEFB$) would then be equal, and the six equations of equilibrium would suffice for a determination of all the unknowns in the resulting structure. The system is therefore statically indeterminate to the first degree, and Fig. 6.12b shows an appropriate base structure. We may conclude that, in general, a single-cell structure having n stiffeners is indeterminate to the degree $n - 3$.

Let us now assume that the degree of indeterminacy R of a particular structure has been determined and that an appropriate set of unknown forces has been designated as redundant. The remaining $m - R = s$ unknowns may be expressed in terms of the redundants Y_i ($i = 1, 2, \ldots, R$) and the loads applied to the structure Q_i ($i = 1, 2, \ldots, n$) by solving the s equations of statics. It is

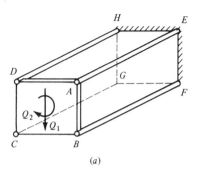 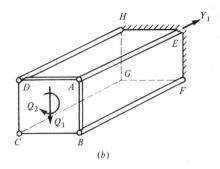

FIGURE 6.12
(a) Statically indeterminate stiffened-panel structure; (b) a base structure.

then possible to write an expression for the complementary strain energy U^* in terms of the Y_i and Q_i forces, that is,

$$U^* = U^*(Q_1, \ldots, Q_n, Y_1, \ldots, Y_R) \qquad (6.3.2)$$

The external complementary energy (5.8.9) is

$$V_E^* = -\sum_{i=1}^n q_i Q_i \qquad (6.3.3)$$

The pmce (5.8.7) requires that the total complementary energy $\Pi^* = U^* + V_E^*$ be a minimum. Thus

$$\delta\Pi^* = \frac{\partial U^*}{\partial Q_1}\delta Q_1 + \cdots + \frac{\partial U^*}{\partial Q_n}\delta Q_n + \frac{\partial U^*}{\partial Y_1}\delta Y_1$$

$$+ \cdots + \frac{\partial U^*}{\partial Y_R}\delta Y_R - q_1\,\delta Q_1 - \cdots - q_n\,\delta Q_n = 0 \qquad (6.3.4)$$

or $\left(\dfrac{\partial U^*}{\partial Q_1} - q_1\right)\delta Q_1 + \cdots + \left(\dfrac{\partial U^*}{\partial Q_n} - q_n\right)\delta Q_n + \dfrac{\partial U^*}{\partial Y_1}\delta Y_1$

$$+ \cdots + \frac{\partial U^*}{\partial Y_R}\delta Y_R = 0 \qquad (6.3.5)$$

Since the variations are arbitrary, Eq. (6.3.5) requires that

$$\frac{\partial U^*}{\partial Q_i} = q_i \qquad i = 1, 2, \ldots, n \qquad (6.3.6)$$

and

$$\frac{\partial U^*}{\partial Y_i} = 0 \qquad i = 1, 2, \ldots, R \qquad (6.3.7)$$

For a linearly elastic structure we recall that the complementary strain energy U^* is equal to the strain energy U, in which case we obtain

$$\frac{\partial U}{\partial Q_i} = q_i \qquad i = 1, 2, \ldots, n \qquad (6.3.8)$$

and

$$\frac{\partial U}{\partial Y_i} = 0 \qquad i = 1, 2, \ldots, R \qquad (6.3.9)$$

Equation (6.3.8) expresses Castigliano's second theorem, and (6.3.9) is a mathematical statement of *Castigliano's theorem of least work*.

> *Castigliano's theorem of least work* If the strain energy U in a linearly elastic, statically indeterminate structure is expressed as a function of the applied loads Q_i and redundants Y_i then the first partial derivative of U with respect to each redundant Y_i is equal to zero.

In physical terms the theorem states that the values of the redundant forces must be such that the strain energy stored in the structure is a minimum.

It should be mentioned here that in using expression (6.3.3) for the potential function V_E^*, we have, in effect, assumed that the redundant forces do no work during the deformation process. If, however, one of the redundants represents a support reaction and the support moves during the deformation, then this redundant does work. Hence, if we wish to account for support movements, Eq. (6.3.3) would have to be written as

$$V_E^* = -\sum_{i=1}^{n} q_i Q_i - \sum_{i=1}^{R} \Delta_i Y_i \qquad (6.3.10)$$

where Δ_i represent the prescribed generalized displacements associated with the redundant generalized forces Y_i. The student may verify that, in this case, the redundants can be found by solving the equations

$$\frac{\partial U^*}{\partial Y_i} = \Delta_i \qquad i = 1, 2, \ldots, R \qquad (6.3.11)$$

or, if the structure is linearly elastic

$$\frac{\partial U}{\partial Y_i} = \Delta_i \qquad i = 1, 2, \ldots, R \qquad (6.3.12)$$

A few examples will illustrate how the above approach can be used to compute unknown forces and displacements in statically indeterminate structures.

Example 6.3   Force method for a statically indeterminate space truss

Let us apply the force method to the truss shown in Fig. 6.13a; the same structure was analyzed using the displacement method in Example 6.1.

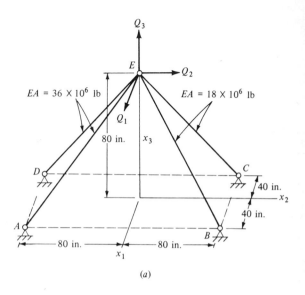

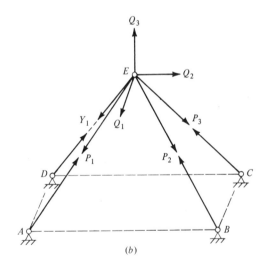

FIGURE 6.13
(*a*) Statically indeterminate space truss; (*b*) a base structure.

The number of unknown axial forces (four) exceeds the number of independent equations of statics for a concurrent force system (three). The structure is therefore indeterminate to the first degree, and we shall arbitrarily choose the axial force in bar DE as the redundant Y_1. The remaining bar forces, designated as P_1, P_2, and P_3, can be expressed as functions of the applied loads Q_1, Q_2, Q_3, and the redundant Y_1 by solving the equations of equilibrium for the base structure (Fig. 6.13b). Considering joint E as a free body and equating the resultant force components in the x_1, x_2, x_3 directions to zero gives

$$\frac{40}{120}P_1 + \frac{40}{120}P_2 - \frac{40}{120}P_3 - \frac{40}{120}Y_1 + Q_1 = 0$$

$$-\frac{80}{120}P_1 + \frac{80}{120}P_2 + \frac{80}{120}P_3 - \frac{80}{120}Y_1 + Q_2 = 0 \qquad (6.3.13)$$

$$-\frac{80}{120}P_1 - \frac{80}{120}P_2 - \frac{80}{120}P_3 - \frac{80}{120}Y_1 + Q_3 = 0$$

Solving these equations for the internal forces P_i yields

$$P_1 = \tfrac{3}{4}Q_2 + \tfrac{3}{4}Q_3 - Y_1$$

$$P_2 = -\tfrac{3}{2}Q_1 - \tfrac{3}{4}Q_2 + Y_1 \qquad (6.3.14)$$

$$P_3 = \tfrac{3}{2}Q_1 + \tfrac{3}{4}Q_3 - Y_1$$

Next the total strain energy U in the structure is computed by summing the energy stored in the individual components of the truss. For a bar subject to an axial force P_i, the strain energy is

$$U = \frac{L}{2EA}P_i^{\,2} \qquad (6.3.15)$$

If the units of the forces Q_1, Q_2, Q_3, and Y_1 are pounds, then for bar AE we obtain

$$U^{(AE)} = \frac{120}{2 \times 36 \times 10^6}\left(\frac{3}{4}Q_2 + \frac{3}{4}Q_3 - Y_1\right)^2$$

$$= \frac{1}{96 \times 10^5}(9Q_2^{\,2} + 9Q_3^{\,2} + 18Q_2Q_3$$

$$- 24Q_2Y_1 - 24Q_3Y_1 + 16Y_1^{\,2}) \qquad \text{in.-lb} \qquad (6.3.16)$$

and similarly for the other bars

$$U^{(BE)} = \frac{1}{48 \times 10^5}(36Q_1^{\,2} + 9Q_2^{\,2} + 36Q_1Q_2$$

$$- 48Q_1Y_1 - 24Q_2Y_1 + 16Y_1^{\,2}) \qquad \text{in.-lb}$$

$$U^{(CE)} = \frac{1}{48 \times 10^5}(36Q_1^{\,2} + 9Q_3^{\,2} + 36Q_1Q_3 \qquad\qquad (6.3.17)$$

$$- 48Q_1Y_1 - 24Q_3Y_1 + 16Y_1^{\,2}) \qquad \text{in.-lb}$$

$$U^{(DE)} = \frac{1}{6 \times 10^5}Y_1^{\,2} \qquad \text{in.-lb}$$

The total strain energy is then given by

$$U = \frac{1}{32 \times 10^5}(48Q_1{}^2 + 9Q_2{}^2 + 9Q_3{}^2 + 24Q_1Q_2 + 24Q_1Q_3$$
$$+ 6Q_2Q_3 - 64Q_1Y_1 - 24Q_2Y_1 - 24Q_3Y_1 + 32Y_1{}^2) \qquad \text{in.-lb} \qquad (6.3.18)$$

Applying the theorem of least work (6.3.9) gives

$$\frac{\partial U}{\partial Y_1} = \frac{1}{32 \times 10^5}(-64Q_1 - 24Q_2 - 24Q_3 + 64Y_1) = 0 \qquad (6.3.19)$$

and solving for Y_1 we obtain

$$Y_1 = Q_1 + \tfrac{3}{8}Q_2 + \tfrac{3}{8}Q_3 \qquad (6.3.20)$$

Substituting Eq. (6.3.20) into (6.3.14) gives the remaining axial forces

$$P_1 = -Q_1 + \tfrac{3}{8}Q_2 + \tfrac{3}{8}Q_3$$
$$P_2 = -\tfrac{1}{2}Q_1 - \tfrac{3}{8}Q_2 + \tfrac{3}{8}Q_3 \qquad (6.3.21)$$
$$P_3 = \tfrac{1}{2}Q_1 - \tfrac{3}{8}Q_2 + \tfrac{3}{8}Q_3$$

Hence, we have succeeded in determining the force in each bar of the indeterminate truss. Naturally these results are the same as those which we obtained using the displacement method.

The displacements q_i corresponding to the applied forces Q_i may be computed using Castigliano's second theorem. First we express the strain energy U in terms of just the applied forces by substituting Eq. (6.3.20) into (6.3.18). The resulting expression may be simplified to read

$$U = \frac{1}{32 \times 10^5}\left(16Q_1{}^2 + \frac{9}{2}Q_2{}^2 + \frac{9}{2}Q_3{}^2 - 3Q_2Q_3\right) \qquad \text{in.-lb} \qquad (6.3.22)$$

Castigliano's theorem then yields

$$q_1 = \frac{\partial U}{\partial Q_1} = \frac{1}{10^5}Q_1 \qquad \text{in.}$$

$$q_2 = \frac{\partial U}{\partial Q_2} = \frac{3}{32 \times 10^5}(3Q_2 - Q_3) \qquad \text{in.} \qquad (6.3.23)$$

$$q_3 = \frac{\partial U}{\partial Q_3} = \frac{3}{32 \times 10^5}(-Q_2 + 3Q_3) \qquad \text{in.}$$

in agreement with our previous result (6.2.12).   ////

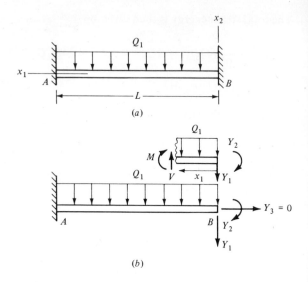

FI‹ URE 6.14
(a) Statically indeterminate beam; (b) a base structure.

Example 6.4   Force method for the bending stress in a statically indeterminate beam

We wish to compute the bending stress distribution in a beam which is clamped at both ends and subject to a uniformly distributed load Q_1 (Fig. 6.14a). If the beam is linearly elastic, the bending stress σ_{11} is given by

$$\sigma_{11} = -\frac{Mx_2}{I} \quad (6.3.24)$$

In order to find the bending moment M, we must first compute the support reactions at the ends of the beam. Recognizing that the beam is statically indeterminate to the third degree (there are three unknown reactive forces at each end of the beam, but only three independent equations of equilibrium), we arbitrarily designate the reactions at end B as the redundants. The corresponding base structure is shown in Fig. 6.14b. It is clear that the transverse load Q_1 will not induce an axial force in the beam (assuming that the deformations are small), and we conclude therefore that the reaction Y_3 is zero. The strain energy U produced by the redundants Y_1 and Y_2 and the applied force Q_1 is (see Prob. 5.6)

$$U = \frac{1}{2EI}\left(\frac{L^5}{20} Q_1{}^2 + \frac{L^4}{4} Q_1 Y_1 + \frac{L^3}{3} Q_1 Y_2 + \frac{L^3}{3} Y_1{}^2 + L^2 Y_1 Y_2 + L Y_2{}^2\right) \quad (6.3.25)$$

Application of Castigliano's theorem of least work (6.3.9) yields

$$\frac{\partial U}{\partial Y_1} = \frac{1}{2EI}\left(\frac{L^4}{4}Q_1 + \frac{2L^3}{3}Y_1 + L^2 Y_2\right) = 0$$

$$\frac{\partial U}{\partial Y_2} = \frac{1}{2EI}\left(\frac{L^3}{3}Q_1 + L^2 Y_1 + 2L Y_2\right) = 0 \qquad (6.3.26)$$

Solving Eqs. (6.3.26) for the redundant forces gives

$$Y_1 = -\frac{L}{2}Q_1$$

$$Y_2 = \frac{L^2}{12}Q_1 \qquad (6.3.27)$$

The value of the reactive force $Y_1 = -LQ_1/2$ could have been anticipated from symmetry arguments.

The bending moment at a distance x_1 along the beam is (see Fig. 6.14b)

$$M = -\frac{x_1^2}{2}Q_1 - x_1 Y_1 - Y_2 = \left(-\frac{x_1^2}{2} + \frac{x_1 L}{2} - \frac{L^2}{12}\right)Q_1 \qquad (6.3.28)$$

Substituting Eq. (6.3.28) into (6.3.24) gives the required bending stress σ_{11}.   ////

Example 6.5   Force method for the determination of a displacement in a statically indeterminate beam   We now shall use Castigliano's second and least work theorems to find the rotation Δ at the right end of the beam shown in Fig. 6.15a. In order to compute the rotation, we must differentiate the strain energy U with respect to the corresponding generalized force. Hence, we shall imagine that a couple Q_2 acts at point B, and after the rotation has been found, we shall set $Q_2 = 0$.

Noting that the degree of indeterminacy of the structure is 1, we arbitrarily select the vertical reaction at B as the redundant. The bending moment at a distance x_1 from the right end of the beam is then (see Fig. 6.15b)

$$M = -\frac{x_1^2}{2}Q_1 + Q_2 + x_1 Y_1 \qquad (6.3.29)$$

We recall that the strain energy U in a symmetric beam is

$$U = \int_0^L \frac{M^2}{2EI}\,dx_1 \qquad (6.3.30)$$

The theorem of least work (6.3.9) then gives

$$\frac{\partial U}{\partial Y_1} = \int_0^L \frac{M}{EI}\frac{\partial M}{\partial Y_1}\,dx_1 = 0 \qquad (6.3.31)$$

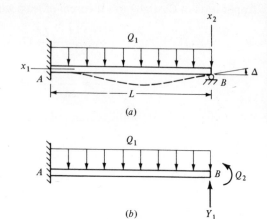

FIGURE 6.15
(a) Statically indeterminate beam with an unknown rotation Δ; (b) base structure with a fictitious couple Q_2.

Substituting Eq. (6.3.29) into (6.3.31) and integrating yields

$$\frac{1}{EI}\left(-\frac{L^4}{8}Q_1 + \frac{L^2}{2}Q_2 + \frac{L^3}{3}Y_1\right) = 0 \quad (6.3.32)$$

and therefore

$$Y_1 = \frac{3L}{8}Q_1 - \frac{3}{2L}Q_2 \quad (6.3.33)$$

Castigliano's second theorem may now be used to compute the rotation q_2

$$q_2 = \frac{\partial U}{\partial Q_2} = \int_0^L \frac{M}{EI}\frac{\partial M}{\partial Q_2}\,dx_1 \quad (6.3.34)$$

By differentiating Eq. (6.3.29) with respect to Q_2, we obtain

$$\frac{\partial M}{\partial Q_2} = 1 + x_1 \frac{\partial Y_1}{\partial Q_2} = 1 - \frac{3x_1}{2L} \quad (6.3.35)$$

Substituting Eqs. (6.3.29), (6.3.33), and (6.3.35) into (6.3.34) and performing the integration gives

$$q_2 = \frac{Q_1 L^3}{48EI} + \frac{Q_2 L}{4EI} \quad (6.3.36)$$

Finally, setting $Q_2 = 0$ yields the desired rotation

$$\Delta = q_2\Big|_{Q_2 = 0} = \frac{Q_1 L^3}{48EI} \quad (6.3.37)$$

////

PROBLEMS

6.1   Use the displacement method to find the axial force in each bar of the truss shown. Assume that all bars have the same longitudinal stiffness EA.

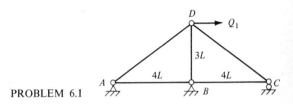

PROBLEM 6.1

6.2   Solve Prob. 6.1 using the force method.

6.3   Find the axial force in each member of the truss shown. Also compute the vertical displacement of joint B. All bars have the same value of EA.

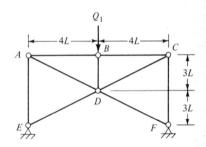

PROBLEM 6.3

6.4   A uniform beam is subjected to the couple Q_1 shown. Use the force method to find (*a*) the external reactions at B, C, and D; (*b*) the maximum bending moment in the beam; and (*c*) the rotation at A.

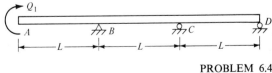

PROBLEM 6.4

6.5   Repeat Prob. 6.4 assuming that the support point B moves upward a distance $\Delta_B = 0.5Q_1L^2/EI$ during application of the load Q_1.

6.6   Compute the maximum displacement in a beam which is clamped at both ends and subject to the uniformly distributed force Q_1 shown.

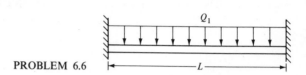

PROBLEM 6.6

6.7   Use the Rayleigh-Ritz method to obtain an approximate value for the maximum displacement of the beam in Prob. 6.6.

6.8   Compute the maximum bending stress in the beam AC shown.

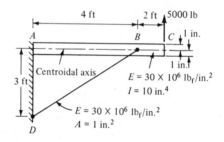

PROBLEM 6.8

6.9   A frame of constant cross section is acted upon by a uniform lateral load Q_1 as shown. Determine the location and the magnitude of the maximum bending moment in the frame. Neglect the strain energy due to axial and shear forces.

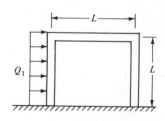

PROBLEM 6.9

6.10–6.11   Find the displacement q_1 for the arch shown.   Neglect strain energy associated with the shear and axial forces, and assume that the cross-sectional dimensions of the arch are small compared with the radius R.

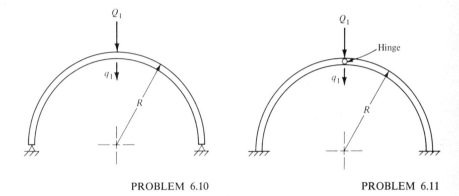

PROBLEM 6.10   PROBLEM 6.11

6.12   Compute the maximum bending stress in a circular ring subject to diametral forces.   Neglect strain energy associated with the shear and axial forces, and assume that the diameter d of the ring's cross section is small compared with R.

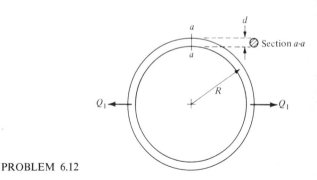

PROBLEM 6.12

6.13   Compute the displacement q_1 of the stiffened cantilever beam shown on page 128. Make the same assumptions as those made in Example 5.16, and let $t = 4EAh/GL^2$, where E and G have the same values for all members.

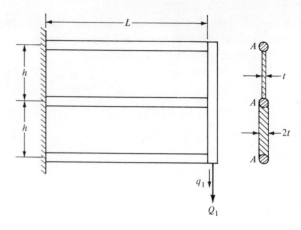

PROBLEM 6.13

REFERENCES

For an introductory account of statically indeterminate structures:

6.1   WILBUR, J. B., and C. H. NORRIS: " Elementary Structural Analysis," McGraw-Hill, New York, 1948.

6.2   TIMOSHENKO, S. P., and D. H. YOUNG: " Theory of Structures," 2d ed., McGraw-Hill, New York, 1965.

For a more advanced account of indeterminate structures:

6.3   HOFF, N. J.: " The Analysis of Structures," Wiley, New York, 1956.

References to several modern treatments of statically indeterminate structures which emphasize matrix methods are given at the end of Chap. 7.

7

MATRIX METHODS

7.1 INTRODUCTION

We have seen that the deformation and stress distributions in a statically indeterminate structure can be computed using either of the dual energy approaches. In the displacement approach the unknown quantities consist of a set of arbitrary generalized displacements which satisfy prescribed constraint conditions; the actual state of deformation is found by applying the pmpe. In the force method a system of arbitrary generalized forces which satisfy equilibrium conditions are taken as the unknowns; the correct state of stress is determined by enforcing the pmce. Although the two approaches differ in their physical concepts, their mathematical formulations are very similar. The governing equations in either case are linear, providing the structural behavior is linearly elastic and the deformations are small. It is convenient to formulate the analysis of such a structure in matrix notation. The mathematical calculations then involve simple matrix operations, which can be carried out on a digital computer.[1]

---

[1] Matrix methods, in conjunction with digital computers, have been used to analyze extremely complicated structures. The techniques which have been employed differ in detail, and the literature in this area is large. Several references on the subject are listed at the end of this chapter.

In this chapter we shall examine the so-called *matrix displacement* and the *matrix force methods*, which represent reformulations of the methods discussed in Chap. 6. Either approach is suitable for the analysis of a relatively simple structure, and the choice is usually a matter of personal taste. For more complicated indeterminate systems, the selection of one method over the other may be more critical. It is generally advisable to use that approach which involves the fewer number of unknowns. We recall that the degree of kinematic indeterminacy of a structure is, in general, different from the degree of static indeterminacy. The indeterminate quantities, i.e., the displacements y_i $(i = 1, 2, \ldots, r)$ or forces $Y_i (i = 1, 2, \ldots, R)$, must be computed by solving the corresponding number of simultaneous equations. Hence, if there are fewer indeterminate displacements than redundant forces $(r < R)$, the displacement method is more efficient, and vice versa.

Another factor to be considered in selecting a method is the type of quantities which are to be computed. For example, if stiffness influence coefficients are the quantities of primary interest, these can be obtained in a more direct way using the displacement method; on the other hand, the flexibility coefficients are computed more easily using the force approach.

We are now ready to reorganize some of our previous results in matrix form. If the student is not familiar with the elements of matrix algebra, he may refer to Appendix B for a brief introduction to the subject.

7.2   FLEXIBILITY AND STIFFNESS MATRICES

In a linearly elastic structure the generalized displacements q_i $(i = 1, 2, \ldots, n)$ are proportional to the corresponding generalized forces Q_i. Following the notation introduced in Chap. 5, we write

$$\{q\} = [c]\{Q\} \qquad (7.2.1)$$

where $\{q\} = \{q_1, q_2, \ldots, q_n\}$ and $\{Q\} = \{Q_1, Q_2, \ldots, Q_n\}$ are column matrices of the displacements and forces, respectively, and $[c]$ is a square matrix called the *flexibility matrix*. The elements of $[c]$ are the flexibility influence coefficients c_{ij} discussed earlier; that is, c_{ij} represents the displacement at point i in the direction of Q_i produced by a unit force Q_j.

Likewise we may express the generalized forces in terms of the corresponding generalized displacements as

$$\{Q\} = [k]\{q\} \qquad (7.2.2)$$

The coefficient k_{ij} of the *stiffness matrix* $[k]$ represents the force required at i for a unit displacement at j, when all the other generalized displacements are zero. Comparing Eqs. (7.2.1) and (7.2.2) we see that the stiffness matrix is equal to the inverse of the flexibility matrix, or

$$[k] = [c]^{-1} \qquad (7.2.3)$$

Recalling Maxwell's reciprocal theorem (5.6.9) which states that $c_{ij} = c_{ji}$, we further note that the flexibility matrix is symmetric. Thus

$$[c]^T = [c] \qquad (7.2.4)$$

where $[c]^T$ denotes the transpose of $[c]$. It follows that

$$[k]^T = [k] \qquad (7.2.5)$$

We also recall from Chap. 5 that the strain energy U stored in a linearly elastic structure is equal to the work done by the externally applied loads during the deformation process, so that

$$U = W_E = \tfrac{1}{2} \sum_{i=1}^{n} Q_i q_i \qquad (7.2.6)$$

Equation (7.2.6) may be written in matrix notation as

$$U = \tfrac{1}{2}\{Q\}^T\{q\} \qquad (7.2.7)$$

Substituting Eq. (7.2.1) into (7.2.7) yields

$$U = \tfrac{1}{2}\{Q\}^T[c]\{Q\} \qquad (7.2.8)$$

Alternatively, we may write

$$U = \tfrac{1}{2}\{q\}^T\{Q\}$$
$$= \tfrac{1}{2}\{q\}^T[k]\{q\} \qquad (7.2.9)$$

Hence if either the flexibility or the stiffness matrix is known, the strain energy can be found easily, and vice versa.

Example 7.1   Flexibility matrix for a beam element   We wish to determine the flexibility matrix for the cantilevered beam loaded as shown in Fig. 7.1. Using the approach described in Sec. 5.2, the strain energy U is found to be (Prob. 5.6)

$$U = \frac{1}{2EI} \left(\frac{L^5}{20} Q_1^{\,2} + \frac{L^3}{3} Q_2^{\,2} + L Q_3^{\,2} + \frac{L^4}{4} Q_1 Q_2 \right.$$
$$\left. + \frac{L^3}{3} Q_1 Q_3 + L^2 Q_2 Q_3 \right) + \frac{L}{2EA} Q_4^{\,2} \qquad (7.2.10)$$

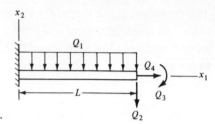

FIGURE 7.1
Cantilevered beam element.

This equation may be written in matrix notation as

$$U = \tfrac{1}{2}[Q_1 Q_2 Q_3 Q_4] \begin{bmatrix} \dfrac{L^5}{20EI} & \dfrac{L^4}{8EI} & \dfrac{L^3}{6EI} & 0 \\[2mm] \dfrac{L^4}{8EI} & \dfrac{L^3}{3EI} & \dfrac{L^2}{2EI} & 0 \\[2mm] \dfrac{L^3}{6EI} & \dfrac{L^2}{2EI} & \dfrac{L}{EI} & 0 \\[2mm] 0 & 0 & 0 & \dfrac{L}{EA} \end{bmatrix} \begin{bmatrix} Q_1 \\[2mm] Q_2 \\[2mm] Q_3 \\[2mm] Q_4 \end{bmatrix} \qquad (7.2.11)$$

The equivalence of these two expressions can be verified by performing the matrix multiplication. The square matrix in Eq. (7.2.11) is the flexibility matrix $[c]$ for the structure, as can be seen by comparing this equation with (7.2.8).   ////

Example 7.2 Stiffness matrix for a beam element As a second example we shall compute the stiffness matrix for a beam element whose ends suffer translations plus rotations. Our procedure will be to consider the deformation of one end of the beam at a time, and then use superposition to obtain the combined effect (Fig. 7.2). First consider the forces Q_i required for equilibrium when the beam's left end undergoes a transverse displacement q_1 and a rotation q_2 while the right end remains stationary (Fig. 7.2b). The strain energy U may be written in terms of the generalized forces Q_1 and Q_2 by substituting the bending moment expression $M = Q_1 x_1 - Q_2$ into the energy formula (5.3.17) for a beam element. Upon integration, this yields

$$U = \frac{1}{2EI}\left(\frac{L^3}{3}Q_1{}^2 - L^2 Q_1 Q_2 + L Q_2{}^2\right) \qquad (7.2.12)$$

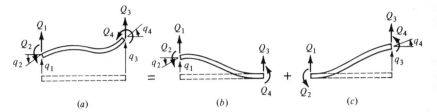

FIGURE 7.2
Deformation of a beam element.

Next, using Castigliano's second theorem (5.9.18) we compute the displacements

$$q_1 = \frac{\partial U}{\partial Q_1} = \frac{L^3}{3EI} Q_1 - \frac{L^2}{2EI} Q_2$$

$$q_2 = \frac{\partial U}{\partial Q_2} = -\frac{L^2}{2EI} Q_1 + \frac{L}{EI} Q_2$$

(7.2.13)

Inverting Eqs. (7.2.13), the forces Q_1 and Q_2 are found to be

$$Q_1 = \frac{12EI}{L^3} q_1 + \frac{6EI}{L^2} q_2$$

$$Q_2 = \frac{6EI}{L^2} q_1 + \frac{4EI}{L} q_2$$

(7.2.14)

If the beam is in equilibrium, the forces at the right end are

$$Q_3 = -Q_1 = -\frac{12EI}{L^3} q_1 - \frac{6EI}{L^2} q_2$$

$$Q_4 = LQ_1 - Q_2 = \frac{6EI}{L^2} q_1 + \frac{2EI}{L} q_2$$

(7.2.15)

Equations (7.2.14) and (7.2.15) give the generalized forces Q_i associated with the generalized displacements q_1 and q_2. Proceeding in a similar manner, the forces produced by the displacements q_3 and q_4 (Fig. 7.2c) are found to be

$$Q_1 = -\frac{12EI}{L^3} q_3 + \frac{6EI}{L^2} q_4$$

$$Q_2 = -\frac{6EI}{L^2} q_3 + \frac{2EI}{L} q_4$$

$$Q_3 = \frac{12EI}{L^3} q_3 - \frac{6EI}{L^2} q_4$$

$$Q_4 = -\frac{6EI}{L^2} q_3 + \frac{4EI}{L} q_4$$

(7.2.16)

By combining the above results, we obtain the matrix equation

$$
\begin{bmatrix} Q_1 \\ Q_2 \\ Q_3 \\ Q_4 \end{bmatrix} =
\begin{bmatrix}
\dfrac{12EI}{L^3} & \dfrac{6EI}{L^2} & -\dfrac{12EI}{L^3} & \dfrac{6EI}{L^2} \\[2ex]
\dfrac{6EI}{L^2} & \dfrac{4EI}{L} & -\dfrac{6EI}{L^2} & \dfrac{2EI}{L} \\[2ex]
-\dfrac{12EI}{L^3} & -\dfrac{6EI}{L^2} & \dfrac{12EI}{L^3} & -\dfrac{6EI}{L^2} \\[2ex]
\dfrac{6EI}{L^2} & \dfrac{2EI}{L} & -\dfrac{6EI}{L^2} & \dfrac{4EI}{L}
\end{bmatrix}
\begin{bmatrix} q_1 \\ q_2 \\ q_3 \\ q_4 \end{bmatrix}
\tag{7.2.17}
$$

The square matrix in this equation is the desired stiffness matrix $[k]$ for the beam element, as can be seen by comparing Eq. (7.2.17) with (7.2.2). Hence the strain energy in the beam is, by virtue of Eq. (7.2.9),

$$
U = \tfrac{1}{2}[q_1 q_2 q_3 q_4]
\begin{bmatrix}
\dfrac{12EI}{L^3} & \dfrac{6EI}{L^2} & -\dfrac{12EI}{L^3} & \dfrac{6EI}{L^2} \\[2ex]
\dfrac{6EI}{L^2} & \dfrac{4EI}{L} & -\dfrac{6EI}{L^2} & \dfrac{2EI}{L} \\[2ex]
-\dfrac{12EI}{L^3} & -\dfrac{6EI}{L^2} & \dfrac{12EI}{L^3} & -\dfrac{6EI}{L^2} \\[2ex]
\dfrac{6EI}{L^2} & \dfrac{2EI}{L} & -\dfrac{6EI}{L^2} & \dfrac{4EI}{L}
\end{bmatrix}
\begin{bmatrix} q_1 \\ q_2 \\ q_3 \\ q_4 \end{bmatrix}
\tag{7.2.18}
$$

////

7.3   MATRIX DISPLACEMENT METHOD

In the analysis of a complicated structure, it is generally necessary to make certain simplifying assumptions with regard to the load-carrying capabilities of the various structural members. Consider, for example, a structure consisting of an assemblage of discrete elements (axially loaded bars, beams, shear webs, etc.) which are either hinged or rigidly joined together. In order to simplify the analysis, we might assume that: the distribution of the tensile stress in each axially loaded bar is uniform; the deformation of each beam element results solely from flexure; and the shear stress is constant throughout each shear web. By making assumptions such as these, we effectively reduce the structure to a simpler idealized model. Naturally the model should retain the essential features of the actual structure. We are not, however, in a position to dwell upon the question of whether or not particular assumptions are justifiable. This depends largely upon the desired accuracy of the final results. At this point we shall merely assume that sufficient approximations have been made so that the strain energy of each individual element of the structure can be readily computed.

It is further assumed that a set of generalized displacements q_i are prescribed at certain points in the structure (Fig. 7.3a). The number and the direction of these prescribed displacements is arbitrary. However, displacements and the corresponding generalized forces will be computed only at those points where a displacement q_i is specified initially; hence the structural engineer must decide in advance which displacements and forces are of primary interest. The chosen set of displacements q_i $(i = 1, 2, \ldots, n)$ are represented by the column matrix $\{q\}$, and the corresponding generalized forces Q_i $(i = 1, 2, \ldots, n)$ are denoted by $\{Q\}$ as before. If displacements q_i are prescribed at some but not all of the joints of the structure, we designate the remaining joint displacements as y_i. These kinematically indeterminate displacements $y_i (i = 1, 2, \ldots, r)$ are represented by the column matrix $\{y\}$.

Next consider the individual members of the structure as separate free bodies. Each element is subject to a system of generalized internal forces arising from the body's contact with the adjacent elements. The internal forces, along with all the applied loads and external reactions, are collectively referred to as *member forces*. These forces are designated as P_i $(i = 1, 2, \ldots, m)$ or $\{P\}$; the corresponding generalized displacements are called *member displacements* and are denoted by v_i $(i = 1, 2, \ldots, m)$ or $\{v\}$. A representation of the structural elements and the associated member displacements is shown in Fig. 7.3b. This assemblage of free bodies shall be referred to as the *unconstrained structure*.

By application of the principles of kinematics, it is possible to compute the member displacements v_i in terms of the prescribed external displacements q_i and the kinematically indeterminate displacements y_i. The resulting *compatibility equations* may be written in matrix notation as

$$\{v\} = [a_q]\{q\} + [a_y]\{y\} \qquad (7.3.1)$$

From Eq. (7.3.1) it is clear that a typical element of the matrix $[a_q]$, say $a_{q_{ij}}$, represents the magnitude of the member displacement v_i which results when the structure experiences a unit displacement q_j, all other joint displacements q_i

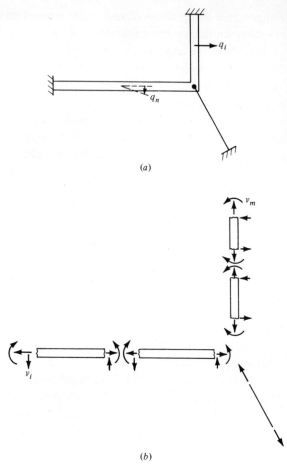

FIGURE 7.3

(a) Prescribed displacements q_i of the indeterminate structure; (b) member displacements v_i of the unconstrained structure.

and x_i being zero. Similarly, an element $a_{y_{ij}}$ of $[a_y]$ represents the value of v_i corresponding to a unit displacement y_j. A procedure for determining the matrices $[a_q]$ and $[a_y]$ for a particular structure will be demonstrated shortly.

The strain energy is computed by summing the energies of the individual members of the unconstrained structure. Since the deformation of each element can be described in terms of its member displacements v_i, the total energy of the structure may be expressed in the form [see Eq. (7.2.9)]

$$U = \tfrac{1}{2}\{v\}^T[K]\{v\} \qquad (7.3.2)$$

where $[K]$ is called the *stiffness matrix of the unconstrained structure*. The strain energy may also be expressed in terms of the prescribed displacements $\{q\}$ and the kinematically indeterminate displacements $\{y\}$. Substituting Eq. (7.3.1) into (7.3.2) and making use of the reversal law for transposition (see Appendix B) gives

$$U = \tfrac{1}{2}(\{q\}^T[a_q]^T + \{y\}^T[a_y]^T)[K]([a_q]\{q\} + [a_y]\{y\}) \qquad (7.3.3)$$

Performing the indicated matrix multiplication yields

$$U = \tfrac{1}{2}\{q\}^T[b_{qq}]\{q\} + \tfrac{1}{2}\{q\}^T[b_{qy}]\{y\}$$
$$+ \tfrac{1}{2}\{y\}^T[b_{yq}]\{q\} + \tfrac{1}{2}\{y\}^T[b_{yy}]\{y\} \qquad (7.3.4)$$

where
$$[b_{qq}] = [a_q]^T[K][a_q]$$
$$[b_{qy}] = [a_q]^T[K][a_y]$$
$$[b_{yq}] = [a_y]^T[K][a_q] = [b_{qy}]^T \qquad (7.3.5)$$
$$[b_{yy}] = [a_y]^T[K][a_y]$$

Next we solve for the kinematically indeterminate displacements by demanding that the total potential energy of the structure be a minimum. It was shown in Sec. 6.2 that this is equivalent to the conditions $\partial U/\partial y_i = 0$ $(i = 1, 2, \ldots, r)$, or in matrix notation

$$\left\{\frac{\partial U}{\partial y}\right\} = \{0\} \qquad (7.3.6)$$

Substitution of Eq. (7.3.4) into (7.3.6) yields the following system of equations

$$[b_{yq}]\{q\} + [b_{yy}]\{y\} = \{0\} \qquad (7.3.7)$$

The details of the differentiation leading to Eq. (7.3.7) are left to the student as an exercise. Solving Eq. (7.3.7) for the indeterminate displacements gives

$$\{y\} = [d]\{q\} \qquad (7.3.8)$$

where
$$[d] = -[b_{yy}]^{-1}[b_{yq}] \qquad (7.3.9)$$

The remaining member displacements are obtained by eliminating the quantities y_i from the compatibility relations. Substituting Eq. (7.3.8) into (7.3.1) gives

$$\{v\} = [a]\{q\} \qquad (7.3.10)$$

where
$$[a] = [a_q] + [a_y][d] \qquad (7.3.11)$$

A typical element a_{ij} of matrix $[a]$ represents the magnitude of the member displacement v_i caused by a unit displacement q_j, all other q_i being zero. [Note, however, that the joint displacements y_i are no longer assumed to be zero; their values are now given by Eq. (7.3.8).] We shall refer to $[a]$ as the *unit-displacement distribution matrix*.

The generalized external forces Q_i corresponding to the displacements q_i may be found using Castigliano's first theorem, $Q_i = \partial U/\partial q_i$ ($i = 1, 2, \ldots, n$), or

$$\{Q\} = \left\{\frac{\partial U}{\partial q}\right\} \qquad (7.3.12)$$

Substituting Eq. (7.3.4) into (7.3.12) yields

$$\{Q\} = [b_{qq}]\{q\} + [b_{qy}]\{y\} \qquad (7.3.13)$$

Eliminating the kinematically indeterminate displacements by using Eq. (7.3.8) gives

$$\{Q\} = [k]\{q\} \qquad (7.3.14)$$

where

$$[k] = [b_{qq}] + [b_{qy}][d] \qquad (7.3.15)$$

Since $[k]$ is the matrix of the coefficients which relate the forces $\{Q\}$ to the displacements $\{q\}$ it is, by definition, the stiffness matrix for the structure.

The member forces P_i ($i = 1, 2, \ldots, m$) may be computed as follows. By definition, the stiffness matrix $[K]$ for the unconstrained structure is

$$\{P\} = [K]\{v\} \qquad (7.3.16)$$

Substituting Eq. (7.3.10) into (7.3.16) gives

$$\{P\} = [K][a]\{q\} \qquad (7.3.17)$$

Equation (7.3.14) can be inverted to give

$$\{q\} = [k]^{-1}\{Q\} \qquad (7.3.18)$$

and Eq. (7.3.17) may therefore be written as

$$\{P\} = [A]\{Q\} \qquad (7.3.19)$$

where

$$[A] = [K][a][k]^{-1} \qquad (7.3.20)$$

Note that a typical element A_{ij} of matrix $[A]$ represents the value of the member force P_i due to the unit applied load Q_j. Therefore $[A]$ will be referred to as the *unit-load distribution matrix*.

Two examples are now presented to illustrate the application of the matrix displacement method. Relatively simple problems have been chosen in order to simplify the presentation. It should be evident, however, that matrix methods in

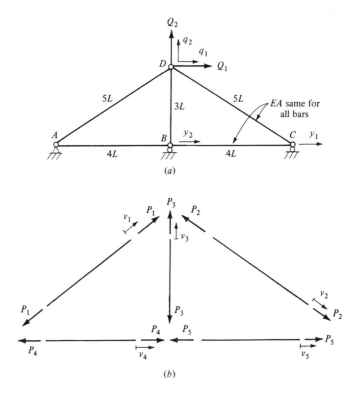

FIGURE 7.4
(a) Prescribed displacements q_i and kinematically indeterminate displacements y_i of a plane truss; (b) member displacements v_i of the unconstrained system.

conjunction with digital computers are indispensable for the analysis of complicated structures.

Example 7.3   Matrix displacement method for a plane truss
As a first example, the analysis of the plane truss considered in Example 6.2 will now be reformulated in matrix notation. The structure is acted upon by the applied loads Q_1 and Q_2 as shown in Fig. 7.4a; the corresponding displacements q_1 and q_2 are therefore taken as the prescribed displacements. This leaves two nonzero joint displacements unspecified, and these are designated as y_1 and y_2 as before.

The unconstrained structure is depicted in Fig. 7.4b. Since the internal forces at the opposite ends of an axial bar are equal in magnitude, it is unnecessary to give a different member load designation to each one. Similarly it is not necessary to define different member displacements at each end of a bar; instead the relative displacement (of one end of a bar with respect to the other) is taken as the member displacement. Thus, there are five member forces P_i ($i = 1, 2, \ldots, 5$) and five member displacements v_i ($i = 1, 2, \ldots, 5$). The latter quantities represent the elongations Δ considered in Example 6.2 (that is, $v_1 = \Delta^{(AD)}$, $v_2 = \Delta^{(CD)}$, etc.) We computed the elongations of each member of the truss in terms of the four degrees of freedom q_1, q_2, y_1, and y_2. Writing the results (6.2.21) in our present notation, we have

$$v_1 = \tfrac{4}{5}q_1 + \tfrac{3}{5}q_2$$
$$v_2 = -\tfrac{4}{5}q_1 + \tfrac{3}{5}q_2 + \tfrac{4}{5}y_1$$
$$v_3 = q_2 \tag{7.3.21}$$
$$v_4 = y_2$$
$$v_5 = y_1 - y_2$$

In matrix form Eq. (7.3.21) becomes

$$\{v\} = [a_q]\{q\} + [a_y]\{y\} \tag{7.3.22}$$

where
$$[a_q] = \begin{bmatrix} \tfrac{4}{5} & \tfrac{3}{5} \\ -\tfrac{4}{5} & \tfrac{3}{5} \\ 0 & 1 \\ 0 & 0 \\ 0 & 0 \end{bmatrix} \quad [a_y] = \begin{bmatrix} 0 & 0 \\ \tfrac{4}{5} & 0 \\ 0 & 0 \\ 0 & 1 \\ 1 & -1 \end{bmatrix} \tag{7.3.23}$$

The strain energy for the structure is

$$U = \frac{1}{2}\left(\frac{EA}{5L}v_1{}^2 + \frac{EA}{5L}v_2{}^2 + \frac{EA}{3L}v_3{}^2 + \frac{EA}{4L}v_4{}^2 + \frac{EA}{4L}v_5{}^2\right) \tag{7.3.24}$$

or, in matrix notation

$$U = \tfrac{1}{2}\{v\}^T[K]\{v\} \tag{7.3.25}$$

where
$$[K] = \begin{bmatrix} \tfrac{1}{5} & 0 & 0 & 0 & 0 \\ 0 & \tfrac{1}{5} & 0 & 0 & 0 \\ 0 & 0 & \tfrac{1}{3} & 0 & 0 \\ 0 & 0 & 0 & \tfrac{1}{4} & 0 \\ 0 & 0 & 0 & 0 & \tfrac{1}{4} \end{bmatrix} \frac{EA}{L} \tag{7.3.26}$$

The remainder of the analysis is based upon the matrices $[a_q]$, $[a_y]$, and $[K]$. Performing the matrix operations described above yields the stiffness matrix

$$[k] = \begin{bmatrix} 0.191 & 0.049 \\ 0.049 & 0.441 \end{bmatrix} \frac{EA}{L} \tag{7.3.27}$$

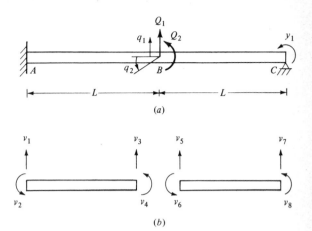

FIGURE 7.5
(a) Indeterminate beam; (b) the unconstrained structure.

and the unit-load distribution matrix

$$[A] = \begin{bmatrix} 0.790 & 0.185 \\ -0.460 & 0.185 \\ -0.198 & 0.778 \\ 0.368 & -0.148 \\ 0.368 & -0.148 \end{bmatrix} \qquad (7.3.28)$$

Multiplication of $[A]$ by the applied forces $\{Q\}$ yields the magnitudes of the internal member forces $\{P\}$. ////

Example 7.4 Matrix displacement method for a statically indeterminate beam The matrix formulation will now be used to compute the stiffness matrix for the beam shown in Fig. 7.5a. We can simplify our analysis considerably by considering each half of the beam separately. The unknown joint displacements in this case include the prescribed displacements q_1 and q_2, and the kinematically indeterminate rotation y_1 at the hinge support C.

From kinematic considerations, it is clear that the member displacements $v_1, v_2, \ldots, v_8$ (Fig. 7.5b) are related to the joint displacement as follows

$$v_1 = v_2 = 0$$
$$v_3 = v_5 = q_1$$
$$v_4 = v_6 = q_2 \qquad (7.3.29)$$
$$v_7 = 0$$
$$v_8 = y_1$$

These compatibility equations may be written in matrix form as

$$\{v\} = [a_q]\{q\} + [a_y]\{y\} \qquad (7.3.30)$$

where

$$[a_q] = \begin{bmatrix} 0 & 0 \\ 0 & 0 \\ 1 & 0 \\ 0 & 1 \\ 1 & 0 \\ 0 & 1 \\ 0 & 0 \\ 0 & 0 \end{bmatrix} \qquad [a_y] = \begin{bmatrix} 0 \\ 0 \\ 0 \\ 0 \\ 0 \\ 0 \\ 0 \\ 1 \end{bmatrix} \qquad (7.3.31)$$

The strain energy in a beam element subject to translations and rotations at its end points is given by Eq. (7.2.18). For element AB we have

$$U^{(AB)} = \tfrac{1}{2}[v_1 v_2 v_3 v_4] \begin{bmatrix} \dfrac{12EI}{L^3} & \dfrac{6EI}{L^2} & -\dfrac{12EI}{L^3} & \dfrac{6EI}{L^2} \\[2ex] \dfrac{6EI}{L^2} & \dfrac{4EI}{L} & -\dfrac{6EI}{L^2} & \dfrac{2EI}{L} \\[2ex] -\dfrac{12EI}{L^3} & -\dfrac{6EI}{L^2} & \dfrac{12EI}{L^3} & -\dfrac{6EI}{L^2} \\[2ex] \dfrac{6EI}{L^2} & \dfrac{2EI}{L} & -\dfrac{6EI}{L^2} & \dfrac{4EI}{L} \end{bmatrix} \begin{bmatrix} v_1 \\[2ex] v_2 \\[2ex] v_3 \\[2ex] v_4 \end{bmatrix} \qquad (7.3.32)$$

and for element BC

$$U^{(BC)} = \tfrac{1}{2}[v_5 v_6 v_7 v_8] \begin{bmatrix} \dfrac{12EI}{L^3} & \dfrac{6EI}{L^2} & -\dfrac{12EI}{L^3} & \dfrac{6EI}{L^2} \\[2ex] \dfrac{6EI}{L^2} & \dfrac{4EI}{L} & -\dfrac{6EI}{L^2} & \dfrac{2EI}{L} \\[2ex] -\dfrac{12EI}{L^3} & -\dfrac{6EI}{L^2} & \dfrac{12EI}{L^3} & -\dfrac{6EI}{L^2} \\[2ex] \dfrac{6EI}{L^2} & \dfrac{2EI}{L} & -\dfrac{6EI}{L^2} & \dfrac{4EI}{L} \end{bmatrix} \begin{bmatrix} v_5 \\[2ex] v_6 \\[2ex] v_7 \\[2ex] v_8 \end{bmatrix} \qquad (7.3.33)$$

The total strain energy of the structure is therefore

$$U = \tfrac{1}{2}\{v\}^T[K]\{v\} \qquad (7.3.34)$$

where

$$\{v\} = \{v_1, v_2, \ldots, v_8\} \qquad (7.3.35)$$

and

$$[K] = \begin{bmatrix} \dfrac{12EI}{L^3} & \dfrac{6EI}{L^2} & -\dfrac{12EI}{L^3} & \dfrac{6EI}{L^2} & 0 & 0 & 0 & 0 \\[2ex] \dfrac{6EI}{L^2} & \dfrac{4EI}{L} & -\dfrac{6EI}{L^2} & \dfrac{2EI}{L} & 0 & 0 & 0 & 0 \\[2ex] -\dfrac{12EI}{L^3} & -\dfrac{6EI}{L^2} & \dfrac{12EI}{L^3} & -\dfrac{6EI}{L^2} & 0 & 0 & 0 & 0 \\[2ex] \dfrac{6EI}{L^2} & \dfrac{2EI}{L} & -\dfrac{6EI}{L^2} & \dfrac{4EI}{L} & 0 & 0 & 0 & 0 \\[2ex] 0 & 0 & 0 & 0 & \dfrac{12EI}{L^3} & \dfrac{6EI}{L^2} & -\dfrac{12EI}{L^3} & \dfrac{6EI}{L^2} \\[2ex] 0 & 0 & 0 & 0 & \dfrac{6EI}{L^2} & \dfrac{4EI}{L} & -\dfrac{6EI}{L^2} & \dfrac{2EI}{L} \\[2ex] 0 & 0 & 0 & 0 & -\dfrac{12EI}{L^3} & -\dfrac{6EI}{L^2} & \dfrac{12EI}{L^3} & -\dfrac{6EI}{L^2} \\[2ex] 0 & 0 & 0 & 0 & \dfrac{6EI}{L^2} & \dfrac{2EI}{L} & -\dfrac{6EI}{L^2} & \dfrac{4EI}{L} \end{bmatrix}$$

(7.3.36)

By performing the necessary matrix computations we obtain the following stiffness matrix for the indeterminate beam

$$[k] = \begin{bmatrix} \dfrac{15EI}{L^3} & -\dfrac{3EI}{L^2} \\[2ex] -\dfrac{3EI}{L^2} & \dfrac{7EI}{L} \end{bmatrix}$$

(7.3.37)

////

7.4 MATRIX FORCE METHOD

The force method of analysis for statically indeterminate structures will now be formulated in matrix notation. Again, we shall assume that the structure to be analyzed can be represented by an assemblage of discrete elements. It is further supposed that the individual elements are subject to a system of internal forces which are to be computed as part of the analysis. As in the matrix displacement

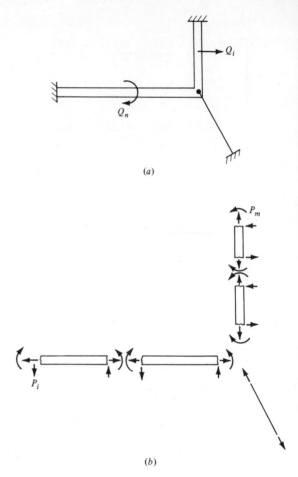

FIGURE 7.6
(a) Applied loads Q_i acting on the statically indeterminate structure; (b) member forces P_i of the unconstrained structure.

method, it is generally necessary to make certain simplifying assumptions concerning the stress distributions in the structural members. Once such assumptions have been made, the degree of static indeterminacy of the idealized structure is determined, and the appropriate number of unknown forces are designated as redundant. Following the notation adopted in Chap. 6, we let Q_i $(i = 1, 2, \ldots, n)$ and Y_i $(i = 1, 2, \ldots, R)$ represent the applied and redundant forces, respectively, and P_i $(i = 1, 2, \ldots, m)$ denote those forces which act on the elements of the unconstrained structure (Fig. 7.6). If an applied force or a redundant acts directly on one of the elements, it will have a P_i designation in addition to its

Q_i or Y_i symbol. In general, therefore, the member forces include all the internal forces, applied loads, and redundants.

Now consider the corresponding base structure, or the structure which would remain if the redundant forces were removed. Since the base structure is statically determinate, it is possible to write m equations of statics which express the member forces in terms of the externally applied loads and redundants. These equilibrium equations can be written in matrix form as

$$\{P\} = [A_Q]\{Q\} + [A_Y]\{Y\} \qquad (7.4.1)$$

From Eq. (7.4.1) it is evident that a typical element of the matrix $[A_Q]$, say $A_{Q_{ij}}$, represents the magnitude of the member force P_i in the base structure resulting from a unit applied load Q_j. Similarly, the element $A_{Y_{ij}}$ of matrix $[A_Y]$ represents the magnitude of P_i when the redundant force Y_j has a magnitude of unity.

Next the strain energy U is computed by summing the energies of the individual members of the unconstrained structure. Since each element is subject to certain of the member forces P_i, the total energy of the structure may be written in the following form [see Eq. (7.2.8)]

$$U = \tfrac{1}{2}\{P\}^T[C]\{P\} \qquad (7.4.2)$$

The square matrix $[C]$ is referred to as the *flexibility matrix of the unconstrained structure*. An alternate form of U is obtained by substituting Eq. (7.4.1) into (7.4.2), which gives

$$U = \tfrac{1}{2}(\{Q\}^T[A_Q]^T + \{Y\}^T[A_Y]^T)[C]([A_Q]\{Q\} + [A_Y]\{Y\})$$
$$= \tfrac{1}{2}\{Q\}^T[B_{QQ}]\{Q\} + \tfrac{1}{2}\{Q\}^T[B_{QY}]\{Y\}$$
$$+ \tfrac{1}{2}\{Y\}^T[B_{YQ}]\{Q\} + \tfrac{1}{2}\{Y\}^T[B_{YY}]\{Y\} \qquad (7.4.3)$$

where
$$[B_{QQ}] = [A_Q]^T[C][A_Q]$$
$$[B_{QY}] = [A_Q]^T[C][A_Y]$$
$$[B_{YQ}] = [A_Y]^T[C][A_Q] = [B_{QY}]^T \qquad (7.4.4)$$
$$[B_{YY}] = [A_Y]^T[C][A_Y]$$

Examining Eq. (7.4.3) we see that the strain energy U has now been expressed solely in terms of the applied forces and the redundants. According to Castigliano's theorem of least work (6.3.9), the redundant forces must be such that U is a minimum. Thus, $\partial U/\partial Y_i = 0 \ (i = 1, 2, \ldots, R)$, or in matrix notation

$$\left\{\frac{\partial U}{\partial Y}\right\} = \{0\} \qquad (7.4.5)$$

Differentiating the expression (7.4.3) according to (7.4.5) leads to the following matrix equation

$$[B_{YQ}]\{Q\} + [B_{YY}]\{Y\} = \{0\} \qquad (7.4.6)$$

Solving for the redundant forces, we obtain

$$\{Y\} = [D]\{Q\} \qquad (7.4.7)$$

with

$$[D] = -[B_{YY}]^{-1}[B_{YQ}] \qquad (7.4.8)$$

The member forces in the indeterminate structure are now obtained by substituting Eq. (7.4.7) into (7.4.1), which gives

$$\{P\} = [A]\{Q\} \qquad (7.4.9)$$

where

$$[A] = [A_Q] + [A_Y][D] \qquad (7.4.10)$$

represents the unit-load distribution matrix for the statically indeterminate structure, the same matrix which was obtained in a different manner using the matrix displacement method.

The generalized displacements q_i corresponding to the applied forces Q_i may be found using Castigliano's second theorem, $q_i = \partial U / \partial Q_i$ ($i = 1, 2, \ldots, n$), or in matrix form

$$\{q\} = \left\{ \frac{\partial U}{\partial Q} \right\} \qquad (7.4.11)$$

Differentiating Eq. (7.4.3) according to (7.4.11) gives

$$\{q\} = [B_{QQ}]\{Q\} + [B_{QY}]\{Y\} \qquad (7.4.12)$$

Eliminating the redundants $\{Y\}$ by using Eq. (7.4.7) yields

$$\{q\} = [c]\{Q\} \qquad (7.4.13)$$

where

$$[c] = [B_{QQ}] + [B_{QY}][D] \qquad (7.4.14)$$

Since Eq. (7.4.13) represents a set of linear relationships between the displacements and the applied forces, the matrix $[c]$ is by definition the flexibility matrix for the indeterminate structure. If the stiffness matrix $[k]$ is desired, it may be found by inverting $[c]$. At this point we may also compute the relationships between the displacements v_i ($i = 1, 2, \ldots, m$) of the individual elements of the structure and the structural displacements q_i. Since $[C]$ is the flexibility matrix of the unconstrained structure,

$$\{v\} = [C]\{P\} \qquad (7.4.15)$$

Substituting Eq. (7.4.9) into (7.4.15) gives

$$\{v\} = [C][A]\{Q\} \qquad (7.4.16)$$

Finally, inverting Eq. (7.4.13) and substituting the forces $\{Q\}$ into Eq. (7.4.16) yields

$$\{v\} = [a]\{q\} \qquad (7.4.17)$$

where

$$[a] = [C][A][c]^{-1} \qquad (7.4.18)$$

Note that $[a]$ is the unit-displacement distribution matrix. The same matrix was obtained in a more direct way using the displacement method.

Example 7.5   Matrix force method for a space truss   The matrix force method will be used to compute the internal forces and the flexibility matrix for the space truss considered in Example 6.3 (Fig. 7.7a). The axial force in bar DE is again designated as redundant. From equilibrium considerations, the member forces P_1, P_2, P_3, and P_4 (Fig. 7.7b) may be expressed in terms of the applied forces Q_1, Q_2, Q_3, and the redundant Y_1 as [see Eq. (6.3.14)]

$$\{P\} = [A_Q]\{Q\} + [A_Y]\{Y\} \qquad (7.4.19)$$

where

$$[A_Q] = \begin{bmatrix} 0 & \frac{3}{4} & \frac{3}{4} \\ -\frac{3}{2} & -\frac{3}{4} & 0 \\ \frac{3}{2} & 0 & \frac{3}{4} \\ 0 & 0 & 0 \end{bmatrix} \qquad [A_Y] = \begin{bmatrix} -1 \\ 1 \\ -1 \\ 1 \end{bmatrix} \qquad (7.4.20)$$

The strain energy of each element of the unconstrained structure is given by Eq. (6.3.15). If the units of the forces P_i are pounds, then the structure's total strain energy may be written as

$$U = \frac{1}{2}\left(\frac{120}{36 \times 10^6} P_1^{\,2} + \frac{120}{18 \times 10^6} P_2^{\,2} + \frac{120}{18 \times 10^6} P_3^{\,2} + \frac{120}{36 \times 10^6} P_4^{\,2} \right)$$

$$= \tfrac{1}{2}(3.33P_1^{\,2} + 6.67P_2^{\,2} + 6.67P_3^{\,2} + 3.33P_4^{\,2})10^{-6} \qquad \text{in.-lb} \qquad (7.4.21)$$

or

$$U = \tfrac{1}{2}\{P\}^T[C]\{P\} \qquad (7.4.22)$$

where the flexibility matrix of the unconstrained structure is

$$[C] = \begin{bmatrix} 3.33 & 0 & 0 & 0 \\ 0 & 6.67 & 0 & 0 \\ 0 & 0 & 6.67 & 0 \\ 0 & 0 & 0 & 3.33 \end{bmatrix} 10^{-6} \qquad \text{in./lb} \qquad (7.4.23)$$

The remainder of the analysis is based upon the matrices $[A_Q]$, $[A_Y]$, and $[C]$. Performing the matrix algebra described above, the axial forces are

$$\{P\} = [A]\{Q\} \qquad (7.4.24)$$

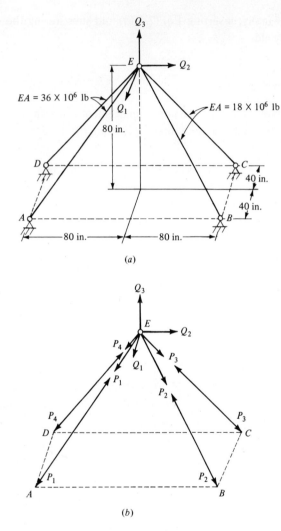

FIGURE 7.7
(a) Statically indeterminate space truss; (b) the unconstrained structure.

where the unit-load distribution matrix is

$$[A] = \begin{bmatrix} -1.0 & 0.375 & 0.375 \\ -0.5 & -0.375 & 0.375 \\ 0.5 & -0.375 & 0.375 \\ 1.0 & 0.375 & 0.375 \end{bmatrix} \qquad (7.4.25)$$

Finally, the flexibility matrix for the indeterminate structure is found to be

$$[c] = \begin{bmatrix} 10.0 & 0 & 0 \\ 0 & 2.81 & -0.938 \\ 0 & -0.938 & 2.81 \end{bmatrix} 10^{-6} \quad \text{in./lb} \quad (7.4.26)$$

////

Example 7.6   Matrix force method for a stiffened-panel structure   The matrix force formulation will now be used to compute the internal forces and the flexibility matrix for the stiffened-panel structure shown in Fig. 7.8a.† In order to simplify the analysis, the following assumptions are introduced:

1   The horizontal flanges or "stringers" have zero bending stiffness. They transmit purely axial forces, and the axial stress in each flange is uniform over its cross section.
2   The webs transmit pure shear, and the shear flow is constant throughout each web.
3   The bulkhead to which the forces Q_1 and Q_2 are applied is rigid.

Based upon these assumptions, the idealized structure is statically indeterminate to the first degree (see Sec. 6.3). The axial force at the fixed end of flange AE is arbitrarily chosen as the redundant Y_1, leaving the base structure shown in Fig. 7.8b. The corresponding unconstrained structure is shown in Fig. 7.8c, in which a system of member forces has been designated. P_1 and P_2 represent loads applied to the bulkhead, P_3 to P_6 are axial forces at the fixed end of the structure, and P_7 to P_{10} denote shear flows in the webs.

As a first step we must relate these member forces P_i to the applied loads Q_1 and Q_2 and the redundant Y_1. In addition to the obvious identities

$$P_1 = Q_1 \qquad P_2 = Q_2 \qquad P_3 = Y_1 \qquad (7.4.27)$$

we may write the following six equations of equilibrium (obtained by considering the entire structure as a free body):

$$\begin{aligned}
\sum F_{x_1} &= 0 & dP_7 - dP_9 &= 0 \\
\sum F_{x_2} &= 0 & P_1 - hP_8 + hP_{10} &= 0 \\
\sum F_{x_3} &= 0 & -P_3 - P_4 - P_5 - P_6 &= 0 \\
\sum M_{x_1} &= 0 & -LP_1 - hP_3 - hP_6 &= 0 \\
\sum M_{x_2} &= 0 & dP_3 + dP_4 &= 0 \\
\sum M_{x_3} &= 0 & \frac{d}{2}P_1 - P_2 - dhP_7 - dhP_8 &= 0
\end{aligned} \qquad (7.4.28)$$

†The displacement method can also be applied to stiffened-panel structures. However, the resulting number of member displacements is generally greater than the number of member forces required in the force method. (See Ref. 7.7.)

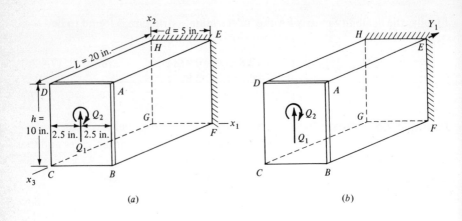

Note: $E = 10^7 \; lb_f/in.^2$
$G = 3.75 \times 10^6 \; lb_f/in.^2$
Stringer areas, $A = 0.1 \; in.^2$
Web thickness, $t = 0.04$ in.

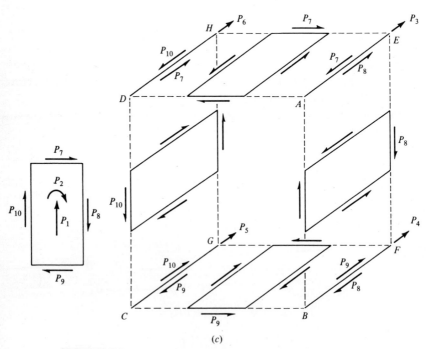

FIGURE 7.8
(a) Statically indeterminate stiffened-panel structure; (b) a base structure;
(c) the unconstrained system.

The equilibrium of a stringer, say AE, provides a 10th independent equation

$$P_3 - LP_7 + LP_8 = 0 \qquad (7.4.29)$$

Solving the 10 simultaneous equations (7.4.27 to 7.4.29) for the member loads in terms of the applied loads Q_1, Q_2, and the redundant Y_1 gives

$$\begin{aligned}
P_1 &= Q_1 \\
P_2 &= Q_2 \\
P_3 &= Y_1 \\
P_4 &= -Y_1 \\
P_5 &= 2Q_1 + Y_1 \\
P_6 &= -2Q_1 - Y_1 \\
P_7 &= 0.025Q_1 - 0.01Q_2 + 0.025Y_1 \\
P_8 &= 0.025Q_1 - 0.01Q_2 - 0.025Y_1 \\
P_9 &= 0.025Q_1 - 0.01Q_2 + 0.025Y_1 \\
P_{10} &= -0.075Q_1 - 0.01Q_2 - 0.025Y_1
\end{aligned} \qquad (7.4.30)$$

in which the units used are pounds and inches. These results may be written in matrix form as

$$\{P\} = [A_Q]\{Q\} + [A_Y]\{Y\} \qquad (7.4.31)$$

where

$$[A_Q] = \begin{bmatrix}
1 & 0 \\
0 & 1 \\
0 & 0 \\
0 & 0 \\
2 & 0 \\
-2 & 0 \\
0.025 & -0.01 \\
0.025 & -0.01 \\
0.025 & -0.01 \\
-0.075 & -0.01
\end{bmatrix} \qquad
[A_Y] = \begin{bmatrix}
0 \\
0 \\
1 \\
-1 \\
1 \\
-1 \\
0.025 \\
-0.025 \\
0.025 \\
-0.025
\end{bmatrix} \qquad (7.4.32)$$

To obtain the strain energy U, we consider the energy stored in the individual elements of the unconstrained structure. For a typical stringer, say AE, we have (see Example 5.16)

$$\begin{aligned}
U^{(AE)} &= \int_{\mathscr{V}} \frac{1}{2E}\sigma_a^2 \, d\mathscr{V} = \int_0^L \int_A \frac{1}{2E}\left[\frac{(P_7 - P_8)x}{A}\right]^2 dA \, dx \\
&= \frac{1}{2}\frac{(P_7 - P_8)^2 L^3}{3EA}
\end{aligned} \qquad (7.4.33)$$

According to Eq. (7.4.29) $(P_7 - P_8)L = P_3$, and therefore

$$U^{(AE)} = \frac{1}{2}\frac{P_3{}^2 L}{3EA} = \frac{1}{2}\left(\frac{20}{3 \times 10^7 \times 0.1}\right)P_3{}^2 = \frac{1}{2}(6.67 \times 10^{-6})P_3{}^2 \qquad \text{in.-lb}$$

$$(7.4.34)$$

Similarly, the energies of the other stringers are

$$U^{(BF)} = \tfrac{1}{2}(6.67 \times 10^{-6})P_4{}^2 \qquad \text{in.-lb}$$
$$U^{(CG)} = \tfrac{1}{2}(6.67 \times 10^{-6})P_5{}^2 \qquad \text{in.-lb} \qquad (7.4.35)$$
$$U^{(DH)} = \tfrac{1}{2}(6.67 \times 10^{-6})P_6{}^2 \qquad \text{in.-lb}$$

The strain energy in web $AEHD$ is (again see Example 5.16)

$$U^{(AEHD)} = \frac{1}{2}\left(\frac{dL}{Gt}\right)P_7{}^2$$

$$= \frac{1}{2}\left(\frac{5 \times 20}{3.75 \times 10^6 \times 0.04}\right)P_7{}^2 = \frac{1}{2}(667 \times 10^{-6})P_7{}^2 \qquad \text{in.-lb} \qquad (7.4.36)$$

and similarly for the other webs

$$U^{(BFEA)} = \tfrac{1}{2}(1333 \times 10^{-6})P_8{}^2 \qquad \text{in.-lb}$$
$$U^{(CGFB)} = \tfrac{1}{2}(667 \times 10^{-6})P_9{}^2 \qquad \text{in.-lb} \qquad (7.4.37)$$
$$U^{(DHGC)} = \tfrac{1}{2}(1333 \times 10^{-6})P_{10}{}^2 \qquad \text{in.-lb}$$

There is no energy stored in the bulkhead $ABCD$ since it is assumed to be rigid. The total strain energy of the structure is therefore

$$U = \tfrac{1}{2}\{P\}^T[C]\{P\} \qquad (7.4.38)$$

where

$$[C] = \begin{bmatrix} 0 & 0 & 0 & 0 & 0 & 0 & 0 & 0 & 0 & 0 \\ 0 & 0 & 0 & 0 & 0 & 0 & 0 & 0 & 0 & 0 \\ 0 & 0 & 6.67 & 0 & 0 & 0 & 0 & 0 & 0 & 0 \\ 0 & 0 & 0 & 6.67 & 0 & 0 & 0 & 0 & 0 & 0 \\ 0 & 0 & 0 & 0 & 6.67 & 0 & 0 & 0 & 0 & 0 \\ 0 & 0 & 0 & 0 & 0 & 6.67 & 0 & 0 & 0 & 0 \\ 0 & 0 & 0 & 0 & 0 & 0 & 667 & 0 & 0 & 0 \\ 0 & 0 & 0 & 0 & 0 & 0 & 0 & 1333 & 0 & 0 \\ 0 & 0 & 0 & 0 & 0 & 0 & 0 & 0 & 667 & 0 \\ 0 & 0 & 0 & 0 & 0 & 0 & 0 & 0 & 0 & 1333 \end{bmatrix} 10^{-5}$$

$$(7.4.39)$$

Performing the matrix algebra described above, the member loads P_i are found to be

$$\{P\} = [A]\{Q\} \qquad (7.4.40)$$

where
$$[A] = \begin{bmatrix} 1.0 & 0 \\ 0 & 1.0 \\ -1.0 & -0.0114 \\ 1.0 & 0.0114 \\ 1.0 & -0.0114 \\ -1.0 & 0.0114 \\ 0 & -0.0103 \\ 0.05 & -0.0097 \\ 0 & -0.0103 \\ -0.05 & -0.0097 \end{bmatrix} \qquad (7.4.41)$$

Thus, the axial force at the fixed end of each stringer and the shear flow in each web have been expressed in terms of the applied loads. (The units for these quantities are pounds and pounds per inch, providing the applied forces Q_1 and Q_2 are given in units of pounds and inch-pounds, respectively.)

The flexibility matrix for the structure is

$$[c] = \begin{bmatrix} 33.3 \text{ in./lb} & 0 \\ 0 & 0.396 \dfrac{1}{\text{in.-lb}} \end{bmatrix} 10^{-6} \qquad (7.4.42)$$

Note that $c_{12} = c_{21} = 0$, which means that the vertical force Q_1 produces no rotation q_2, and the couple Q_2 produces no vertical displacement q_1 of the bulkhead. This result could have been anticipated from a consideration of the symmetry of the structure.   ////

7.5   SUMMARY OF THE MATRIX METHODS

A summary of the matrix methods described in this chapter is given below. The duality which exists between the displacement and the force formulations is evident. Both techniques involve three basic input matrices $[a_q]$, $[a_y]$, and $[K]$ or $[A_Q]$, $[A_Y]$, and $[C]$. Once these arrays have been established, the remainder of the analysis consists of simple matrix algebra. The quantities calculated at any step in one method are the conjugates to those calculated in the other method.

Displacement method	Force method
q_i = Prescribed displacements $(i = 1, 2, \ldots, n)$	Q_i = Prescribed forces $(i = 1, 2, \ldots, n)$
y_i = Kinematically indeterminate displacements $(i = 1, 2, \ldots, r)$	Y_i = Statically indeterminate forces $(i = 1, 2, \ldots, R)$
v_i = Member displacements $(i = 1, 2, \ldots, m)$	P_i = Member forces $(i = 1, 2, \ldots, m)$
Compatibility equations:	Equilibrium equations:
$\{v\} = [a_q]\{q\} + [a_y]\{y\}$	$\{P\} = [A_Q]\{Q\} + [A_Y]\{Y\}$
Strain energy:	Strain energy:
$U = \tfrac{1}{2}\{v\}^T[K]\{v\}$	$U = \tfrac{1}{2}\{P\}^T[C]\{P\}$
Member displacements:	Member forces:
$\{v\} = [a]\{q\}$	$\{P\} = [A]\{Q\}$
Force-displacement relation:	Displacement-force relation:
$\{Q\} = [k]\{q\}$	$\{q\} = [c]\{Q\}$
Member forces:	Member displacements:
$\{P\} = [A]\{Q\}$	$\{v\} = [a]\{q\}$
Matrix operations:	Matrix operations:
$[b_{qq}] = [a_q]^T[K][a_q]$	$[B_{QQ}] = [A_Q]^T[C][A_Q]$
$[b_{qy}] = [a_q]^T[K][a_y]$	$[B_{QY}] = [A_Q]^T[C][A_Y]$
$[b_{yq}] = [a_y]^T[K][a_q]$	$[B_{YQ}] = [A_Y]^T[C][A_Q]$
$[b_{yy}] = [a_y]^T[K][a_y]$	$[B_{YY}] = [A_Y]^T[C][A_Y]$
$[d] = -[b_{yy}]^{-1}[b_{yq}]$	$[D] = -[B_{YY}]^{-1}[B_{YQ}]$
$[a] = [a_q] + [a_y][d]$	$[A] = [A_Q] + [A_Y][D]$
$[k] = [b_{qq}] + [b_{qy}][d]$	$[c] = [B_{QQ}] + [B_{QY}][D]$
$[A] = [K][a][k]^{-1}$	$[a] = [C][A][c]^{-1}$

PROBLEMS

7.1   Joint A of the truss shown experiences the displacements q_1 and q_2. Use the matrix displacement method to determine the resulting elongations of the members of the truss. Also compute the stiffness matrix corresponding to the prescribed displacements.

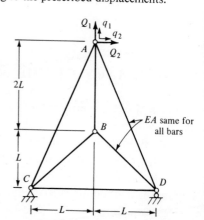

PROBLEM 7.1

7.2 Use the matrix displacement method to determine the stiffness matrix for the beam shown.

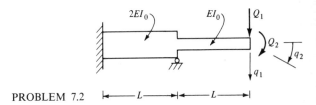

PROBLEM 7.2 |←——L——→|←——L——→|

7.3 Use the matrix force method to determine the internal forces produced by the applied loads Q_1 and Q_2 for the truss considered in Prob. 7.1. Also compute the corresponding flexibility matrix.

7.4 Find the flexibility matrix for the beam in Prob. 7.2 using the matrix force method.

7.5 Solve Prob. 6.13 using the matrix force method. Let $L = 20$ in., $h = 10$ in., $t = 0.025$ in., $A = 0.1$ in.^{2}, $E = 10^7$ lb$_f$/in.^{2}, $G = 4 \times 10^6$ lb$_f$/in.^{2}, and $Q_1 = 1,000$ lb.

Problems *7.6* to *7.11* can be solved most efficiently using a digital computer. Write a computer program which will perform the matrix operations summarized in Sec. 7.5. The input data for the program include the numerical values of the elements of $[a_q]$, $[a_y]$, $[K]$ (or $[A_Q]$, $[A_Y]$, $[C]$) and the order of these matrices. Make use of appropriate matrix subroutines if they are available at your computer facility.

7.6 Use the matrix displacement method to find the stiffness and flexibility matrices and the unit-displacement and unit-load distribution matrices, for the section of the two-story frame building shown. If $Q_1 = Q_3 = 275,000$ lb and $Q_2 = Q_4 = 2,750$ lb, what is the maximum normal stress in the vertical columns? Let $E = 30 \times 10^6$ lb$_f$/in.^{2}, $A_0 = 10$ in.^{2} and $I_0 = 250$ in.^{4}.

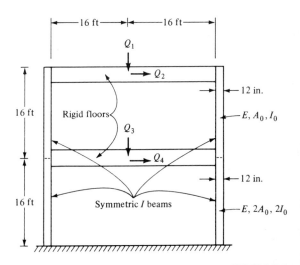

PROBLEM 7.6

7.7   A railroad bridge is to be constructed from two plane trusses of the type shown.

(a) Using the matrix displacement method, calculate the stiffness, flexibility, and unit-load distribution matrices for the truss.

(b) When a long-haul locomotive is on the bridge, the truss is subject to the following loads:

$$Q_1 = Q_3 = Q_5 = Q_7 = 30 \text{ tons}$$
$$Q_2 = Q_4 = Q_6 = Q_8 = 0$$

Determine the minimum bar area A for which the truss will not fail if the ultimate tensile (or compressive) stress is 30,000 $\text{lb}_f/\text{in.}^2$, and the buckling load for a pinned bar in compression is $Q_{\text{critical}} = \pi^2 EI/L^2$ (see Chap. 9). Assume all bars have a circular cross section, let $L = 10$ ft, and take $E = 30 \times 10^6 \text{ lb}_f/\text{in.}^2$.

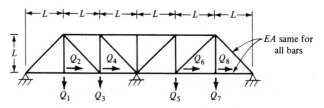

PROBLEM 7.7

7.8   A helicopter's rescue hoist is supported by the space truss shown.

(a) Use the matrix displacement method to find the stiffness matrix, the flexibility matrix, and the axial force in each member of the truss. Assume EA is the same for every bar.

(b) The load conditions for which the hoist is designed are

$$Q_1 = +5,000 \text{ lb}$$
$$Q_2 = \pm 1,000 \text{ lb}$$
$$Q_3 = \pm 1,000 \text{ lb}$$

If the members of the truss are 24S aluminum alloy rods of circular cross section, find the minimum area A for which the truss will not fail when subjected to any combination of the above loads. Assume that the material's ultimate tensile (or compressive) stress is 60,000 $\text{lb}_f/\text{in.}^2$ and $E = 10 \times 10^6 \text{ lb}_f/\text{in.}^2$. The buckling load for a pinned bar in compression is $Q_{\text{critical}} = \pi^2 EI/L^2$ (see Chap. 9).

(c) Optional: Attempt to improve upon the design of the truss (i.e., try to decrease its total weight) by using bars having different cross-sectional areas.

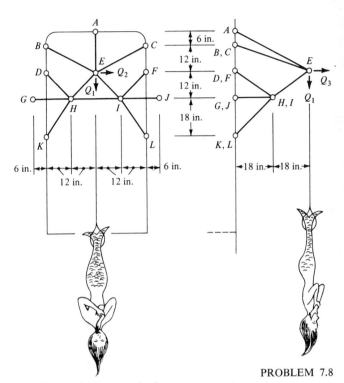

PROBLEM 7.8

7.9 Solve Prob. 7.6 using the matrix force method.

7.10 Use the matrix force method to find the axial force in each stringer and the shear flow in each panel (in terms of Q_1, Q_2, and Q_3) for the structure shown. Make the same assumptions as those introduced in Example 7.6.

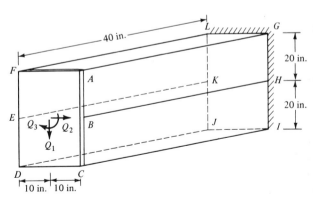

Note: $E = 10^7$ psi
$G = 3.75 \times 10^6$ psi
Stringer areas = 0.2 in.^{2}
Panel thickness = 0.040 in.

PROBLEM 7.10

7.11 The Ranger spacecraft was designed to support TV cameras and associated electronic equipment, for taking pictures of the moon prior to impact. The electronic components are housed inside a *thermal shroud* or outer shell. The approximate geometry and loads for the shroud are as shown.

(*a*) Use the matrix force method to find the unit-load distribution matrix and the flexibility matrix for the shroud. Make the same assumptions as those introduced in Example 7.6.

(*b*) Calculate the maximum axial stress in the stringers and the maximum shear stress in the panels for the design loads

$$Q_1 = 1,900 \text{ lb} \qquad Q_2 = 2,200 \text{ lb}$$

(Due to alignment requirements for the Ranger cameras, rigidity was the controlling factor in the design; hence the stresses are relatively small.)

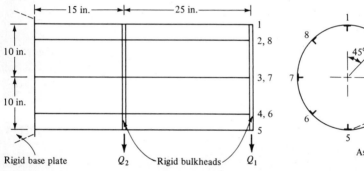

Note: Stringer areas 1, 3, 5, 7 = 0.44 in.^{2}
2, 4, 6, 8 = 0.20 in.^{2}
Panel thickness = 0.032 in.
Material: 7075 Aluminum
($E = 10^7$ psi, $G = 3.75 \times 10^6$ psi)

PROBLEM 7.11

REFERENCES

For a detailed discussion of the use of matrix methods for analyzing framed structures:

7.1 GERE, J. M., and W. WEAVER, JR.: "Analysis of Framed Structures," Van Nostrand, Princeton, N.J., 1965.

7.2 HALL, A. S., and R. W. WOODHEAD: "Frame Analysis," 2d ed., Wiley, New York, 1967.

For a general description of matrix methods and their applications to various types of structures:

7.3   ARGYRIS, J. H., and S. KELSEY: " Energy Theorems and Structural Analysis," Butterworth, London, 1960.

7.4   MARTIN, H. C.: " Introduction to Matrix Methods of Structural Analysis," McGraw-Hill, New York, 1966.

7.5   ROBINSON, J.: "Structural Matrix Analysis for the Engineer," Wiley, New York, 1966.

7.6   RUBINSTEIN, M. F.: " Matrix Computer Analysis of Structures," Prentice-Hall, Englewood Cliffs, N.J., 1966.

7.7   PRZEMIENIECKI, J. S.: " Theory of Matrix Structural Analysis," McGraw-Hill, New York, 1968.

7.8   WILLEMS, N., and W. M. LUCAS, JR.: " Matrix Analysis for Structural Engineers," Prentice-Hall, Englewood Cliffs, N.J., 1968.

For a treatment of the so-called *finite-element technique*, which represents an extension of the matrix methods discussed here to problems in elasticity, plasticity, etc.:

7.9   ZIENKIEWICZ, O. C., and Y. K. CHEUNG: " The Finite Element Method in Structural and Continuum Mechanics," McGraw-Hill, New York, 1967.

7.10   ZIENKIEWICZ, O. C.: "The Finite Element Method in Engineering Science," 2d ed., McGraw-Hill, New York, 1971.

8

THERMAL STRESSES AND DISPLACEMENTS IN STRUCTURES

8.1 INTRODUCTION

Thus far we have been concerned with the behavior of structures under the action of applied forces. Another important source of deformation and stress is heating. When a body is subject to variations in temperature, its elements undergo volume changes; if the temperature distribution is nonuniform, then contiguous elements experience unequal expansion, and stresses develop. Stresses may also result from uniform temperature changes if geometric constraints prevent free expansion of the body. A few examples of situations in which severe *thermal stresses* may be encountered include: atmospheric reentry of space vehicles, heat generation in nuclear reactors, and storage of cryogenic fuels.

The basic equations which govern nonisothermal behavior of elastic bodies are developed in the following article. A strength-of-materials approach for analyzing thermal stresses in beams is illustrated in Sec. 8.3. The remaining sections of the chapter are devoted to applications of energy methods; the virtual work and potential energy principles introduced in Chap. 5 are used to study the response of structures subject to combined thermal and mechanical loadings.

8.2 THERMOELASTIC BEHAVIOR

Before developing the equations which govern the deformation and stress in a heated body, it will be helpful to review the equations of isothermal elasticity and the assumptions implicit in their derivation. The basic equations include, respectively, the equilibrium conditions, the strain-displacement relations, and the strain-stress laws:

$$\sigma_{ij,j} + f_i = 0 \tag{8.2.1}$$

$$e_{ij} = \tfrac{1}{2}(u_{i,j} + u_{j,i}) \tag{8.2.2}$$

$$e_{ij} = \frac{1}{E}[(1 + v)\sigma_{ij} - v\delta_{ij}\sigma_{kk}] \tag{8.2.3}$$

Equation (8.2.1) expresses the fact that every element of the structure is in equilibrium, a condition which must be fulfilled whether the loads are mechanical or thermal in origin. The strain-displacement equations (8.2.2) are based upon purely geometric considerations, and hence they too apply to nonisothermal conditions. The strain-stress equations (8.2.3) require modification, however, since strains may be induced in a structure by temperature changes as well as by applied stresses.

The relationship between stress, strain, and temperature for a linearly elastic material may be obtained by superimposing the strain components induced by the temperature change (in the absence of stress) upon those produced by the stresses. Denoting the stress-induced strains by e'_{ij}, we have according to Eq. (8.2.3)

$$e'_{ij} = \frac{1}{E}[(1 + v)\sigma_{ij} - v\,\delta_{ij}\sigma_{kk}] \tag{8.2.4}$$

Now consider the deformation of a volume element which is stress free but subject to a temperature change Θ. Assuming that the material is isotropic, the element undergoes a pure expansion or contraction, such that each edge of the element experiences a fractional elongation which is proportional to the temperature change. The thermal strains e''_{ij} can therefore be expressed as

$$e''_{ij} = \alpha\,\delta_{ij}\Theta \tag{8.2.5}$$

where the quantity α is the *coefficient of linear thermal expansion*. The total strain components $e_{ij} = e'_{ij} + e''_{ij}$ are then given by

$$e_{ij} = \frac{1}{E}[(1 + v)\sigma_{ij} - v\,\delta_{ij}\sigma_{kk}] + \alpha\,\delta_{ij}\Theta \tag{8.2.6}$$

or, in expanded notation

$$e_{11} = \frac{1}{E}[\sigma_{11} - v(\sigma_{22} + \sigma_{33})] + \alpha\Theta$$

$$e_{22} = \frac{1}{E}[\sigma_{22} - v(\sigma_{33} + \sigma_{11})] + \alpha\Theta$$

$$e_{33} = \frac{1}{E}[\sigma_{33} - v(\sigma_{11} + \sigma_{22})] + \alpha\Theta$$

$$e_{12} = \frac{1+v}{E}\sigma_{12}$$

$$e_{23} = \frac{1+v}{E}\sigma_{23}$$

$$e_{31} = \frac{1+v}{E}\sigma_{31}$$

(8.2.7)

It is sometimes convenient to express the stresses explicitly in terms of the strains and temperature. Inverting Eq. (8.2.6) yields

$$\sigma_{ij} = 2\mu e_{ij} + \lambda\,\delta_{ij}e_{kk} - \alpha(2\mu + 3\lambda)\,\delta_{ij}\Theta \qquad (8.2.8)$$

Relationships between the Lamé constants μ, λ and the engineering material constants E, v are given in Eqs. (4.4.4).

The differential equations (8.2.1), (8.2.2), and (8.2.6) or (8.2.8), together with appropriate boundary conditions consitute the *classical theory of thermo-elasticity*. Providing the temperature distribution Θ is known, these 15 equations suffice for a determination of the 15 unknown quantities σ_{ij}, e_{ij}, and u_i. Unfortunately, computing the temperature field in a structure is often a rather difficult task;[1] this problem is the subject matter of courses in heat transfer and will not be treated here. We shall assume that the temperature distribution is known and concern ourselves only with finding the resulting stresses and displacements.

Due to the linearity of the equations of thermoelasticity, the principle of superposition is valid. Thus if a structure is subject to both mechanical and thermal loadings, it is possible to compute the response to each effect separately and use superposition to obtain the combined result. As we shall see, however, it

---

[1] The temperature distribution in an isotropic elastic solid is governed by the differential equation $\kappa\Theta,_{ii} - \partial\Theta/\partial t = m\Theta_a\,\partial e_{ii}/\partial t$ (see Ref. 8.1), where κ and m represent material properties which may be assumed constant for small temperature changes, Θ_a is the absolute temperature, and t denotes time. In the majority of engineering problems, the effect of strain rate upon the temperature is very small and the term $m\Theta_a(\partial e_{ii}/\partial t)$ may be neglected. Solutions to the resulting *uncoupled heat conduction equation* corresponding to various initial and boundary conditions are given in Ref. 8.6.

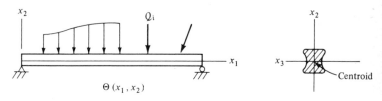

FIGURE 8.1
Symmetric beam.

is often just as simple to consider all the prescribed loads simultaneously. The resulting quantities will be referred to as thermal stresses, thermal strains, and thermal displacements, whether or not they are induced by heating alone.

As in isothermal problems, a direct solution to the governing differential equations is generally difficult to obtain. We shall see that energy methods again provide a useful alternative approach.

8.3   THERMAL STRESSES AND DISPLACEMENTS IN BEAMS

Let us now investigate the response of beams to combined mechanical and thermal disturbances. Since exact solutions to the equations of thermoelasticity are known for only the most trivial problems of bending, our analysis will necessarily be based upon the so-called "classical" or *Bernoulli-Euler* beam theory.

Consider a slender prismatic bar having a longitudinal plane of symmetry (the $x_1 x_2$ plane).[1] As shown in Fig. 8.1, x_1 is the axial coordinate, and x_2 and x_3 represent centroidal axes. We assume that the beam is subject to a system of external forces Q_i which act in the plane of symmetry, as well as to a temperature change $\Theta(x_1, x_2)$.

We begin by introducing the well-known Bernoulli-Euler assumption that plane cross sections of the beam remain plane and perpendicular to the x_1 axis after the deformation. An element of undeformed length dx_1 of the beam thus has the geometry shown in Fig. 8.2, in which ρ represents the radius of curvature of the x_1 axis. If the normal strain of a line element AB situated along the x_1 axis is denoted by $e_{11}{}^0$, then the length of this element after deformation is

$$AB = (1 + e_{11}{}^0)\, dx_1 = \rho\, d\phi \qquad (8.3.1)$$

---

[1]The more general case of unsymmetrical bending is left as an exercise for the student (Prob. 8.2).

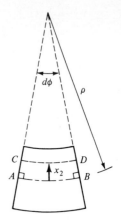

FIGURE 8.2
Assumed deformation of a beam element.

Similarly, a line element CD located at an arbitrary distance x_2 above the x_1 axis suffers a strain e_{11} and has a final length

$$CD = (1 + e_{11}) \, dx_1 = (\rho - x_2) \, d\phi \qquad (8.3.2)$$

Eliminating the angle $d\phi$ from Eqs. (8.3.1) and (8.3.2), and solving for e_{11} in the case of infinitesimal strain ($e_{11}{}^0 \ll 1$), gives

$$e_{11} = e_{11}{}^0 - \frac{x_2}{\rho} \qquad (8.3.3)$$

We now introduce the second important assumption of classical beam theory; namely, we assume that the transverse stresses σ_{22} and σ_{33} are zero throughout the beam. This assumption is made on the grounds that σ_{22} and σ_{33} are zero on the lateral surfaces of the beam, and that owing to the slenderness of the beam, they are negligibly small throughout the interior as well. We are then left with one nonzero normal stress component, σ_{11}. Substitution of Eq. (8.3.3) into the first of Eqs. (8.2.7) gives

$$\sigma_{11} = E(e_{11} - \alpha\Theta) = E\left(e_{11}{}^0 - \frac{x_2}{\rho} - \alpha\Theta\right) \qquad (8.3.4)$$

Because we have introduced assumptions with regard to the deformation and stresses in the beam, the expression (8.3.4) for σ_{11} does not represent an exact solution to the equations of thermoelasticity. However, we can insure that the resultant internal forces are such that every element of length dx_1 is in equilibrium. The axial force N at an arbitrary distance x_1 along the beam is found by integrating the stress σ_{11} over the corresponding cross-sectional area A; thus,

$$N = \int_A \sigma_{11} \, dA = \int_A E\left(e_{11}{}^0 - \frac{x_2}{\rho} - \alpha\Theta\right) dA \qquad (8.3.5)$$

Since $e_{11}{}^0$ and ρ are functions of x_1 only, and $\int_A x_2\, dA = 0$ because x_3 is a centroidal axis, the integration indicated in Eq. (8.3.5) yields

$$N = EAe_{11}{}^0 - \alpha E \int_A \Theta\, dA \qquad (8.3.6)$$

The integral term in Eq. (8.3.6) is sometimes referred to as the *thermal force* since it has the dimensions of a force. We denote this quantity by N_Θ; that is,

$$N_\Theta = \alpha E \int_A \Theta\, dA \qquad (8.3.7)$$

Solving Eq. (8.3.6) for the strain along the x_1 axis gives

$$e_{11}{}^0 = \frac{N + N_\Theta}{EA} \qquad (8.3.8)$$

The resultant bending moment at an arbitrary cross section of the beam is given by

$$M = -\int_A \sigma_{11} x_2\, dA = -\int_A E\left(e_{11}{}^0 x_2 - \frac{x_2{}^2}{\rho} - \alpha\Theta x_2\right) dA \qquad (8.3.9)$$

or, after integration

$$M = \frac{EI}{\rho} + \alpha E \int_A \Theta x_2\, dA \qquad (8.3.10)$$

where I represents the moment of inertia of the area A about the x_3 axis ($I = \int_A x_2{}^2\, dA$). Since the last term in Eq. (8.3.10) has the dimensions of a bending moment, it is called the *thermal moment* and is denoted by M_Θ; that is,

$$M_\Theta = \alpha E \int_A \Theta x_2\, dA \qquad (8.3.11)$$

The curvature of the beam may therefore be written as

$$\frac{1}{\rho} = \frac{M - M_\Theta}{EI} \qquad (8.3.12)$$

Substituting Eqs. (8.3.8) and (8.3.12) into (8.3.4) gives the flexure stress

$$\sigma_{11} = \frac{N + N_\Theta}{A} - \frac{(M - M_\Theta)x_2}{I} - \alpha E\Theta \qquad (8.3.13)$$

In using Eq. (8.3.13) it should be kept in mind that the internal force N and the bending moment M may result from thermal expansion or contraction as well as from the applied forces. This will be the case whenever free expansion of the beam is inhibited.

In general, the normal bending stress σ_{11} is accompanied by a shear stress σ_{12}. As in the isothermal beam theory, the shear stress is often negligibly small compared with the normal stress; this is especially true for relatively long, slender members. In this chapter we will presume that shear does not significantly influence the deformation.

Finally, thermal displacements in the beam can be computed by simple integration of the appropriate strain-displacement relations. To obtain the longitudinal displacement of a point on the x_1 axis, for example, we would integrate the following expression

$$\frac{du_1{}^0}{dx_1} = e_{11}{}^0 = \frac{N + N_\Theta}{EA} \qquad (8.3.14)$$

where the superscript ^{0} again refers to quantities evaluated along the x_1 axis.

The transverse displacement can be computed using a procedure analogous to that normally employed in isothermal problems. Restricting our analysis to the case of small deformations so that the approximate curvature relationship $(1/\rho \cong d^2u_2{}^0/dx_1{}^2)$ is valid, we can obtain $u_2{}^0$ by double integration of the equation

$$\frac{d^2u_2{}^0}{dx_1{}^2} = \frac{1}{\rho} = \frac{M - M_\Theta}{EI} \qquad (8.3.15)$$

For convenience we shall hereafter omit the superscripts ^{0} in the above equations. Thus, the thermal displacements may be computed by integrating the following differential equations

$$\frac{du_1}{dx_1} = \frac{N + N_\Theta}{EA} \qquad (8.3.16)$$

and

$$\frac{d^2u_2}{dx_1{}^2} = \frac{M - M_\Theta}{EI} \qquad (8.3.17)$$

where we must keep in mind that the resulting displacements apply strictly to points located along the x_1 axis.

Example 8.1   Stress and deformation in a beam subject to a linear temperature distribution

We shall now investigate the stress and deflections in a symmetric cantilever beam (Fig. 8.3) resulting from the linear temperature distribution

$$\Theta = ax_1 + bx_2 + c \qquad (8.3.18)$$

where a, b, and c are specified constants.

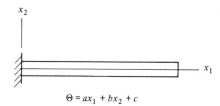

FIGURE 8.3
Beam subject to a linear temperature
distribution.

$$\Theta = ax_1 + bx_2 + c$$

From equilibrium considerations it follows that the resultant axial force N and bending moment M are zero at every cross section of the beam. The thermal force and moment, found by substituting Eq. (8.3.18) into (8.3.7) and (8.3.11), are

$$N_\Theta = \alpha E \int_A \Theta \, dA = \alpha E A(ax_1 + c)$$

$$(8.3.19)$$

$$M_\Theta = \alpha E \int_A \Theta x_2 \, dA = \alpha E I b$$

where use has been made of the relations $\int_A x_2 \, dA = 0$ and $\int_A x_2{}^2 \, dA = I$. Substituting Eqs. (8.3.19) into (8.3.13) gives the axial thermal stress

$$\sigma_{11} = \frac{N + N_\Theta}{A} - \frac{(M - M_\Theta)x_2}{I} - \alpha E \Theta$$

$$= \alpha E(ax_1 + c) + \alpha E b x_2 - \alpha E(ax_1 + bx_2 + c) = 0 \quad (8.3.20)$$

Hence, no thermal stresses are induced in the beam by a linear temperature field. In fact, it can be proven (see Ref. 8.1) that the stress components are identically zero throughout any thermoelastic body with stress-free boundaries, providing the temperature variation is linear with respect to a rectangular cartesian coordinate system.

The axial displacement u_1 in the beam is governed by the equation

$$\frac{du_1}{dx_1} = \frac{N + N_\Theta}{EA} = \alpha(ax_1 + c) \quad (8.3.21)$$

Integrating Eq. (8.3.21) gives

$$u_1 = \alpha\left(\frac{ax_1{}^2}{2} + cx_1\right) + C_1 \quad (8.3.22)$$

Since the longitudinal displacement at the left end of the beam is zero, the constant of integration C_1 must be zero.

The transverse displacement u_2 is obtained by integrating twice the differential equation

$$\frac{d^2u_2}{dx_1{}^2} = \frac{M - M_\Theta}{EI} = -\alpha b \quad (8.3.23)$$

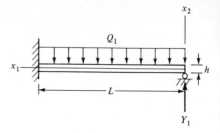

FIGURE 8.4
Beam subject to mechanical and thermal loads.

$$\Theta = \Theta_0 \left(1 + a\frac{x_2}{h}\right)$$

This gives

$$u_2 = -\alpha b \frac{x_1{}^2}{2} + C_2 x_1 + C_3 \qquad (8.3.24)$$

Applying the boundary conditions that the displacement and slope are zero at the beam's clamped end, the constants C_2 and C_3 are found to be zero.

Thus, the longitudinal and transverse displacements of a point on the x_1 axis of the beam are, respectively,

$$u_1 = \alpha \left(\frac{ax_1{}^2}{2} + cx_1\right) \qquad u_2 = -\frac{\alpha b x_1{}^2}{2} \qquad (8.3.25)$$

////

Example 8.2   Analysis of a statically indeterminate beam by simple integration   We wish to determine the stress and deformation in a symmetric, elastic beam supported and loaded as shown in Fig. 8.4 and subject to the temperature change

$$\Theta = \Theta_0\left(1 + a\frac{x_2}{h}\right) \qquad (8.3.26)$$

where Θ_0 and a are specified constants.

Designating the unknown external reaction at the right end of the beam as Y_1, the normal force and bending moment at a distance x_1 along the beam become

$$N = 0$$

$$M = -\frac{Q_1 x_1{}^2}{2} + Y_1 x_1 \qquad (8.3.27)$$

The corresponding thermal force and moment are

$$N_\Theta = \alpha E \int_A \Theta \, dA = \alpha E A \Theta_0$$

$$M_\Theta = \alpha E \int_A \Theta x_2 \, dA = \frac{\alpha E I a \Theta_0}{h} \qquad (8.3.28)$$

Since the beam is statically indeterminate, it is not possible to compute the stresses without first considering the deformation. We shall therefore compute the transverse displacement u_2 by integrating the governing differential equation

$$\frac{d^2 u_2}{dx_1^2} = \frac{M - M_\Theta}{EI} = -\frac{Q_1 x_1^2}{2EI} + \frac{Y_1 x_1}{EI} - \frac{\alpha a \Theta_0}{h} \qquad (8.3.29)$$

Consecutive integrations with respect to x_1 give

$$\frac{du_2}{dx_1} = -\frac{Q_1 x_1^3}{6EI} + \frac{Y_1 x_1^2}{2EI} - \frac{\alpha a \Theta_0 x_1}{h} + C_1 \qquad (8.3.30)$$

and

$$u_2 = -\frac{Q_1 x_1^4}{24EI} + \frac{Y_1 x_1^3}{6EI} - \frac{\alpha a \Theta_0 x_1^2}{2h} + C_1 x_1 + C_2 \qquad (8.3.31)$$

The external reaction Y_1 and the constants of integration C_1 and C_2 are found by applying the three boundary conditions

$$u_2(0) = u_2(L) = \frac{du_2(L)}{dx_1} = 0 \qquad (8.3.32)$$

which yield the results

$$Y_1 = \frac{3Q_1 L}{8} + \frac{3\alpha EIa\Theta_0}{2hL}$$

$$C_1 = -\frac{Q_1 L^3}{48EI} + \frac{\alpha a L \Theta_0}{4h} \qquad (8.3.33)$$

$$C_2 = 0$$

The transverse displacement of a point on the x_1 axis is, therefore,

$$u_2 = \frac{Q_1}{48EI}(-L^3 x_1 + 3Lx_1^3 - 2x_1^4) + \frac{\alpha a \Theta_0}{4hL}(L^2 x_1 - 2Lx_1^2 + x_1^3) \qquad (8.3.34)$$

and the thermal stress, found by substituting Eqs. (8.3.27), (8.3.28), and (8.3.33) into (8.3.13), is

$$\sigma_{11} = \frac{N + N_\Theta}{A} - \frac{(M - M_\Theta)x_2}{I} - \alpha E\Theta$$

$$= -\frac{Q_1(3L - 4x_1)x_1 x_2}{8I} - \frac{3\alpha Ea\Theta_0 x_1 x_2}{2hL} \qquad (8.3.35)$$

As noted earlier, the resulting displacement and stress expressions could have been obtained by considering the applied force and the temperature change separately and then superimposing the two solutions.   ////

8.4   THERMOELASTIC STRAIN ENERGY AND COMPLEMENTARY STRAIN ENERGY

Thermal stresses and displacements can often be computed more easily using an energy approach than by integrating directly the governing differential equations for the structure. In order for the energy principles developed in Chap. 5 to be valid in nonisothermal situations, it is necessary to generalize our previous definitions of strain energy and complementary strain energy. With this in mind, we again consider separately the stress-induced strain components e'_{ij} and the temperature-induced components $e''_{ij} = \alpha \, \delta_{ij} \, \Theta$. The *thermoelastic strain energy* is now defined as the energy associated with the stressing only. In particular, if we replace e_{ij} by e'_{ij} in expression (5.2.5) for U, we obtain

$$U = \int_{\mathscr{V}} \left(\int_0^{e'_{ij}} \sigma_{ij} \, de'_{ij} \right) d\mathscr{V} \qquad (8.4.1)$$

Since $e_{ij} = e'_{ij} + e''_{ij}$ and therefore $de'_{ij} = de_{ij} - \alpha \, \delta_{ij} \, d\Theta$, Eq. (8.4.1) may be written as

$$U = \int_{\mathscr{V}} \left(\int_0^{e_{ij}} \sigma_{ij} \, de_{ij} - \int_0^{\Theta} \alpha \sigma_{kk} \, d\Theta \right) d\mathscr{V} \qquad (8.4.2)$$

Furthermore, we define the *thermoelastic complementary strain energy* as

$$U^* = \int_{\mathscr{V}} \left(\int_0^{\sigma_{ij}} e'_{ij} \, d\sigma_{ij} + e''_{ij} \sigma_{ij} \right) d\mathscr{V} \qquad (8.4.3)$$

so that

$$U^* = \int_{\mathscr{V}} \left[\int_0^{\sigma_{ij}} (e_{ij} - \alpha \, \delta_{ij} \Theta) \, d\sigma_{ij} + \alpha \sigma_{kk} \Theta \right] d\mathscr{V} \qquad (8.4.4)$$

The physical meaning of U and U^* can be seen by evaluating these quantities for the case of a nonlinearly elastic bar subject to a uniaxial stress σ_{11} and a uniform temperature rise Θ. For this combination of stressing and heating, Eqs. (8.4.1) and (8.4.3) yield, respectively,

$$U = \mathscr{V} \int_0^{e'_{11}} \sigma_{11} \, de'_{11} \qquad U^* = \mathscr{V} \left(\int_0^{\sigma_{11}} e'_{11} \, d\sigma_{11} + e''_{11} \sigma_{11} \right) \qquad (8.4.5)$$

Next consider the bar's stress-strain curve corresponding to two separate loading histories. If the material were first stressed and then heated, the stress-strain curve would be as shown in Fig. 8.5a, while if the heating were to precede the stressing, the curve would be that of Fig. 8.5b. Note that in either case, the area ABC under the stress-strain curve is equal to the integral $\int_0^{e'_{11}} \sigma_{11} \, de'_{11}$, which according to Eq. (8.4.5) is equal to $U/\mathscr{V}$. Thus the thermoelastic strain energy per unit volume in the bar represents that portion of the area underneath the stress-strain curve associated with stressing alone. It is also seen that the complementary strain energy per unit volume $U^*/\mathscr{V}$ represents the area which

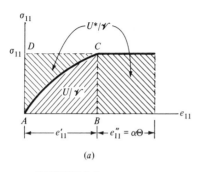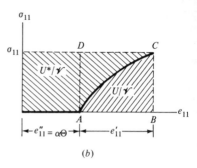

FIGURE 8.5
Stress-strain diagrams for a bar subject to two different loading histories:
(a) tensile loading followed by a temperature increase; (b) temperature increase
followed by a tensile loading.

complements ABC, that is, the area ADC above the stress-strain curve plus the
area $e''_{11}\sigma_{11}$ associated with heating alone.

Expressions (8.4.2) and (8.4.4) define U and U^* for an arbitrary elastic
body. In the special case of linear, isotropic material behavior the stress-strain
law is

$$\sigma_{ij} = 2\mu e_{ij} + \lambda\, \delta_{ij} e_{kk} - \alpha(2\mu + 3\lambda)\, \delta_{ij}\, \Theta \qquad (8.4.6)$$

Substituting Eq. (8.4.6) into (8.4.2) and integrating gives

$$U = \int_{\mathscr{V}} \left[\mu e_{ij} e_{ij} + \frac{\lambda}{2} e_{kk}{}^2 - \alpha(2\mu + 3\lambda)e_{kk}\,\Theta + \frac{3\alpha^2}{2}(2\mu + 3\lambda)\Theta^2 \right] d\mathscr{V} \qquad (8.4.7)$$

or, in expanded notation

$$U = \int_{\mathscr{V}} \left[\frac{Ev}{2(1 + v)(1 - 2v)}(e_{11} + e_{22} + e_{33})^2 \right.$$
$$+ G(e_{11}{}^2 + e_{22}{}^2 + e_{33}{}^2 + 2e_{12}{}^2 + 2e_{23}{}^2 + 2e_{31}{}^2)$$
$$\left. - \frac{\alpha E}{(1 - 2v)}(e_{11} + e_{22} + e_{33})\Theta + \frac{3}{2}\frac{\alpha^2 E}{(1 - 2v)}\,\Theta^2 \right] d\mathscr{V} \qquad (8.4.8)$$

where the Lamé constants have been replaced by E, v, and G using Eq. (4.4.4).
It is sometimes more convenient to have U expressed in terms of stresses rather
than strains, in which case

$$U = \int_{\mathscr{V}} \left[\frac{1}{2E}(\sigma_{11}{}^2 + \sigma_{22}{}^2 + \sigma_{33}{}^2) - \frac{v}{E}(\sigma_{11}\sigma_{22} + \sigma_{22}\sigma_{33} + \sigma_{33}\sigma_{11}) \right.$$
$$\left. + \frac{1}{2G}(\sigma_{12}{}^2 + \sigma_{23}{}^2 + \sigma_{31}{}^2) \right] d\mathscr{V} \qquad (8.4.9)$$

which is identical to our previous (isothermal) expression (5.2.10).

Similarly the thermoelastic complementary strain energy for a linearly elastic, isotropic solid, found by substituting Eq. (8.2.6) into (8.4.4), is

$$U^* = \int_{\mathcal{V}} \left(\frac{1+\nu}{2E} \sigma_{ij}\sigma_{ij} - \frac{\nu}{2E} \sigma_{kk}^2 + \alpha\sigma_{kk}\Theta \right) d\mathcal{V} \qquad (8.4.10)$$

or, in standard notation

$$U^* = \int_{\mathcal{V}} \left[\frac{1}{2E} (\sigma_{11}^2 + \sigma_{22}^2 + \sigma_{33}^2) - \frac{\nu}{E} (\sigma_{11}\sigma_{22} + \sigma_{22}\sigma_{33} + \sigma_{33}\sigma_{11}) \right.$$
$$\left. + \frac{1}{2G} (\sigma_{12}^2 + \sigma_{23}^2 + \sigma_{31}^2) + \alpha(\sigma_{11} + \sigma_{22} + \sigma_{33})\Theta \right] d\mathcal{V} \qquad (8.4.11)$$

In particular, note that

$$U^* = U + \int_{\mathcal{V}} \alpha(\sigma_{11} + \sigma_{22} + \sigma_{33})\Theta \, d\mathcal{V} \qquad (8.4.12)$$

Example 8.3   Thermoelastic strain energy and complementary strain energy for a beam element   Expressions for U and U^* will now be developed for a thermoelastic beam element. The beam has a cross-sectional area A, and a flexural rigidity EI. We assume that the line of action of the resultant normal force N coincides with the centroidal axis x_1, and that the flexure produced by the bending moment M occurs about the principal axis x_3. When the thermal force and moment associated with the temperature $\Theta(x_1,x_2)$ are $N_\Theta = \alpha E \int_A \Theta \, dA$ and $M_\Theta = \alpha E \int_A \Theta x_2 \, dA$, respectively, then the axial stress σ_{11} is

$$\sigma_{11} = \frac{N + N_\Theta}{A} - \frac{(M - M_\Theta)x_2}{I} - \alpha E\Theta \qquad (8.4.13)$$

Assuming that the beam is sufficiently slender that deformation due to shear may be neglected, the strain energy expression (8.4.9) yields

$$U = \int_{\mathcal{V}} \frac{1}{2E} \sigma_{11}^2 \, d\mathcal{V}$$
$$= \int_0^L \int_A \frac{1}{2E} \left[\frac{N + N_\Theta}{A} - \frac{(M - M_\Theta)x_2}{I} - \alpha E\Theta \right]^2 dA \, dx_1$$
$$= \int_0^L \left(\frac{N^2 - N_\Theta^2}{2EA} + \frac{M^2 - M_\Theta^2}{2EI} + \frac{\alpha^2 E}{2} \int_A \Theta^2 \, dA \right) dx_1 \qquad (8.4.14)$$

where use has been made of the definitions of N_Θ, M_Θ, and of the relations $\int_A x_2 \, dA = 0$ and $\int_A x_2{}^2 \, dA = I$ for principal axes. An alternate expression for U, obtained using Eqs. (8.3.16), (8.3.17), and (8.4.14) is

$$U = \int_0^L \left[\frac{EA}{2} \left(\frac{du_1}{dx_1}\right)^2 - N_\Theta \frac{du_1}{dx_1} + \frac{EI}{2} \left(\frac{d^2u_2}{dx_1{}^2}\right)^2 + M_\Theta \frac{d^2u_2}{dx_1{}^2} + \frac{\alpha^2 E}{2} \int_A \Theta^2 \, dA \right] dx_1$$

(8.4.15)

The following expressions for the beam's complementary strain energy can be obtained in a similar fashion:

$$U^* = \int_0^L \left[\frac{(N + N_\Theta)^2}{2EA} + \frac{(M - M_\Theta)^2}{2EI} - \frac{\alpha^2 E}{2} \int_A \Theta^2 \, dA \right] dx_1 \qquad (8.4.16)$$

and

$$U^* = \int_0^L \left[\frac{EA}{2} \left(\frac{du_1}{dx_1}\right)^2 + \frac{EI}{2} \left(\frac{d^2u_2}{dx_1{}^2}\right)^2 - \frac{\alpha^2 E}{2} \int_A \Theta^2 \, dA \right] dx_1 \qquad (8.4.17)$$

////

8.5 APPLICATIONS OF THE VIRTUAL WORK AND COMPLEMENTARY VIRTUAL WORK PRINCIPLES

A review of the derivations of the pvw and the pcvw given earlier will indicate that these principles were developed without reference to a particular stress-strain relation. The results are therefore applicable to structures having arbitrary material behavior, and in particular they are valid for thermoelastic solids. In their most general form these principles are as follows:

pvw:
$$\int_{\mathscr{S}} T_i \delta u_i \, d\mathscr{S} + \int_{\mathscr{V}} f_i \, \delta u_i \, d\mathscr{V} = \int_{\mathscr{V}} \sigma_{ij} \, \delta e_{ij} \, d\mathscr{V} \qquad (8.5.1)$$

pcvw:
$$\int_{\mathscr{S}} \delta T_i u_i \, d\mathscr{S} + \int_{\mathscr{V}} \delta f_i u_i \, d\mathscr{V} = \int_{\mathscr{V}} \delta \sigma_{ij} e_{ij} \, d\mathscr{V} \qquad (8.5.2)$$

Equation (8.5.1) implies that the virtual work done by the surface tractions T_i and the body forces f_i during a virtual distortion δu_i is equal to the virtual work done by the internal stresses σ_{ij} under the virtual strains δe_{ij}. The complementary principle (8.5.2), on the other hand, says that the work done by a system of virtual external forces under the actual displacements is equal to the work done by the virtual stresses during the actual strains.

We have already seen that the pvw in conjunction with an appropriate dummy-displacement system provides a convenient method for computing an unknown force; similarly the pcvw together with a dummy-force system can be used to find an unknown displacement. Consider, for example, a thermoelastic

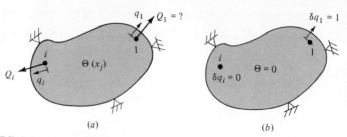

FIGURE 8.6
(a) Thermoelastic body with an unknown force Q_1; (b) dummy displacement $\delta q_1 = 1$.

body subject to a temperature change Θ in addition to a set of generalized forces Q_i (Fig. 8.6a). In order to compute one of the external forces, say Q_1, we consider a virtual distortion consisting of a dummy displacement $\delta q_1 = 1$ (Fig. 8.6b). Application of the pvw then yields

$$Q_1 = \int_{\mathscr{V}} \sigma_{ij}\, \delta e_{ij}\, d\mathscr{V} \qquad (8.5.3)$$

where σ_{ij} now represents the actual thermal stresses and δe_{ij} are the virtual strains associated with the dummy displacement $\delta q_1 = 1$.

Similarly, to compute an unknown thermal displacement Δ in a structure (Fig. 8.7a), we introduce a unit dummy force at the point where the displacement is desired (Fig. 8.7b). For this virtual-force system the pcvw gives

$$\Delta = \int_{\mathscr{V}} \delta\sigma_{ij}\, e_{ij}\, d\mathscr{V} \qquad (8.5.4)$$

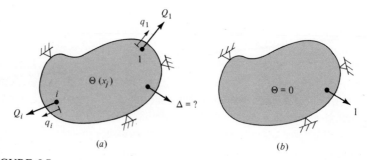

FIGURE 8.7
(a) Thermoelastic body with an unknown displacement Δ; (b) dummy-force system.

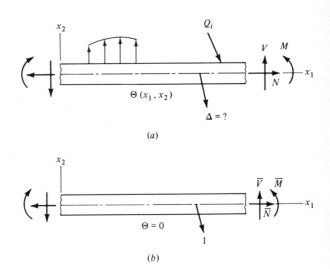

FIGURE 8.8
(a) Thermoelastic bar with an unknown displacement Δ; (b) dummy-force system.

In this case e_{ij} are the strains produced by the actual applied loads and the temperature change, while $\delta\sigma_{ij}$ are the stresses caused by the dummy force.

Example 8.4   Dummy-force method for a beam element   We shall now derive an expression for the generalized displacement Δ in a symmetrical beam element subject to a temperature change $\Theta(x_1,x_2)$ and to external loads Q_i, as shown in Fig. 8.8a.

If we neglect the effect of shear, the nonzero strain components at a distance x_1 along the beam are, according to Eqs. (8.2.7) and (8.3.13),

$$e_{11} = \frac{\sigma_{11}}{E} + \alpha\Theta = \frac{N + N_\Theta}{EA} - \frac{(M - M_\Theta)x_2}{EI}$$

$$e_{22} = e_{33} = -\frac{v\sigma_{11}}{E} + \alpha\Theta = -\frac{v(N + N_\Theta)}{EA} + \frac{v(M - M_\Theta)x_2}{EI} + (1 + v)\alpha\Theta$$

(8.5.5)

Next consider the stresses $\delta\sigma_{ij}$ associated with the dummy-force system shown in Fig. 8.8b. Denoting the axial force, shear force, and bending moment by $\overline{N}, \overline{V},$ and $\overline{M}$, respectively, and again neglecting shear effects, the only nonzero virtual stress component is

$$\delta\sigma_{11} = \frac{\overline{N}}{A} - \frac{\overline{M}x_2}{I}$$

(8.5.6)

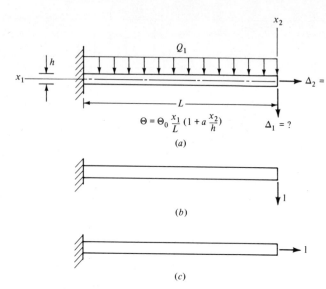

FIGURE 8.9
(a) Cantilever beam subject to distributed force and a temperature change; (b) dummy-force system for computing Δ_1; (c) dummy-force system for computing Δ_2.

Application of the pcvw in the form of Eq. (8.5.4) gives

$$\Delta = \int_{\mathscr{V}} \delta\sigma_{11}e_{11} \, d\mathscr{V}$$

$$= \int_0^L \int_A \left(\frac{\overline{N}}{A} - \frac{\overline{M}x_2}{I}\right)\left[\frac{N + N_\Theta}{EA} - \frac{(M - M_\Theta)x_2}{EI}\right] dA \, dx_1 \qquad (8.5.7)$$

Because $\int_A x_2 \, dA = 0$ and $\int_A x_2{}^2 \, dA = I$ for centroidal axes, Eq. (8.5.7) may be simplified to

$$\Delta = \int_0^L \left[\frac{(N + N_\Theta)\overline{N}}{EA} + \frac{(M - M_\Theta)\overline{M}}{EI}\right] dx_1 \qquad (8.5.8)$$

Note that Eq. (8.5.8) reduces to our previous result (5.5.11) when the thermal force N_Θ and moment M_Θ are zero.   ////

Example 8.5   Dummy-force method for a cantilever beam   A cantilever beam of uniform cross section is acted upon by a distributed load Q_1 (Fig. 8.9a), and is also subject to the temperature distribution

$$\Theta = \Theta_0 \frac{x_1}{L}\left(1 + a\frac{x_2}{h}\right) \qquad (8.5.9)$$

where Θ_0 and a are specified constants. We shall make use of the results of Example 8.4 to compute the resulting displacements Δ_1 and Δ_2.

The axial force and bending moment at a distance x_1 from the beam's free end are

$$N = 0 \qquad M = \frac{-Q_1 x_1^2}{2} \qquad (8.5.10)$$

and the thermal force and moment are

$$N_\Theta = \alpha E \int_A \Theta \, dA = \frac{\alpha E A \Theta_0 x_1}{L}$$

$$M_\Theta = \alpha E \int_A \Theta x_2 \, dA = \frac{\alpha E I a \Theta_0 x_1}{hL} \qquad (8.5.11)$$

To compute the displacement Δ_1, we introduce the unit force shown in Fig. 8.9b, in which case

$$\bar{N} = 0 \qquad \bar{M} = -x_1 \qquad (8.5.12)$$

Equation (8.5.8) then gives

$$\begin{aligned}
\Delta_1 &= \int_0^L \left[\frac{(N + N_\Theta)\bar{N}}{EA} + \frac{(M - M_\Theta)\bar{M}}{EI} \right] dx_1 \\
&= \int_0^L \left(-\frac{Q_1 x_1^2}{2EI} - \frac{\alpha a \Theta_0 x_1}{hL} \right)(-x_1) \, dx_1 \\
&= \frac{Q_1 L^4}{8EI} + \frac{\alpha a L^2 \Theta_0}{3h} \qquad (8.5.13)
\end{aligned}$$

To determine the horizontal displacement Δ_2, we apply the axial dummy force shown in Fig. 8.9c. Thus

$$\bar{N} = 1 \qquad \bar{M} = 0 \qquad (8.5.14)$$

in which case Eq. (8.5.8) yields

$$\Delta_2 = \int_0^L \frac{\alpha \Theta_0 x_1}{L} (1) \, dx_1 = \frac{\alpha L \Theta_0}{2} \qquad (8.5.15)$$

$////$

Example 8.6 Dummy-force method for a curved bar   A slender quarter-circle bar of rectangular cross section (Fig. 8.10a) experiences a temperature change which varies in the radial direction according to

$$\Theta = \Theta_0 \left[1 + a_1 \frac{x_2}{h} + a_2 \left(\frac{x_2}{h} \right)^2 \right] \qquad (8.5.16)$$

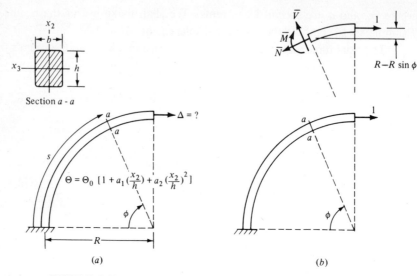

FIGURE 8.10
(a) Curved bar with an unknown displacement Δ; (b) dummy-force system.

where Θ_0, a_1, and a_2 are specified constants. We shall use the dummy-force method to compute the horizontal displacement Δ of the bar's free end; Fig. 8.10b shows the appropriate dummy-force system.

Assuming that the cross-sectional dimensions b and h are small compared with the radius of curvature R, the displacement Δ can be found using

$$\Delta = \int \left[\frac{(N + N_\Theta)\bar{N}}{EA} + \frac{(M - M_\Theta)\bar{M}}{EI} \right] ds \quad (8.5.17)$$

From equilibrium considerations, it is evident that the axial force N and bending moment M are zero at every cross section of the bar. The temperature integrals N_Θ and M_Θ are given by

$$N_\Theta = \alpha E \int_A \Theta \, dA = \alpha E \Theta_0 \left(A + \frac{a_2 I}{h^2} \right)$$

$$M_\Theta = \alpha E \int_A \Theta x_2 \, dA = \alpha E \Theta_0 \frac{a_1 I}{h} \qquad (8.5.18)$$

From Fig. 8.10b we note that the virtual normal force $\bar{N}$ and bending moment $\bar{M}$ at section aa are

$$\bar{N} = \sin \phi$$

$$\bar{M} = -R(1 - \sin \phi) \qquad (8.5.19)$$

Substituting Eqs. (8.5.18) and (8.5.19) into (8.5.17), and letting $ds = R\,d\phi$, we obtain

$$\Delta = \int_0^{\pi/2} \left[\alpha\Theta_0 \left(1 + \frac{a_2 I}{Ah^2} \right) \sin\phi + \alpha\Theta_0 \frac{a_1}{h} R(1 - \sin\phi) \right] R\,d\phi \qquad (8.5.20)$$

Integrating Eq. (8.5.20) and letting $A = bh$ and $I = bh^3/12$, it is found that

$$\Delta = \alpha R\Theta_0 \left[1 + \left(\frac{\pi}{2} - 1 \right) \frac{a_1 R}{h} + \frac{a_2}{12} \right] \qquad (8.5.21)$$

////

8.6   APPLICATIONS OF THE PRINCIPLES OF MINIMUM POTENTIAL ENERGY AND MINIMUM COMPLEMENTARY ENERGY

The principles of minimum potential energy and minimum complementary energy are basically restatements of the virtual work concepts as they apply to elastic bodies subject to potential force fields. Hence they too can be applied in non-isothermal situations. Mathematical statements of these conjugate principles are:

pmpe: $\qquad\qquad\qquad\qquad \delta\Pi = \delta U + \delta V_E = 0 \qquad (8.6.1)$

pmce: $\qquad\qquad\qquad\qquad \delta\Pi^* = \delta U^* + \delta V_E^* = 0 \qquad (8.6.2)$

From a physical point of view, the pmpe implies that of all geometrically possible states of deformation of a structure, the correct state is that for which the total potential energy Π is a minimum. On the other hand, the pmce says that of all states of stress which satisfy the equations of equilibrium, the correct state is that which makes the total complementary energy Π^* a minimum.

To prove that the pmpe is applicable in nonisothermal problems, we need only to prove that it is mathematically equivalent to the pvw, the validity of which has already been discussed. By the definition of the potential of the external forces, Eq. (5.7.9), we have

$$\delta V_E = -\int_{\mathscr{S}} T_i \,\delta u_i \,d\mathscr{S} - \int_{\mathscr{V}} f_i \,\delta u_i \,d\mathscr{V} \qquad (8.6.3)$$

A comparison of Eqs. (8.6.1) and (8.5.1) then indicates that the two expressions are identical providing the variation of the thermoelastic strain energy is given by

$$\delta U = \int_{\mathscr{V}} \sigma_{ij} \,\delta e_{ij} \,d\mathscr{V} \qquad (8.6.4)$$

The verification of Eq. (8.6.4) is left to the student as an exercise (Prob. 8.12).

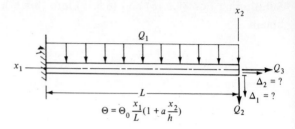

FIGURE 8.11
Cantilever beam subject to a distributed force and a temperature change.

The equivalence of the pmce and the pcvw for thermoelastic bodies can be established in a similar fashion. Consequently, each of our variational principles can be used in the analysis of thermal stresses and displacements. In particular, unknown forces Q_i or displacements q_i can be computed using the following relations:

$$\frac{\partial U}{\partial q_i} = Q_i \qquad i = 1, 2, \ldots, n \qquad (8.6.5)$$

and

$$\frac{\partial U^*}{\partial Q_i} = q_i \qquad i = 1, 2, \ldots, n \qquad (8.6.6)$$

Castigliano's first theorem (8.6.5) and *Engesser's theorem* (8.6.6) follow directly from the pmpe and the pmce; their derivations were given in Chap. 5. In applying Eqs. (8.6.5) and (8.6.6) to nonisothermal problems, however, one must be careful to use the thermoelastic energies U and U^* given by Eqs. (8.4.2) and (8.4.4). Remember that U^* is not equal to U even for linearly elastic materials unless the temperature change Θ is everywhere zero.

Example 8.7 Engesser's theorem for a cantilever beam   The cantilever beam problem treated in Example 8.5 using the dummy-force method will now be solved using Engesser's theorem. The loading consists of a uniformly distributed force Q_1 and the temperature field $\Theta = \Theta_0 x_1(1 + a x_2/h)/L$. Since there are no generalized forces applied to the structure at the points where the displacements Δ_1 and Δ_2 are required, we shall introduce the forces Q_2 and Q_3 (Fig. 8.11) and later set them equal to zero. The axial force and bending moment at an arbitrary cross section of the beam are seen to be

$$N = Q_3 \qquad M = \frac{-Q_1 x_1{}^2}{2} - Q_2 x_1 \qquad (8.6.7)$$

and the temperature integrals are (see Example 8.5)

$$N_\Theta = \frac{\alpha E A \Theta_0 x_1}{L} \qquad M_\Theta = \frac{\alpha E I a \Theta_0 x_1}{hL} \qquad (8.6.8)$$

Recall that the complementary strain energy U^* in a thermoelastic beam is given by

$$U^* = \int_0^L \left[\frac{(N + N_\Theta)^2}{2EA} + \frac{(M - M_\Theta)^2}{2EI} - \frac{\alpha^2 E}{2} \int_A \Theta^2 \, dA \right] dx_1 \qquad (8.6.9)$$

Applying Engesser's theorem (8.6.6) gives the generalized displacements

$$
\begin{aligned}
q_2 &= \frac{\partial U^*}{\partial Q_2} = \int_0^L \frac{M - M_\Theta}{EI} \frac{\partial M}{\partial Q_2} \, dx_1 \\
&= \int_0^L \left(-\frac{Q_1 x_1{}^2}{2EI} - \frac{Q_2 x_1}{EI} - \frac{\alpha a \, \Theta_0 x_1}{hL} \right) (-x_1) \, dx_1 \\
&= \frac{Q_1 L^4}{8EI} + \frac{Q_2 L^3}{3EI} + \frac{\alpha a L^2 \Theta_0}{3h} \qquad (8.6.10)
\end{aligned}
$$

and

$$
\begin{aligned}
q_3 &= \frac{\partial U^*}{\partial Q_3} = \int_0^L \frac{N + N_\Theta}{EA} \frac{\partial N}{\partial Q_3} \, dx_1 \\
&= \int_0^L \left(\frac{Q_3}{EA} + \frac{\alpha \Theta_0 x_1}{L} \right) (1) \, dx_1 \\
&= \frac{Q_3 L}{EA} + \frac{\alpha L \Theta_0}{2} \qquad (8.6.11)
\end{aligned}
$$

The unknown displacements Δ_1 and Δ_2 are now found by setting $Q_2 = Q_3 = 0$ in the above expressions. This gives

$$\Delta_1 = q_2 \Big|_{Q_2 = Q_3 = 0} = \frac{Q_1 L^4}{8EI} + \frac{\alpha a L^2 \Theta_0}{3h} \qquad (8.6.12)$$

and

$$\Delta_2 = q_3 \Big|_{Q_2 = Q_3 = 0} = \frac{\alpha L \Theta_0}{2} \qquad (8.6.13)$$

////

Besides providing a means for computing unknown displacements and forces in a structure, the minimum energy principles can also be used to derive the differential equations governing a body's deformation and/or equilibrium. While the same equations can sometimes be obtained in a more direct way (i.e., from a consideration of the equilibrium of a typical element of the body), the energy approach offers one distinct advantage. Namely, it yields a system of physically

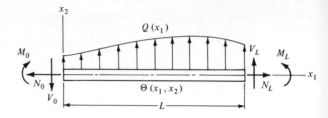

FIGURE 8.12
Bar subject to applied forces and a temperature change.

admissible or *natural* boundary conditions for the structure. To illustrate this we shall now use the pmpe to derive the displacement equations and boundary conditions for a thermoelastic bar. Variational techniques will also be employed in Chap. 9 to derive the equations governing the equilibrium of a beam column, and in Chap. 12 to formulate the equations describing the dynamic behavior of various continuous structures.

Example 8.8   Derivation of the differential equations and boundary conditions governing the deformation of a thermoelastic bar   Consider a symmetric beam subject to a temperature change $\Theta(x_1,x_2)$ in addition to the transverse and axial applied loads shown in Fig. 8.12. No assumptions will be made with regard to the beam's boundary conditions. We shall see that if the end conditions are left unspecified, application of the pmpe yields the beam's natural boundary conditions. Thus, the ends of the beam are presumed to undergo arbitrary displacements u_1, u_2 and rotations du_2/dx_1; they are also subject to arbitrary end loads N_0, V_0, M_0 and N_L, V_L, M_L as shown.

In accordance with the classical beam theory assumptions introduced earlier, the strain energy is

$$U = \int_0^L \left[\frac{EA}{2} (u_1')^2 - N_\Theta u_1' + \frac{EI}{2} (u_2'')^2 + M_\Theta u_2'' + \frac{\alpha^2 E}{2} \int_A \Theta^2 \, dA \right] dx_1 \qquad (8.6.14)$$

where the primes (') denote differentiation with respect to the axial coordinate x_1. The potential of the external forces is

$$V_E = -\int_0^L Qu_2 \, dx_1 + N_0 u_1(0) + V_0 u_2(0) + M_0 u_2'(0)$$

$$- N_L u_1(L) - V_L u_2(L) - M_L u_2'(L) \qquad (8.6.15)$$

Requiring that the total potential energy be a minimum, $\delta\Pi = \delta(U + V_E) = 0$, we obtain

$$\int_0^L [EAu_1' \, \delta(u_1') - N_\Theta \, \delta(u_1') + EIu_2'' \, \delta(u_2'') + M_\Theta \, \delta(u_2'') - Q \, \delta u_2] \, dx_1$$
$$+ N_0 \, \delta u_1(0) + V_0 \, \delta u_2(0) + M_0 \, \delta u_2'(0)$$
$$- N_L \, \delta u_1(L) - V_L \, \delta u_2(L) - M_L \, \delta u_2'(L) = 0 \qquad (8.6.16)$$

Noting that $\delta(u_1') = (\delta u_1)'$ and $\delta(u_2'') = (\delta u_2')' = (\delta u_2)''$, the first four terms of Eq. (8.6.16) may be integrated by parts as follows

$$\int_0^L EAu_1' \, \delta(u_1') \, dx_1 = [EAu_1' \, \delta u_1]_0^L - \int_0^L (EAu_1')' \, \delta u_1 \, dx_1$$
$$-\int_0^L N_\Theta \, \delta(u_1') \, dx_1 = -[N_\Theta \, \delta u_1]_0^L + \int_0^L N_\Theta' \, \delta u_1 \, dx_1$$
$$\int_0^L EIu_2'' \, \delta(u_2'') \, dx_1 = [EIu_2'' \, \delta u_2']_0^L - [(EIu_2'')' \, \delta u_2]_0^L + \int_0^L (EIu_2'')'' \, \delta u_2 \, dx_1 \qquad (8.6.17)$$
$$\int_0^L M_\Theta \, \delta(u_2'') \, dx_1 = [M_\Theta \, \delta u_2']_0^L - [M_\Theta' \, \delta u_2]_0^L + \int_0^L M_\Theta'' \, \delta u_2 \, dx_1$$

Substituting Eqs. (8.6.17) into (8.6.16) and rearranging terms gives

$$\int_0^L [-(EAu_1')' + N_\Theta'] \, \delta u_1 \, dx_1 + \int_0^L [(EIu_2'')'' + M_\Theta'' - Q] \, \delta u_2 \, dx_1$$
$$+ [N_0 + N_\Theta - EAu_1']_{x_1 = 0} \, \delta u_1(0) + [-N_L - N_\Theta + EAu_1']_{x_1 = L} \, \delta u_1(L)$$
$$+ [V_0 + M_\Theta' + (EIu_2'')']_{x_1 = 0} \, \delta u_2(0) + [-V_L - M_\Theta' - (EIu_2'')']_{x_1 = L} \, \delta u_2(L)$$
$$+ [M_0 - M_\Theta - EIu_2'']_{x_1 = 0} \, \delta u_2'(0) + [-M_L + M_\Theta + EIu_2'']_{x_1 = L} \, \delta u_2'(L) = 0$$
$$(8.6.18)$$

Since the displacements and rotation at each end of the beam were left unspecified, the corresponding variations $\delta u_1(0)$, $\delta u_1(L)$, ..., $\delta u_2'(L)$ are arbitrary. Furthermore the variations δu_1 and δu_2 are arbitrary for $0 < x_1 < L$. Consequently each of the eight terms in Eq. (8.6.18) must vanish independently. The differential equations for the bar are, accordingly,

$$-(EAu_1')' + N_\Theta' = 0 \qquad (8.6.19)$$

and

$$(EIu_2'')'' + M_\Theta'' - Q = 0 \qquad (8.6.20)$$

The remaining terms in Eq. (8.6.18) are equal to zero providing the following end conditions are satisfied

$$N_0 = [EAu'_1 - N_\Theta]_{x_1=0} \quad \text{or} \quad \delta u_1(0) = 0$$
$$N_L = [EAu'_1 - N_\Theta]_{x_1=L} \quad \text{or} \quad \delta u_1(L) = 0 \tag{8.6.21}$$

and

$$V_0 = -[(EIu''_2)' + M'_\Theta]_{x_1=0} \quad \text{or} \quad \delta u_2(0) = 0$$
$$V_L = -[(EIu''_2)' + M'_\Theta]_{x_1=L} \quad \text{or} \quad \delta u_2(L) = 0 \tag{8.6.22}$$

and also

$$M_0 = [EIu''_2 + M_\Theta]_{x_1=0} \quad \text{or} \quad \delta u'_2(0) = 0$$
$$M_L = [EIu''_2 + M_\Theta]_{x_1=L} \quad \text{or} \quad \delta u'_2(L) = 0 \tag{8.6.23}$$

Observe that the differential equation (8.6.19) and the boundary conditions (8.6.21) involve the axial displacement u_1 but are independent of the transverse displacement u_2; these equations thus define a purely *extensional* or *stretching* *problem*. If the displacement u_1 is prescribed at $x_1 = 0$, then $\delta u_1(0) = 0$. On the other hand if $u_1(0)$ is unspecified, then $\delta u_1(0)$ is arbitrary and Eq. (8.6.21) requires that $N_0 = [EAu'_1 - N_\Theta]_{x_1=0}$. We showed earlier that the resultant normal force at a cross section of a beam is given by $N = EAu'_1 - N_\Theta$. Hence, for the equilibrium of a beam we must prescribe either the axial displacement $u_1(0)$ or the corresponding normal force $N(0) = N_0$. Similarly the displacement $u_1(L)$ or the force $N(L) = N_L$ must be specified at the boundary $x_1 = L$. The first type of condition (prescribed displacement) is referred to as a kinematic boundary condition (see Sec. 5.10); the second type is called a static boundary condition. The latter, obtained as a natural consequence of the variational principle, is also referred to as one of the system's natural boundary conditions.

The problem of *flexure* is defined by the differential equation (8.6.20) and the boundary conditions (8.6.22) and (8.6.23). Since these equations are independent of the displacement u_1 we see that the flexure and stretching problems are uncoupled in this theory.[1]

From Eq. (8.6.22) it is evident that at $x_1 = 0$ either the displacement $u_2(0)$ or the shear force $V(0) = V_0$ must be specified; and from Eq. (8.6.23) we see that either the slope $u'_2(0)$ or the bending moment $M(0) = M_0$ must be prescribed. Analogous conditions apply to the boundary $x_1 = L$.

The four types of end conditions most often considered in engineering beam theory are shown in Fig. 8.13. Note that each corresponding pair of end conditions satisfies the Eqs. (8.6.22) and (8.6.23) which apply to the boundary $x_1 = 0$.

---

[1]The flexure and stretching problems are uncoupled providing the transverse displacement $u_2(x_1)$ is everywhere small compared with the depth of the beam. In Chap. 9 we shall study the deformation of long, slender beams in which case this restriction is not always met.

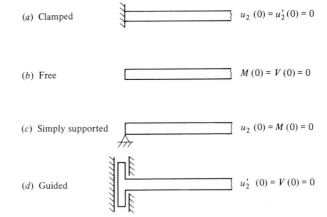

$\longmapsto x_1$

(a) Clamped — $u_2(0) = u_2'(0) = 0$

(b) Free — $M(0) = V(0) = 0$

(c) Simply supported — $u_2(0) = M(0) = 0$

(d) Guided — $u_2'(0) = V(0) = 0$

FIGURE 8.13
Ordinary boundary conditions for the classical beam theory.

While there are other possible combinations which one might construct mathematically, only those satisfying both (8.6.22) and (8.6.23) insure equilibrium of the beam. The boundary conditions $u_2(0) = 0$ and $V(0) = 0$, for example, do not necessarily satisfy Eq. (8.6.23) and are therefore inadmissible. Furthermore this set of conditions is unreasonable from a physical viewpoint. (To convince yourself of this, try to design an end support for which the beam's displacement and shear force are zero for arbitrary loadings.)   ////

8.7   THERMAL STRESSES AND DISPLACEMENTS IN INDETERMINATE STRUCTURES

Two dual approaches for analyzing indeterminate structures were described in Chap. 6. In the approach known as the displacement method, compatibility equations are written which insure that the deformation is consistent with all kinematic constraints. The correct state of deformation is then found by invoking the pmpe. If the structure is kinematically indeterminate to the rth degree, for example, minimization of the total potential energy is tantamount to minimizing the strain energy U with respect to each of the r kinematically indeterminate displacements y_i; that is,

$$\frac{\partial U}{\partial y_i} = 0 \qquad i = 1, 2, \ldots, r \qquad (8.7.1)$$

Equations (8.7.1) together with the structure's compatibility equations suffice for a determination of the unknown displacements.

In the force method, on the other hand, equations of equilibrium are written in terms of unknown generalized forces, and the pmce is applied to obtain the true force distribution. For a structure which is statically indeterminate to the Rth degree, minimization of the total complementary energy is equivalent to minimizing the complementary strain energy with respect to the redundant forces Y_i; that is,

$$\frac{\partial U^*}{\partial Y_i} = 0 \qquad i = 1, 2, \ldots, R \qquad (8.7.2)$$

Equation (8.7.2) is a mathematical statement of *Engesser's theorem of least work*, also known as the generalized form of Castigliano's theorem of least work [see Eq. (6.3.9)]. These relations together with the equations of statics are sufficient to provide a complete solution to the problem.

Equations (8.7.1) and (8.7.2) are direct consequences of the pmpe and the pmce. They can be used in the analysis of thermal stress and deformation, providing that U and U^* represent the thermoelastic strain energy (8.4.2) and complementary strain energy (8.4.4).

Example 8.9 Force method for a statically indeterminate axially loaded bar   The longitudinal expansion of a heated bar is inhibited by hinge supports at A and B as shown in Fig. 8.14a. We shall use the force method to investigate the thermal stresses induced in the bar by the following temperature field:

$$\Theta = \Theta_0 \sin \frac{\pi x_1}{L} \cos \frac{\pi x_2}{h} \qquad (8.7.3)$$

Note that Θ is zero at each end of the beam ($x_1 = 0, L$) and along the top and bottom surfaces ($x_2 = \pm h/2$), is a maximum at the center of the beam ($x_1 = L/2$, $x_2 = 0$), and varies harmonically in the x_1 and x_2 directions as shown.

Due to the symmetry of the applied loading and the temperature distribution about the $x_1 x_3$ plane, no flexure occurs; that is,

$$M = 0 \qquad M_\Theta = \alpha E \int_A \Theta x_2 \, dA = 0 \qquad (8.7.4)$$

Noting that the bar is statically indeterminate to the first degree, we arbitrarily select the force at B as the redundant Y_1 (Fig. 8.14b), Then

$$N = -Y_1$$

$$N_\Theta = \alpha E \int_A \Theta_0 \sin \frac{\pi x_1}{L} \cos \frac{\pi x_2}{h} \, dA = \frac{2\alpha EA\Theta_0}{\pi} \sin \frac{\pi x_1}{L} \qquad (8.7.5)$$

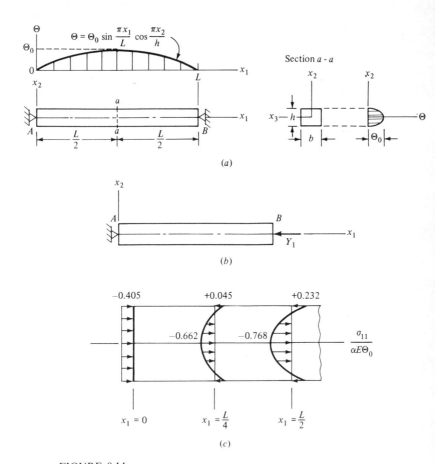

FIGURE 8.14
(a) Temperature distribution in a constrained bar; (b) a base structure; (c) the resulting thermal stress distribution.

The complementary strain energy for the thermoelastic bar is, according to Eq. (8.4.16),

$$U^* = \int_0^L \left[\frac{(N + N_\Theta)^2}{2EA} - \frac{\alpha^2 E}{2} \int_A \Theta^2 \, dA \right] dx_1 \qquad (8.7.6)$$

Application of the theorem of least work (8.7.2) gives

$$\frac{\partial U^*}{\partial Y_1} = \int_0^L \frac{N + N_\Theta}{EA} \frac{\partial N}{\partial Y_1} \, dx_1 = 0 \qquad (8.7.7)$$

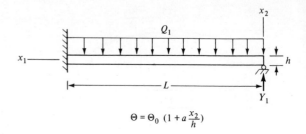

$$\Theta = \Theta_0 \left(1 + a\frac{x_2}{h}\right)$$

FIGURE 8.15
Statically indeterminate beam.

Substituting Eqs. (8.7.5) into (8.7.7) and integrating yields

$$\frac{Y_1 L}{EA} - \frac{4\alpha L\Theta_0}{\pi^2} = 0 \qquad (8.7.8)$$

so that the redundant force has the value

$$Y_1 = \frac{4\alpha EA\Theta_0}{\pi^2} \qquad (8.7.9)$$

The thermal stress σ_{11} is, according to Eq. (8.3.13),

$$\sigma_{11} = \frac{N + N_\Theta}{A} - \alpha E\Theta$$

$$= -\frac{4\alpha E\Theta_0}{\pi^2} + \frac{2\alpha E\Theta_0}{\pi} \sin\frac{\pi x_1}{L} - \alpha E\Theta_0 \sin\frac{\pi x_1}{L} \cos\frac{\pi x_2}{h}$$

$$= \alpha E\Theta_0 \left[-\frac{4}{\pi^2} + \left(\frac{2}{\pi} - \cos\frac{\pi x_2}{h}\right) \sin\frac{\pi x_1}{L}\right] \qquad (8.7.10)$$

From Eq. (8.7.10) we note that the maximum compressive stress occurs at the center of the bar ($x_1 = L/2$, $x_2 = 0$), and the maximum tensile stress is located at $x_1 = L/2$, $x_2 = \pm h/2$. The stress distributions at three different cross sections of the beam are illustrated in Fig. 8.14c.   ////

Example 8.10   Force method for a statically indeterminate beam   We shall now apply the force method to determine the thermal stress distribution and deformation in the statically indeterminate beam shown in Fig. 8.15; the same problem was solved using direct integration in Example 8.2.

Recognizing that the degree of indeterminacy of the structure is 1, we arbitrarily select the vertical reaction at the right end of the beam as the redundant Y_1. The normal force and bending moment at a distance x_1 along the beam are then given by

$$N = 0 \qquad M = -\frac{Q_1 x_1^2}{2} + Y_1 x_1 \qquad (8.7.11)$$

For the prescribed temperature field $\Theta = \Theta_0(1 + a x_2/h)$, the thermal force and moment are

$$N_\Theta = \alpha E A \Theta_0 \qquad M_\Theta = \frac{\alpha E I a \Theta_0}{h} \qquad (8.7.12)$$

According to Eq. (8.4.16) the complementary strain energy in a thermoelastic beam is

$$U^* = \int_0^L \left[\frac{(N + N_\Theta)^2}{2EA} + \frac{(M - M_\Theta)^2}{2EI} - \frac{\alpha^2 E}{2} \int_A \Theta^2 \, dA \right] dx_1 \qquad (8.7.13)$$

Engesser's theorem of least work (8.7.2) requires that U^* be a minimum. Since M is the only quantity in Eq. (8.7.13) which is a function of Y_1, we have

$$\frac{\partial U^*}{\partial Y_1} = \int_0^L \frac{M - M_\Theta}{EI} \frac{\partial M}{\partial Y_1} \, dx_1 = 0 \qquad (8.7.14)$$

Substituting Eqs. (8.7.11) and (8.7.12) into (8.7.14) and integrating gives

$$-\frac{Q_1 L^4}{8EI} + \frac{Y_1 L^3}{3EI} - \frac{\alpha a L^2 \Theta_0}{2h} = 0 \qquad (8.7.15)$$

so that

$$Y_1 = \frac{3 Q_1 L}{8} + \frac{3 \alpha E I a \Theta_0}{2hL} \qquad (8.7.16)$$

The bending moment expression (8.7.11) then becomes

$$M = -\frac{Q_1 x_1^2}{2} + \frac{3 Q_1 L x_1}{8} + \frac{3 \alpha E I a \Theta_0 x_1}{2hL} \qquad (8.7.17)$$

and the resulting thermal stress is

$$\sigma_{11} = \frac{N + N_\Theta}{A} - \frac{(M - M_\Theta) x_2}{I} - \alpha E \Theta$$

$$= -\frac{Q_1(3L - 4x_1) x_1 x_2}{8I} - \frac{3 \alpha E a \Theta_0 x_1 x_2}{2hL} \qquad (8.7.18)$$

which agrees with our previous result (8.3.35).

The displacement curve $u_2(x_1)$ can be obtained by integrating the equation $d^2 u_2/dx_1^2 = (M - M_\Theta)/EI$ twice and choosing the constants of integration so that they satisfy the boundary conditions $du_2(L)/dx_1 = u_2(L) = 0$. [Note, the third kinematic boundary condition $u_2(0) = 0$ is satisfied automatically since we have set $\partial U^*/\partial Y_1 = 0$.] The results of this computation were given in Example 8.2 and will not be repeated here.

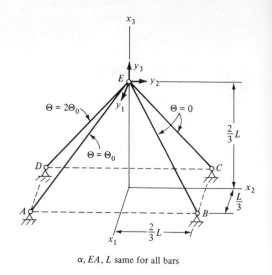

α, EA, L same for all bars

FIGURE 8.16
Space truss subject to temperature changes.

Example 8.11   Displacement method for a statically indeterminate space truss   We wish to determine the stress in each bar of the pin-jointed space truss shown in Fig. 8.16 when members AE and DE experience uniform temperature rises of Θ_0 and $2\Theta_0$, respectively. The displacement approach will be used in this example; the same problem will be solved using the force method in Example 8.12.

We observe that the structure has three kinematically indeterminate joint displacements, namely, the three degrees of freedom of joint E; these displacements are denoted by y_1, y_2, and y_3 as shown. Following the procedure used in Chap. 6, the extension Δ of each bar may be expressed in terms of the kinematically indeterminate displacements as follows (see Example 6.1):

$$\Delta^{(AE)} = -\tfrac{1}{3}y_1 + \tfrac{2}{3}y_2 + \tfrac{2}{3}y_3$$
$$\Delta^{(BE)} = -\tfrac{1}{3}y_1 - \tfrac{2}{3}y_2 + \tfrac{2}{3}y_3$$
$$\Delta^{(CE)} = \tfrac{1}{3}y_1 - \tfrac{2}{3}y_2 + \tfrac{2}{3}y_3 \qquad (8.7.19)$$
$$\Delta^{(DE)} = \tfrac{1}{3}y_1 + \tfrac{2}{3}y_2 + \tfrac{2}{3}y_3$$

Next, the thermoelastic strain energy U in the truss is computed by summing the energy stored in each member. For an axially loaded bar $(M = M_\Theta = 0)$

subject to a uniform normal strain $e_{11} = du_1/dx_1 = \Delta/L$ and a uniform temperature rise Θ, Eq. (8.4.15) gives

$$U = \int_0^L \left[\frac{EA}{2} \left(\frac{du_1}{dx_1}\right)^2 - N_\Theta \frac{du_1}{dx_1} + \frac{\alpha^2 E}{2} \int_A \Theta^2 \, dA \right] dx_1$$

$$= \int_0^L \left[\frac{EA}{2} \left(\frac{\Delta}{L}\right)^2 - \alpha EA\Theta \frac{\Delta}{L} + \frac{\alpha^2 EA\Theta^2}{2} \right] dx_1$$

$$= \frac{EA}{2L} (\Delta - \alpha L\Theta)^2 \tag{8.7.20}$$

Thus, for the members of truss we have

$$U^{(AE)} = \frac{EA}{2L} (\Delta^{(AE)} - \alpha L\Theta_0)^2$$

$$U^{(BE)} = \frac{EA}{2L} (\Delta^{(BE)} - 0)^2$$

$$\tag{8.7.21}$$

$$U^{(CE)} = \frac{EA}{2L} (\Delta^{(CE)} - 0)^2$$

$$U^{(DE)} = \frac{EA}{2L} (\Delta^{(DE)} - 2\alpha L\Theta_0)^2$$

Substituting the expressions (8.7.19) for the elongations into Eqs. (8.7.21) and adding, we obtain the total strain energy

$$U = \frac{EA}{18L} [4y_1^2 + 16y_2^2 + 16y_3^2 - 6\alpha L\Theta_0 y_1 - 36\alpha L\Theta_0 y_2$$

$$- 36\alpha L\Theta_0 y_3 + 45(\alpha L\Theta_0)^2] \tag{8.7.22}$$

Application of Eq. (8.7.1) then gives

$$\frac{\partial U}{\partial y_1} = \frac{EA}{18L} (8y_1 - 6\alpha L\Theta_0) = 0$$

$$\frac{\partial U}{\partial y_2} = \frac{EA}{18L} (32y_2 - 36\alpha L\Theta_0) = 0 \tag{8.7.23}$$

$$\frac{\partial U}{\partial y_3} = \frac{EA}{18L} (32y_3 - 36\alpha L\Theta_0) = 0$$

Solving Eqs. (8.7.23) for the kinematically indeterminate displacements yields

$$y_1 = \tfrac{3}{4}\alpha L\Theta_0 \qquad y_2 = y_3 = \tfrac{9}{8}\alpha L\Theta_0 \tag{8.7.24}$$

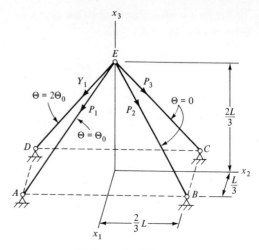

α, EA, L same for all bars

FIGURE 8.17
Space truss subject to temperature changes.

The elongations of the bars, found by substituting the relations (8.7.24) into Eqs. (8.7.19), are

$$\Delta^{(AE)} = \tfrac{5}{4}\alpha L\Theta_0 \qquad \Delta^{(BE)} = -\tfrac{1}{4}\alpha L\Theta_0$$
$$\Delta^{(CE)} = \tfrac{1}{4}\alpha L\Theta_0 \qquad \Delta^{(DE)} = \tfrac{7}{4}\alpha L\Theta_0 \qquad (8.7.25)$$

Computing the axial strain $e_{11} = \Delta/L$ in each member, and then solving for the corresponding stress $\sigma_{11} = E(e_{11} - \alpha\Theta)$ leads to the results

$$\sigma_{11}{}^{(AE)} = E(\tfrac{5}{4}\alpha\Theta_0 - \alpha\Theta_0) = \tfrac{1}{4}\alpha E\Theta_0$$
$$\sigma_{11}{}^{(BE)} = E(-\tfrac{1}{4}\alpha\Theta_0 - 0) = -\tfrac{1}{4}\alpha E\Theta_0$$
$$\sigma_{11}{}^{(CE)} = E(\tfrac{1}{4}\alpha\Theta_0 - 0] = \tfrac{1}{4}\alpha E\Theta_0 \qquad (8.7.26)$$
$$\sigma_{11}{}^{(DE)} = E(\tfrac{7}{4}\alpha\Theta_0 - 2\alpha\Theta_0) = -\tfrac{1}{4}\alpha E\Theta_0 \qquad ////$$

Example 8.12 Force method for a statically indeterminate space truss
The force method will now be used to determine the temperature-induced stresses in the space truss of Example 8.11. Since there are four unknown bar forces in the structure but only three independent equations of statics for the concurrent force system, the truss is statically indeterminate to the first degree; the force in bar DE is taken as the redundant Y_1 (Fig. 8.17). The remaining bar forces, designated as P_1, P_2, and P_3 can be expressed in terms of Y_1 by writing equations of statics. Considering joint E as a free body, and equating the x_1, x_2, and x_3 components of the resultant force to zero gives

$$\tfrac{1}{3}P_1 + \tfrac{1}{3}P_2 - \tfrac{1}{3}P_3 - \tfrac{1}{3}Y_1 = 0$$
$$-\tfrac{2}{3}P_1 + \tfrac{2}{3}P_2 + \tfrac{2}{3}P_3 - \tfrac{2}{3}Y_1 = 0 \qquad (8.7.27)$$
$$-\tfrac{2}{3}P_1 - \tfrac{2}{3}P_2 - \tfrac{2}{3}P_3 - \tfrac{2}{3}Y_1 = 0$$

Solving Eqs. (8.7.27) for the member forces yields

$$P_1 = -P_2 = P_3 = -Y_1 \qquad (8.7.28)$$

The complementary strain energy U^* is computed by adding the energies of the individual members. For a bar subject to an end load P and a uniform temperature rise Θ, Eq. (8.4.16) gives

$$U^* = \int_0^L \left[\frac{(N + N_\Theta)^2}{2EA} - \frac{\alpha^2 E}{2} \int_A \Theta^2 \, dA \right] dx_1$$
$$= \int_0^L \left[\frac{(P + \alpha EA\Theta)^2}{2EA} - \frac{\alpha^2 EA\Theta^2}{2} \right] dx_1$$
$$= \frac{LP^2}{2EA} + \alpha LP\Theta \qquad (8.7.29)$$

Thus, for the members of the truss

$$U^{*(AE)} = \frac{LP_1{}^2}{2EA} + \alpha LP_1 \Theta_0$$

$$U^{*(BE)} = \frac{LP_2{}^2}{2EA}$$

$$U^{*(CE)} = \frac{LP_3{}^2}{2EA} \qquad (8.7.30)$$

$$U^{*(DE)} = \frac{LY_1{}^2}{2EA} + 2\alpha LY_1 \Theta_0$$

Summing the individual energies, and using Eqs. (8.7.28) to eliminate the dependence of U^* upon P_1, P_2, and P_3, we obtain

$$U^* = \frac{2LY_1{}^2}{EA} + \alpha LY_1 \Theta_0 \qquad (8.7.31)$$

Engesser's theorem of least work then yields

$$\frac{\partial U^*}{\partial Y_1} = \frac{4LY_1}{EA} + \alpha L\Theta_0 = 0 \qquad (8.7.32)$$

so that

$$Y_1 = -\frac{\alpha EA\Theta_0}{4} \qquad (8.7.33)$$

The remaining bar forces are

$$P_1 = -P_2 = P_3 = \frac{\alpha EA\Theta_0}{4} \qquad (8.7.34)$$

and the corresponding thermal stresses, computed using Eq. (8.3.13), are

$$\sigma_{11}{}^{(AE)} = -\sigma_{11}{}^{(BE)} = \sigma_{11}{}^{(CE)} = -\sigma_{11}{}^{(DE)} = \frac{\alpha E \Theta_0}{4} \qquad (8.7.35)$$

in agreement with the results of Example 8.11.   ////

PROBLEMS

8.1   The beam shown is free of surface tractions but subject to the temperature change $\Theta = \Theta_0(1 - 4x_2{}^2/h^2)$. Compute the stress σ_{11}.

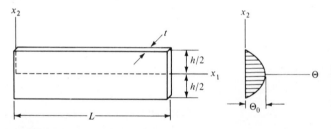

PROBLEM 8.1

8.2   Show that the longitudinal stress σ_{11} in an unsymmetric elastic bar subject to a temperature change $\Theta(x_1, x_2, x_3)$ is given by

$$\sigma_{11} = \frac{N + N_\Theta}{A} - \frac{(M_3 - M_{3\Theta})x_2}{I_{33}} + \frac{(M_2 + M_{2\Theta})x_3}{I_{22}} - \alpha E \Theta$$

providing x_2 and x_3 are principal axes. Here N is the normal force, M_2 and M_3 are components of the resultant bending moment in the x_2 and x_3 directions, $N_\Theta = \alpha E \int_A \Theta \, dA$, $M_{2\Theta} = \alpha E \int_A \Theta x_3 \, dA$, and $M_{3\Theta} = \alpha E \int_A \Theta x_2 \, dA$. The principal moments of inertia of the cross section are $I_{22} = \int_A x_3{}^2 \, dA$ and $I_{33} = \int_A x_2{}^2 \, dA$.

8.3   The statically indeterminate beam shown is acted upon by a linearly varying distributed load and is also subject to the temperature change $\Theta = \Theta_0 x_1 x_2/Lh$. Compute the displacement curve $u_2(x_1)$ by integrating Eq. (8.3.17). Also compute the stress σ_{11}.

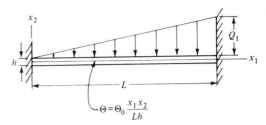

PROBLEM 8.3

8.4 Use direct integration to investigate the deformation and stress for the pin-ended prismatic bar shown. The temperature change is $\Theta = \Theta_0 \sin(\pi x_1/L)\cos(\pi x_2/h)$.

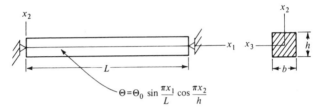

PROBLEM 8.4

8.5 Use direct integration to find the rotation at the right end of a simply supported beam subject to a linearly varying distributed load and a temperature distribution of the form $\Theta = \Theta_0 x_1 x_2/Lh$.

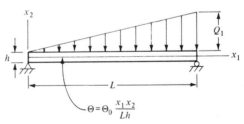

PROBLEM 8.5

8.6 Verify the strain energy expressions (8.4.14) and (8.4.15).

8.7 Verify the complementary strain energy expressions (8.4.16) and (8.4.17).

8.8 The structure shown is loaded by external forces Q_1 and Q_2. In addition, each bar is subject to a linear temperature variation. If the resulting displacements of joint C are q_1 and q_2, compute the corresponding forces Q_1 and Q_2 using the dummy-displacement method.

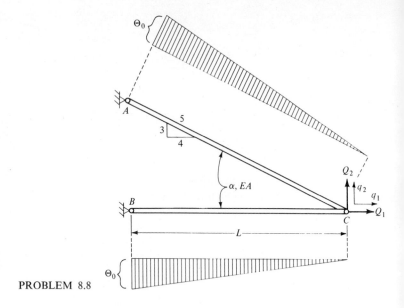

PROBLEM 8.8

8.9   Use the dummy-force method to find the displacements q_1 and q_2 produced by the applied loads and the temperature changes in Prob. 8.8.

8.10   A symmetric cantilever bar is subject to the linear temperature field $\Theta = ax_1 + bx_2 + c$. Use the dummy-force method to compute the horizontal displacement Δ_1 and the vertical displacement Δ_2 of a point located at a distance ξ_1 from the fixed end of the beam. Compare your results with Eq. (8.3.25).

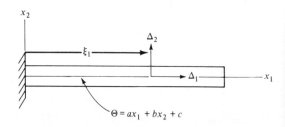

PROBLEM 8.10

8.11   Solve Prob. 8.5 using the dummy-force method.

8.12   Show that the variation of the thermoelastic strain energy U associated with variations in the strain components e_{ij} is given by

$$\delta U = \int_{\mathcal{V}} \sigma_{ij} \, \delta e_{ij} \, d\mathcal{V}.$$

8.13   Show that the variation of the thermoelastic complementary strain energy U^* with respect to the stress components σ_{ij} is given by

$$\delta U^* = \int_{\mathscr{V}} e_{ij}\, \delta\sigma_{ij}\, d\mathscr{V}.$$

8.14   Solve Prob. 8.5 using Engesser's theorem.

8.15   Solve Example 8.6 using Engesser's theorem.

8.16   Joint D of the truss shown suffers a horizontal displacement q_1 due to the applied force Q_1 and the uniform temperature changes indicated. Use the displacement method to compute the force Q_1 in terms of q_1, Θ_0 and α, L, EA.

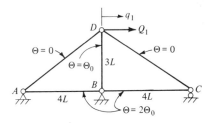

PROBLEM 8.16 α, EA same for all bars

8.17   Use the force method to compute the displacement q_1 in terms of Q_1, Θ_0 and α, L, EA for the truss in Prob. 8.16.

8.18   Use Engesser's theorem of least work to compute the external reactions for the statically indeterminate beam in Prob. 8.3. Also compute the stress σ_{11}.

8.19   Compute the maximum circumferential stress in a circular ring of rectangular cross section which is subject to the radial temperature distribution $\Theta = \Theta_0[1 + a(x_2/h)^3]$, in which x_2 is the radial coordinate and a is a constant.

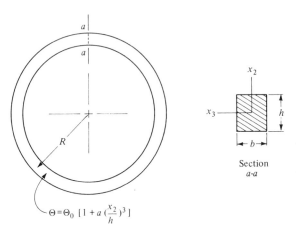

PROBLEM 8.19

REFERENCES

For a comprehensive treatment of the theoretical foundations of thermoelasticity, general methods of solution for two- and three-dimensional problems, thermal-stress analysis of beams (8.1), plates (8.1 and 8.2), and shells (8.2), and an introduction to thermal stresses in inelastic bodies (8.1 and 8.2):

8.1 BOLEY, B. A., and J. H. WEINER: "Theory of Thermal Stresses," Wiley, New York, 1960.

8.2 NOWACKI, W.: "Thermoelasticity," Addison-Wesley, Reading, Mass., 1962.

8.3 PARKUS, H.: "Thermoelasticity," Blaisdell, Waltham, Mass., 1968.

For the analysis of thermal stresses in aerospace structures, including applications of energy methods:

8.4 GATEWOOD, B. E.: "Thermal Stresses: With Applications to Airplanes, Missiles, Turbines and Nuclear Reactors," McGraw-Hill, New York, 1957.

8.5 RIVELLO, R. M.: "Theory and Analysis of Flight Structures," McGraw-Hill, New York, 1969.

For an account of the general theory of heat conduction, including solutions to a large number of special problems:

8.6 CARSLAW, H. S., and J. C. JAEGER: "Conduction of Heat in Solids," 2d ed., Oxford University Press, London, 1959.

STRUCTURAL STABILITY

9.1 INTRODUCTION

Thus far in our study of the behavior of elastic structures, we have assumed that all elements of the structure are in a state of stable equilibrium. We know from experience, however, that for certain types of loadings a given element may become unstable; i.e., it may buckle. This in turn may lead to a total collapse of the structure.

Elements having one dimension much smaller than another are particularly susceptible to buckling instability. For example the slender column, the frame, the circular ring, the thin panel, and the cylindrical shell shown in Fig. 9.1a to e will suffer disproportionately large displacements when the magnitudes of the applied compressive forces reach certain critical values. Buckling may also occur in thin-walled members as a result of torsion or combined torsion and flexure as shown in Fig. 9.1f. All the above examples represent conditions of *primary instability*, wherein the wavelength of the buckled shape is of the order of the member's length.

It is also possible for buckling to occur locally, such that the cross-sectional shape of the affected member experiences significant distortion. If the buckling is confined to a region on the order of the member's thickness, the instability is

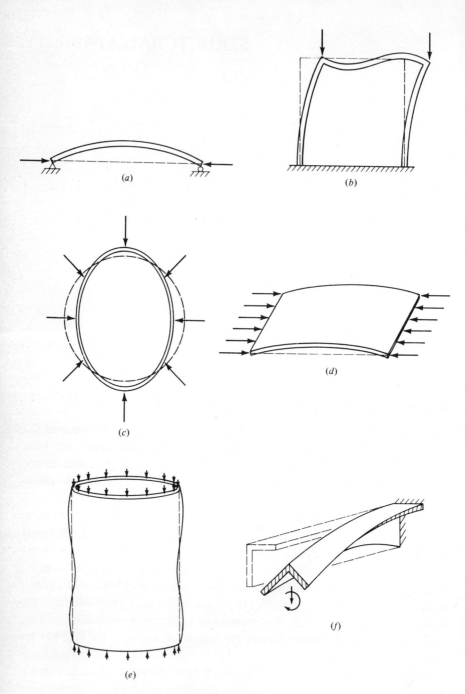

FIGURE 9.1
Examples of structures which are susceptible to buckling.

referred to as *local* or *secondary*. Secondary instabilities are generally less catastrophic than primary ones. While they may affect the structure's appearance, quite often they have no significant effect upon the structure's capacity to support loads.

In this chapter we will concentrate on one of the simpler types of elastic instability, namely, primary instability of a slender beam column. For simplicity the discussion will be restricted to the case of small displacements, i.e., displacements no larger than the beam column's depth. While this problem is rather specialized, it does demonstrate the essential characteristics of elastic stability. The methods of analysis used here are equally applicable in more complicated problems. A list of references covering the stability of various other types of elastic structures, inelastic bodies, large deformations of buckled members, etc., is given at the end of the chapter.

9.2 STABILITY CRITERIA

In order to answer the question of whether a structure is in stable equilibrium under a given set of loads, we must define first explicitly what we mean by stability. Suppose that a statically loaded structure were to undergo a virtual distortion of the type described in Chap. 5 (i.e., an imaginary, infinitesimal deformation consistent with all the geometric constraints). If upon releasing the structure from this virtually deformed state, the system returns to its previous configuration, then we say that the equilibrium configuration is *stable*. On the other hand if the loaded structure does not return to its original configuration following a virtual distortion, the condition is one of *unstable equilibrium*.

Alternatively, stability can be defined in terms of the total potential energy Π of the structure. Recall that Π is the sum of the internal energy U and the potential V_E of the external forces. If the total potential Π increases during a virtual distortion, then the equilibrium configuration is defined to be stable; if Π decreases or remains constant, the configuration is unstable. In other words a relative minimum (maximum) of Π is associated with stable (unstable) equilibrium.

Equivalence of the above two definitions is easily demonstrated. Consider a system which is in stable equilibrium. Our second definition of stability implies that the change in the internal strain energy U is greater than the change in the work $-V_E$ done by the applied loads during a virtual distortion. The strain energy U represents energy associated with the structure's elastic restoring forces. Hence, when the structure is released from a state of virtual deformation, it is subject to a net restoring action. Consequently the system returns to its initial configuration, in accordance with our first definition.

Let us now express the stability criteria in mathematical form. For simplicity it is assumed that the structure's deformation is characterized by a finite number of generalized displacements q_i. If the body is given a virtual distortion δq_i about its equilibrium configuration, then it is possible to write the total potential energy in a Taylor's expansion about q_i. For a two-degree-of-freedom system, for example,

$$\Pi(q_1 + \delta q_1, q_2 + \delta q_2) = \Pi(q_1, q_2) + \frac{\partial \Pi}{\partial q_1} \delta q_1 + \frac{\partial \Pi}{\partial q_2} \delta q_2$$

$$+ \frac{1}{2!} \left[\frac{\partial^2 \Pi}{\partial q_1^2} (\delta q_1)^2 + 2 \frac{\partial^2 \Pi}{\partial q_1 \partial q_2} \delta q_1 \delta q_2 + \frac{\partial^2 \Pi}{\partial q_2^2} (\delta q_2)^2 \right] + \cdots \qquad (9.2.1)$$

The change in potential energy $\Delta \Pi = \Pi(q_1 + \delta q_1, q_2 + \delta q_2) - \Pi(q_1, q_2)$ can therefore be expressed as

$$\Delta \Pi = \delta \Pi + \frac{1}{2!} \delta^2 \Pi + \cdots \qquad (9.2.2)$$

where the first variation is zero by virtue of the pmpe

$$\delta \Pi = \frac{\partial \Pi}{\partial q_1} \delta q_1 + \frac{\partial \Pi}{\partial q_2} \delta q_2 = 0 \qquad (9.2.3)$$

and the *second variation* $\delta^2 \Pi$ is defined as

$$\delta^2 \Pi = \delta(\delta \Pi) = \frac{\partial^2 \Pi}{\partial q_1^2} (\delta q_1)^2 + 2 \frac{\partial^2 \Pi}{\partial q_1 \partial q_2} \delta q_1 \delta q_2 + \frac{\partial^2 \Pi}{\partial q_2^2} (\delta q_2)^2 \qquad (9.2.4)$$

Note that the sign of $\Delta \Pi$ in Eq. (9.2.2) is determined by the first nonvanishing term in the Taylor's expansion. Since $\delta \Pi = 0$, the second variation is generally the relevant term. For instance if $\delta^2 \Pi$ is positive, then $\Delta \Pi$ is positive, Π is a relative minimum, and consequently the equilibrium configuration is stable. If $\delta^2 \Pi$ is negative, Π is a relative maximum, and the configuration is unstable. The special case in which the second and all higher variations of Π are zero corresponds to a state known as *neutral equilibrium*. When a structure which is in neutral equilibrium is released from a virtual distortion, there is no net restoring force present, and the system remains in its virtually deformed state. Hence, by our first definition of stability, neutral equilibrium is a special case of unstable equilibrium. In summary, the criteria for stability are as follows:

$$\Delta \Pi > 0 \qquad \text{stable equilibrium}$$
$$\Delta \Pi = 0 \qquad \text{neutral equilibrium} \qquad (9.2.5)$$
$$\Delta \Pi < 0 \qquad \text{unstable equilbrium}$$

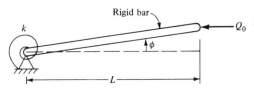

FIGURE 9.2
One-degree-of-freedom system.

If the potential energy Π is quadratic in the displacements q_i, which is the case when the structure is linearly elastic and the deformations are infinitesimal, then all variations higher than the second are necessarily zero. In this case the type of equilibrium is governed by the following conditions:

$$\delta^2\Pi > 0 \qquad \text{stable equilibrium}$$
$$\delta^2\Pi = 0 \qquad \text{neutral equilibrium} \qquad (9.2.6)$$
$$\delta^2\Pi < 0 \qquad \text{unstable equilibrium}$$

Before examining the buckling behavior of continuous elastic structures, it will be helpful to consider a few problems involving idealized one- and two-degree-of-freedom systems.

Example 9.1 Stability of a one-degree-of-freedom system As a first example we shall investigate the stability characteristics of the rigid, weightless bar shown in Fig. 9.2. The pinned end of the bar is fastened to a torsional spring of stiffness k. For the time being we will assume that the bar is subject to a longitudinal force Q_0 alone; in Example 9.2 we shall investigate the effect of adding a small transverse force.

Since the bar is assumed to be rigid, the system's strain energy consists solely of the energy stored in the torsional spring; thus

$$U = \tfrac{1}{2}k\phi^2 \qquad (9.2.7)$$

where ϕ represents the bar's rotation. The potential of the applied force Q_0 is

$$V_E = -Q_0 L(1 - \cos\phi) \qquad (9.2.8)$$

and the total potential energy $\Pi = U + V_E$ is therefore given by

$$\Pi = \tfrac{1}{2}k\phi^2 - Q_0 L(1 - \cos\phi) \qquad (9.2.9)$$

For equilibrium the first variation of the total potential is zero, so that

$$\delta\Pi = (k\phi - Q_0 L \sin\phi)\delta\phi = 0 \qquad (9.2.10)$$

The positions of the bar for which Eq. (9.2.10) is satisfied are $\phi = 0$ and $\phi/\sin \phi = Q_0/(k/L)$. To determine whether these equilibrium configurations are stable or unstable, we evaluate the second variation of Π

$$\delta^2\Pi = (k - Q_0 L \cos \phi)(\delta\phi)^2 \quad (9.2.11)$$

Substituting $\phi = 0$ into Eq. (9.2.11) gives

$$\delta^2\Pi = (k - Q_0 L)(\delta\phi)^2 \quad (9.2.12)$$

Hence the equilibrium position $\phi = 0$ is stable if $Q_0 < k/L$ (in which case $\delta^2\Pi > 0$), and unstable ($\delta^2\Pi < 0$) if $Q_0 > k/L$. If $Q_0 = k/L$, $\delta^2\Pi$ vanishes identically. In order to determine the type of equilibrium in this case, it becomes necessary to examine the higher order terms in the expansion (9.2.2); that is,

$$\delta^3\Pi = Q_0 L \sin \phi(\delta\phi)^3$$
$$\delta^4\Pi = Q_0 L \cos \phi(\delta\phi)^4 \quad (9.2.13)$$
$$\vdots$$

Note that $\delta^3\Pi$ is zero when $\phi = 0$. However, $\delta^4\Pi$ is positive when $\phi = 0$, for arbitrary variations $\delta\phi$. Therefore $\Delta\Pi > 0$, and the equilibrium is stable. In other words, if the bar were given an infinitesimal rotation about $\phi = 0$, it would return to its undisturbed position only if the magnitude of the axial force Q_0 were less than or equal to k/L.

Next consider the type of equilibrium associated with the second possible deformation configuration. Substituting $Q_0 L = k\phi/\sin \phi$ into Eq. (9.2.11) yields

$$\delta^2\Pi = k(1 - \phi \cot \phi)(\delta\phi)^2 \quad (9.2.14)$$

Since the quantity $1 - \phi \cot \phi$ is positive for all values of ϕ between 0 and π, this configuration is stable.

The load-versus-rotation relationship for the bar is shown as curve $0AB$ in Fig. 9.3. Note that every point on the curve represents a state of stable equilibrium.

The results obtained above are valid for arbitrarily large rotations. It is rarely possible to obtain such generality when dealing with more complicated systems. You will recall, for instance, that in analyzing continuous structures, we invariably make use of small deformation theories. Let us now investigate how the assumption of small displacements affects the solution to this problem.

Using the small angle approximation $\cos \phi = 1 - \phi^2/2$,[1] the total potential energy (9.2.9) can be written as

$$\Pi = \tfrac{1}{2}(k - Q_0 L)\phi^2 \quad (9.2.15)$$

---

[1] Based upon the assumption that ϕ is small compared to unity, terms of order higher than quadratic have been neglected in the trigonometric series

$$\cos \phi = 1 - \frac{\phi^2}{2!} + \frac{\phi^4}{4!} - \cdots$$

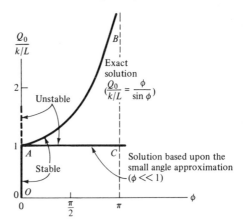

FIGURE 9.3
Load-versus-rotation curve for the one-
degree-of-freedom system.

The first and second variations of Π are

$$\delta\Pi = (k - Q_0 L)\phi \,\delta\phi \qquad (9.2.16)$$

and

$$\delta^2\Pi = (k - Q_0 L)(\delta\phi)^2 \qquad (9.2.17)$$

From Eq. (9.2.16) we note that $\phi = 0$ and $Q_0 = k/L$ represent two independent conditions for equilibrium. Equation (9.2.17) shows that the equilibrium position $\phi = 0$ is stable providing $Q_0 < k/L$ and unstable if $Q_0 \geq k/L$.

The type of equilibrium corresponding to the condition $Q_0 = k/L$ is neutral, since $\delta^2\Pi$ and all higher variations of Π are identically zero in this case. Hence, if the system were subject to a virtual rotation $\delta\phi$ and then released, the rod would not return to its original position but instead would remain in the virtually dis-placed position. Consequently the angle ϕ has arbitrary magnitude when $Q_0 = k/L$.

Having investigated each of the possible equilibrium configurations, we see that stable equilibrium is possible only when the magnitude of the axial force Q_0 is less than k/L, in which case $\phi = 0$. The load at which neutral equilibrium is reached is referred to as the *critical* or *buckling* load. Hence the small displace-ment theory predicts that

$$(Q_0)_{\text{critical}} = \frac{k}{L} \qquad (9.2.18)$$

The results of this analysis are represented by the curve $0AC$ in Fig. 9.3; points along $0A$ represent conditions of stable equilibrium, and points on AC correspond to neutral equilibrium configurations. It is particularly important to note that while the approximate formulation predicts that the bar is unstable when ϕ is greater than zero, the exact theory (curve $0AB$) shows that there is a

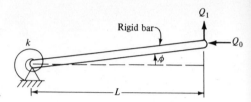

FIGURE 9.4
One-degree-of-freedom system.

position of stable equilibrium for every value of ϕ less than π. Moreover the possibility of ϕ becoming infinite when $Q_0 = k/L$ is ruled out by the exact solution. Thus while the approximate solution yields valid results as long as the magnitude of Q_0 is less than $(Q_0)_{\text{critical}}$, it is not applicable for larger values of Q_0.

////

Example 9.2   Stability of a one-degree-of-freedom system subject to a transverse load   The effect of applying a small transverse force to the structure analyzed in Example 9.1 will now be investigated. The total potential energy of the system in this case (Fig. 9.4) is

$$\Pi = \tfrac{1}{2}k\phi^2 - Q_0 L(1 - \cos \phi) - Q_1 L \sin \phi \qquad (9.2.19)$$

in which the last term represents the potential of the additional force Q_1. Equilibrium of the structure requires that

$$\delta\Pi = (k\phi - Q_0 L \sin \phi - Q_1 L \cos \phi)\,\delta\phi = 0 \qquad (9.2.20)$$

and therefore

$$Q_0 = \frac{k\phi - Q_1 L \cos \phi}{L \sin \phi} \qquad (9.2.21)$$

The type of equilibrium depends upon the sign of the second variation

$$\delta^2\Pi = (k - Q_0 L \cos \phi + Q_1 L \sin \phi)(\delta\phi)^2 \qquad (9.2.22)$$

Evaluating $\delta^2\Pi$ for the equilibrium condition (9.2.21) gives

$$\delta^2\Pi = \left[k(1 - \phi \cot \phi) + Q_1 L \frac{1}{\sin \phi}\right](\delta\phi)^2 \qquad (9.2.23)$$

Since $\delta^2\Pi$ is positive for all values of ϕ between 0 and π the configuration is stable. The axial force-versus-rotation relationship (9.2.21) corresponding to an arbitrarily chosen value of Q_1, namely, $Q_1 = 0.5\ k/L$ is plotted as curve $0D$ in Fig. 9.5. Note that the rotation ϕ is nonzero for all finite values of the axial force Q_0. When ϕ is equal to π, the force Q_0 is theoretically infinite.

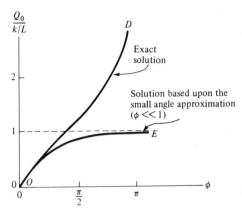

FIGURE 9.5
Axial force-versus-rotation curve for the case $Q_1 = 0.5 \, k/L$.

The way in which the assumption of small displacements affects the solution to this problem can be investigated following the approach described in Example 9.2. In so doing, it is found that equilibrium exists when

$$Q_0 = \frac{k}{L} - \frac{Q_1}{\phi} \quad \text{or} \quad \phi = \frac{Q_1}{k/L - Q_0} \qquad (9.2.24)$$

The equilibrium is stable if $Q_0 < k/L$ and unstable if $Q_0 > k/L$. Neutral equilibrium exists when $Q_0 = k/L$, in which case the rotation ϕ is infinite. These results are represented by the curve $0E$ in Fig. 9.5. Note that the solution (9.2.24) is accurate only for very small values of ϕ. While the rotation is proportional to the lateral force Q_1, it is clearly not proportional to the axial load Q_0. It is further noted that the critical value of the compressive force is $(Q_0)_{\text{critical}} = k/L$, regardless of the magnitude of the transverse force Q_1.   ////

Example 9.3   Stability of a two-degree-of-freedom system   Let us now compute the critical loads and the corresponding deflected shapes for the structure shown in Fig. 9.6. Bars AB and BC are assumed to be weightless and rigid. Two torsional springs of stiffness k constrain the system as shown. The rotations ϕ_1 and ϕ_2, assumed small, will be used to describe the deflected shape of this two-degree-of-freedom system. Accordingly the structure's strain energy is written as

$$U = \tfrac{1}{2}k(\phi_1 - \phi_2)^2 + \tfrac{1}{2}k\phi_2{}^2 \qquad (9.2.25)$$

and the potential of the external force Q_0 is

$$V_E = -Q_0[L(1 - \cos \phi_1) + L(1 - \cos \phi_2)] \qquad (9.2.26)$$

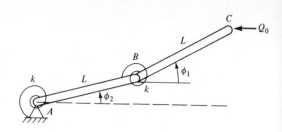

FIGURE 9.6
Two-degree-of-freedom system.

Therefore the total potential energy is given by

$$\Pi = \tfrac{1}{2}k(\phi_1{}^2 - 2\phi_1\phi_2 + 2\phi_2{}^2) - Q_0 L(2 - \cos\phi_1 - \cos\phi_2) \qquad (9.2.27)$$

The condition for equilibrium is

$$\delta\Pi = \frac{\partial\Pi}{\partial\phi_1}\,\delta\phi_1 + \frac{\partial\Pi}{\partial\phi_2}\,\delta\phi_2 = 0 \qquad (9.2.28)$$

or

$$\delta\Pi = [k(\phi_1 - \phi_2) - Q_0 L \sin\phi_1]\,\delta\phi_1$$
$$+ [-k(\phi_1 - 2\phi_2) - Q_0 L \sin\phi_2]\,\delta\phi_2 = 0 \qquad (9.2.29)$$

Since the variations $\delta\phi_1$ and $\delta\phi_2$ are arbitrary, Eq. (9.2.29) requires that

$$k(\phi_1 - \phi_2) - Q_0 L \sin\phi_1 = 0$$
$$-k(\phi_1 - 2\phi_2) - Q_0 L \sin\phi_2 = 0 \qquad (9.2.30)$$

In order to simplify the analysis, we now introduce the small angle approximation $\sin\phi \cong \phi$, in which case Eqs. (9.2.30) can be written as

$$(k - Q_0 L)\phi_1 - k\phi_2 = 0$$
$$-k\phi_1 + (2k - Q_0 L)\phi_2 = 0 \qquad (9.2.31)$$

This system of homogeneous equations represents an *eigenvalue problem* (see Appendix B). For a solution other than the trivial one, $\phi_1 = \phi_2 = 0$, the determinant of the coefficients of ϕ_1 and ϕ_2 must vanish. Thus

$$\begin{vmatrix} k - Q_0 L & -k \\ -k & 2k - Q_0 L \end{vmatrix} = 0 \qquad (9.2.32)$$

Expanding the determinant and dividing by L^2 yields the characteristic equation

$$Q_0{}^2 - 3\frac{k}{L}Q_0 + \left(\frac{k}{L}\right)^2 = 0 \qquad (9.2.33)$$

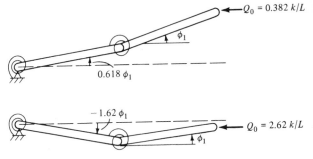

FIGURE 9.7
Modes of buckling.

the roots of which are

$$Q_0 = 0.382 \frac{k}{L} \quad \text{and} \quad 2.62 \frac{k}{L} \quad (9.2.34)$$

Ratios between the angles of rotation of the bars (the eigenvectors) associated with these forces (eigenvalues) are obtained by substituting the values (9.2.34) into Eqs. (9.2.31). This gives

$$\frac{\phi_2}{\phi_1} = 0.618 \quad \left(Q_0 = 0.382 \frac{k}{L}\right)$$

$$\frac{\phi_2}{\phi_1} = -1.62 \quad \left(Q_0 = 2.62 \frac{k}{L}\right) \quad (9.2.35)$$

The corresponding deflected shapes are shown in Fig. 9.7. Since the amplitude ϕ_1 is indeterminate, these shapes represent positions of neutral equilibrium. This conclusion may be verified by examining the second variation of the potential energy Π. In the case of small rotations we obtain from Eq. (9.2.29)

$$\delta^2\Pi = (k - Q_0 L)(\delta\phi_1)^2 - 2k\,\delta\phi_1\,\delta\phi_2 + (2k - Q_0 L)(\delta\phi_2)^2 \quad (9.2.36)$$

The sign of $\delta^2\Pi$ depends upon the value of the determinant given in Eq. (9.2.32). It can be shown (see Ref. 9.4) that $\delta^2\Pi$ will be positive (definite) if the value of the determinant is positive, in which case the equilibrium position is stable. Similarly $\delta^2\Pi$ will be negative if the value of the determinant is negative, in which case the state of equilibrium is unstable. Neutral equilibrium is associated with a zero value of the determinant. Consequently solutions to the eigenvalue problem represent conditions of neutral equilibrium. The eigenvalues (9.2.34) are therefore the critical or buckling forces, and the eigenvectors (9.2.35) represent the corresponding *buckling modes* or shapes.

It must be remembered that our results are applicable only in the case of small rotations. The possibility of large rotations is great when Q_0 is close to the lowest critical value $(0.382\,k/L)$. ////

9.3 EQUILIBRIUM OF A BEAM COLUMN

Bars subjected to combined lateral and compressive axial loads are known as beam columns. The conditions for equilibrium of a beam column are developed in this article. Critical values of the compressive loads and the deflections, for various loadings and support conditions, will be examined in the subsequent sections of this chapter.

Consider a long, slender, symmetric beam column acted upon by the axial forces Q_0 and the distributed transverse load $Q(x_1)$ shown in Fig. 9.8. The lateral deflection $u_2(x_1)$ is governed by the equation

$$\frac{d^2}{dx_1^2}\left(EI\,\frac{d^2u_2}{dx_1^2}\right) + Q_0\,\frac{d^2u_2}{dx_1^2} = Q \qquad (9.3.1)$$

This differential equation is normally derived using a strength-of-materials approach by considering the equilibrium of a differential element of the bar (see Prob. 9.3). The same result may be obtained using the pmpe as demonstrated below. An advantage of using the energy approach is that it yields the system's natural boundary conditions as part of the solution. That is, by leaving the end conditions initially unspecified it is possible to determine the types of boundary conditions which are admissible from an equilibrium standpoint. The following quantities are therefore considered to be arbitrary (see Fig. 9.8): the displacements $u_2(0)$ and $u_2(L)$; the slopes $du_2(0)/dx_1$ and $du_2(L)/dx_1$; the shear forces V_0 and V_L; and the bending moments M_0 and M_L.

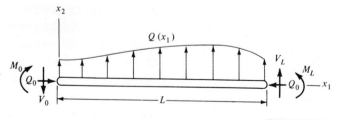

FIGURE 9.8
Beam column.

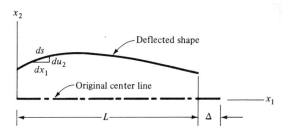

FIGURE 9.9
Deflected shape of the beam column's axis.

Neglecting the strain energy associated with the axial and shear forces in the bar, the total strain energy is that due to bending; thus

$$U = \int_0^L \frac{EI}{2} (u_2'')^2 \, dx_1 \qquad (9.3.2)$$

As before, a prime denotes differentiation with respect to the axial coordinate x_1. The potential of the external forces is

$$V_E = -Q_0 \Delta - \int_0^L Q u_2 \, dx_1 + V_0 u_2(0) + M_0 u_2'(0) - V_L u_2(L) - M_L u_2'(L) \qquad (9.3.3)$$

where Δ represents the displacement through which Q_0 moves due to the change in curvature of the bar (Fig. 9.9). Since the bar is assumed to be inextensible (we have already neglected the strain energy associated with its axial deformation), its total length is

$$\Delta + L = \int ds \qquad (9.3.4)$$

The differential quantity ds represents the length of an element of the deformed beam column. From Fig. 9.9 it is seen that

$$ds = (dx_1{}^2 + du_2{}^2)^{1/2} = [1 + (u_2')^2]^{1/2} \, dx_1 \qquad (9.3.5)$$

If the deflected shape of the bar differs only slightly from the original straight line, then $(u_2')^2$ is small compared to unity. In this case the square-root term in Eq. (9.3.5) may be expanded in a binomial series. Retaining only the first two terms of the series, we obtain

$$ds = [1 + \tfrac{1}{2}(u_2')^2] \, dx_1 \qquad (9.3.6)$$

Equation (9.3.4) then gives

$$\Delta + L = \int_0^L [1 + \tfrac{1}{2}(u_2')^2] \, dx_1 = L + \tfrac{1}{2} \int_0^L (u_2')^2 \, dx_1 \qquad (9.3.7)$$

and the displacement Δ is found to be

$$\Delta = \tfrac{1}{2} \int_0^L (u_2')^2 \, dx_1 \qquad (9.3.8)$$

By substituting this expression for Δ into Eq. (9.3.3), the potential V_E may be expressed as

$$V_E = - \int_0^L \left[\frac{Q_0}{2} (u_2')^2 + Qu_2 \right] dx_1 + V_0 u_2(0) + M_0 u_2'(0) - V_L u_2(L) - M_L u_2'(L) \qquad (9.3.9)$$

and the total potential $\Pi = U + V_E$ is

$$\Pi = \int_0^L \left[\frac{EI}{2} (u_2'')^2 - \frac{Q_0}{2} (u_2')^2 - Qu_2 \right] dx_1$$

$$+ V_0 u_2(0) + M_0 u_2'(0) - V_L u_2(L) - M_L u_2'(L) \qquad (9.3.10)$$

For equilibrium the first variation of the total potential is zero, so that

$$\delta\Pi = \int_0^L \left[EIu_2'' \, \delta(u_2'') - Q_0 u_2' \, \delta(u_2') - Q \, \delta u_2 \, dx_1 \right]$$

$$+ V_0 \, \delta u_2(0) + M_0 \, \delta u_2'(0) - V_L \, \delta u_2(L) - M_L \, \delta u_2'(L) = 0 \qquad (9.3.11)$$

Noting that $\delta(u_2') = (\delta u_2)'$ and $\delta(u_2'') = (\delta u_2)''$, the first two terms of the integrand in Eq. (9.3.11) may be integrated by parts to obtain

$$\int_0^L EI\, u_2'' \, \delta(u_2'') \, dx_1 = \left[EIu_2'' \, \delta u_2' \right]_0^L - \left[(EIu_2'')' \, \delta u_2 \right]_0^L + \int_0^L (EIu_2'')'' \, \delta u_2 \, dx_1$$

$$- \int_0^L Q_0 u_2' \, \delta(u_2') \, dx_1 = -\left[Q_0 u_2' \, \delta u_2 \right]_0^L + \int_0^L Q_0 u_2'' \, \delta u_2 \, dx_1 \qquad (9.3.12)$$

Substituting Eqs. (9.3.12) into (9.3.11) and rearranging terms gives

$$\int_0^L \left[(EIu_2'')'' + Q_0 u_2'' - Q \right] \delta u_2 \, dx_1$$

$$+ \left[V_0 + (EIu_2'')' + Q_0 u_2' \right]_{x_1=0} \delta u_2(0)$$

$$+ \left[-V_L - (EIu_2'')' - Q_0 u_2' \right]_{x_1=L} \delta u_2(L)$$

$$+ \left[M_0 - EIu_2'' \right]_{x_1=0} \delta u_2'(0) + \left[-M_L + EIu_2'' \right]_{x_1=L} \delta u_2'(L) = 0 \qquad (9.3.13)$$

Since the displacement and slope at each end of the beam column were left unspecified, the variations $\delta u_2(0)$, $\delta u_2(L)$, $\delta u_2'(0)$ and $\delta u_2'(L)$ are arbitrary. Moreover $\delta u_2(x_1)$ is arbitrary for $0 < x_1 < L$. Consequently each of the five bracketed terms in Eq. (9.3.13) must vanish independently; thus

$$(EIu_2'')'' + Q_0 u_2'' = Q \tag{9.3.14}$$

and
$$V_0 = -[(EIu_2'')' + Q_0 u_2']_{x_1=0} \quad \text{or} \quad \delta u_2(0) = 0 \tag{9.3.15}$$

$$V_L = -[(EIu_2'')' + Q_0 u_2']_{x_1=L} \quad \text{or} \quad \delta u_2(L) = 0 \tag{9.3.16}$$

$$M_0 = [EIu_2'']_{x_1=0} \quad \text{or} \quad \delta u_2'(0) = 0 \tag{9.3.17}$$

$$M_L = [EIu_2'']_{x_1=L} \quad \text{or} \quad \delta u_2'(L) = 0 \tag{9.3.18}$$

Equation (9.3.14) is the governing differential equation for a beam column, and Eqs. (9.3.15) to (9.3.18) represent admissible boundary conditions for the system. The first of each pair of boundary conditions (the one involving the shear force or the bending moment) is known as a static or natural boundary condition; the second is called a kinematic or rigid end condition. One or the other of each pair must be satisfied in order that $\delta\Pi$ vanish. It can be seen from Eq. (9.3.15), for example, that either the displacement or the shear force must be specified at $x_1 = 0$. If the displacement $u_2(0)$ is prescribed, then $\delta u_2(0) = 0$ and Eq.(9.3.15) is satisfied. On the other hand if the displacement is not specified, then $\delta u_2(0)$ is arbitrary and Eq. (9.3.15) requires that $V_0 = -[(EIu_2'')' + Q_0 u_2']_{x_1=0}$. Since the shear force at an arbitrary cross section of the beam column is given by $V(x_1) = -[(EIu_2'')' + Q_0 u_2']$ (see Prob. 9.3), $V_0 = V(0)$. Thus Eq. (9.3.15) will be satisfied if either the displacement $u_2(0)$ or the shear force $V(0)$ is prescribed. Similarly, Eq. (9.3.17) will be satisfied if either the slope $u_2'(0)$ or the bending moment $M_0 = M(0)$ is specified. Analogous conditions apply to the boundary $x_1 = L$.

Note that each of the four ordinary end conditions for a beam, namely, clamped, free, simply supported, and guided (see Fig. 8.13) represents a special case of the boundary conditions (9.3.15) to (9.3.18) for a beam column.

9.4 BUCKLING OF A PIN-ENDED COLUMN

For a column of uniform flexural rigidity EI in equilibrium under the application of compressive loads Q_0 alone, the differential equation (9.3.14) reduces to

$$u_2{}^{iv} + \beta^2 u_2'' = 0 \tag{9.4.1}$$

in which
$$\beta^2 = \frac{Q_0}{EI} \tag{9.4.2}$$

The general solution of Eq. (9.4.1) is

$$u_2 = C_1 \cos \beta x_1 + C_2 \sin \beta x_1 + C_3 x_1 + C_4 \tag{9.4.3}$$

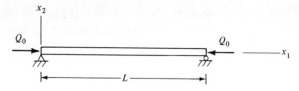

FIGURE 9.10
Pin-ended column.

where C_1, C_2, C_3, and C_4 are independent constants of integration to be determined from known boundary conditions. Considering the pin-ended column shown in Fig. 9.10, for example, we have

$$u_2(0) = u_2(L) = M(0) = M(L) = 0 \qquad (9.4.4)$$

or since $M = EIu_2''$,

$$u_2(0) = u_2(L) = u_2''(0) = u_2''(L) = 0 \qquad (9.4.5)$$

Application of these conditions to the general solution (9.4.3) leads to the following system of homogeneous equations:

$$
\begin{aligned}
C_1 + C_4 &= 0 \\
C_1 \cos \beta L + C_2 \sin \beta L + C_3 L + C_4 &= 0 \\
- C_1 \beta^2 &= 0 \\
- C_1 \beta^2 \cos \beta L - C_2 \beta^2 \sin \beta L &= 0
\end{aligned}
\qquad (9.4.6)
$$

For a nontrivial solution, the determinant of the coefficients in Eqs. (9.4.6) must vanish. Thus

$$
\begin{vmatrix}
1 & 0 & 0 & 1 \\
\cos \beta L & \sin \beta L & L & 1 \\
-\beta^2 & 0 & 0 & 0 \\
-\beta^2 \cos \beta L & -\beta^2 \sin \beta L & 0 & 0
\end{vmatrix} = 0 \qquad (9.4.7)
$$

The evaluation of this determinant gives the characteristic equation

$$-\beta^4 L \sin \beta L = 0 \qquad (9.4.8)$$

Therefore either $\beta = 0$ (which is a trivial solution since this implies $Q_0 = 0$) or $\sin \beta L = 0$. The latter is satisfied when

$$\beta = \frac{n\pi}{L} \qquad n = 1, 2, \ldots \qquad (9.4.9)$$

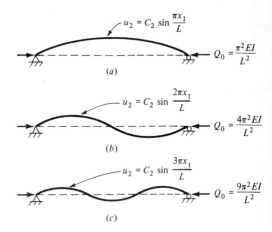

FIGURE 9.11
Lowest three modes of buckling for a pin-ended column.

Substitution of Eq. (9.4.9) into (9.4.2) shows that the column is in equilibrium when the magnitude of the compressive force has one of the following values:

$$Q_0 = \frac{n^2\pi^2 EI}{L^2} \qquad n = 1, 2, \ldots \qquad (9.4.10)$$

The corresponding deflections are determined by substituting Eq. (9.4.9) into (9.4.6). It is found that $C_1 = C_3 = C_4 = 0$ and C_2 is indeterminate, in which case

$$u_2 = C_2 \sin \frac{n\pi x_1}{L} \qquad n = 1, 2, \ldots \qquad (9.4.11)$$

Since these deflections satisfy the differential equation of equilibrium and the boundary conditions, and in addition have arbitrary amplitude, they represent positions of neutral equilibrium.

Figure 9.11 shows the modes of buckling for $n = 1, 2,$ and 3. In order to verify that these modes are indeed positions of neutral equilibrium, let us examine the second variation of the total potential energy Π. From Eq. (9.3.10) we have

$$\Pi = \int_0^L \left[\frac{EI}{2} (u_2'')^2 - \frac{Q_0}{2} (u_2')^2 \right] dx_1 \qquad (9.4.12)$$

For the *fundamental mode* $n = 1$, the deflected shape is of the form (Fig. 9.11a)

$$u_2 = a \sin \frac{\pi x_1}{L} \qquad (9.4.13)$$

where a represents the displacement at the column's midpoint. Substituting Eq. (9.4.13) into (9.4.12) gives

$$\Pi = \int_0^L \left[\frac{EI}{2} \left(-a \frac{\pi^2}{L^2} \sin \frac{\pi x_1}{L} \right)^2 - \frac{Q_0}{2} \left(a \frac{\pi}{L} \cos \frac{\pi x_1}{L} \right)^2 \right] dx_1 \qquad (9.4.14)$$

which upon integration yields

$$\Pi = \frac{\pi^2}{4L} \left(\frac{\pi^2 EI}{L^2} - Q_0 \right) a^2 \qquad (9.4.15)$$

For equilibrium we require that

$$\delta\Pi = \frac{d\Pi}{da} \delta a = \frac{\pi^2}{2L} \left(\frac{\pi^2 EI}{L^2} - Q_0 \right) a\, \delta a = 0 \qquad (9.4.16)$$

There are therefore two conditions for equilibrium. One is $a = 0$, which implies that the original straight-line shape of the column is an equilibrium configuration. The second condition is $Q_0 = \pi^2 EI/L^2$, in agreement with the solution (9.4.10) which we obtained by integrating the differential equation of equilibrium (9.4.1). Evaluating the second variation of Π, we obtain

$$\delta^2\Pi = \frac{\pi^2}{2L} \left(\frac{\pi^2 EI}{L^2} - Q_0 \right) (\delta a)^2 \qquad (9.4.17)$$

The straight-line position $a = 0$ is evidently stable ($\delta^2\Pi > 0$) providing $Q_0 < \pi^2 EI/L^2$, unstable ($\delta^2\Pi < 0$) when $Q_0 > \pi^2 EI/L^2$, and neutral ($\delta^2\Pi = 0$) if $Q_0 = \pi^2 EI/L^2$. On the other hand the sine-curve deflected shape (9.4.13), corresponding to the equilibrium condition $Q_0 = \pi^2 EI/L^2$, represents a position of neutral equilibrium. The amplitude a of the sine curve is indeterminate in this case.

The types of equilibrium corresponding to the higher buckling modes $n > 1$ can be determined in a similar fashion. One finds that each of the infinite number of shapes described by Eq. (9.4.11) also represents a condition of neutral equilibrium. We conclude that stable equilibrium is possible only when the magnitude of the compressive force Q_0 is less than $\pi^2 EI/L^2$, in which case the column remains straight. The structure becomes unstable when Q_0 reaches the critical value

$$(Q_0)_{\text{critical}} = \frac{\pi^2 EI}{L^2} \qquad (9.4.18)$$

known as the *Euler-buckling load*[1] for a pin-ended column. Critical loads for columns having other end conditions can be computed in a similar way.

[1]The buckling of an elastic column was first studied by Leonhard Euler in 1744.

Once the buckling has occurred, the behavior of the column is uncertain. The situation here is analogous to that of the one-degree-of-freedom system described in Example 9.1. Because of the nature of the equations which we used to compute the critical force, we are not able to determine the deflections of the buckled member. Only by solving the exact differential equation, rather than Eq. (9.4.1) which is based upon an approximate expression for the curvature of the buckled bar, is it possible to investigate the structure's post-buckling behavior. (See Ref. 9.2).

The critical load (9.4.18) was derived for an inextensible, initially straight, axially loaded (loads coincident with the bar's centroidal axis) column. It can be shown (see Ref. 9.3), however, that the same buckling load is obtained if axial deformation, initial crookedness, and eccentric loadings are considered. (This is not to say that the deformation will be the same in each case!) The response of a beam column to combined axial and lateral loads is examined in the following section.

9.5   DEFORMATION AND STABILITY OF BEAM COLUMNS

The presence of a large axial force may have a significant influence upon the deformation of a laterally loaded bar, or beam column. When the axial force is compressive, the displacements are magnified and stability becomes a consideration. Furthermore, as we shall see in this section, the deformation is not proportional to the magnitude of the axial force, and hence the principle of superposition is not directly applicable.

The differential equation for the deflection of a beam column was found to be

$$(EIu_2'')'' + Q_0 u_2'' = Q \qquad (9.5.1)$$

When the member has a constant flexural stiffness EI, this equation may be written as

$$u_2{}^{iv} + \beta^2 u_2'' = \frac{Q}{EI} \qquad (9.5.2)$$

where, as before, $\beta^2 = Q_0/EI$. Equation (9.5.2) is a linear, nonhomogeneous differential equation. The general solution for u_2 consists of the complementary solution (the solution to the corresponding homogeneous equation) plus a particular solution. Thus,

$$u_2 = C_1 \cos \beta x_1 + C_2 \sin \beta x_1 + C_3 x_1 + C_4 + u_p \qquad (9.5.3)$$

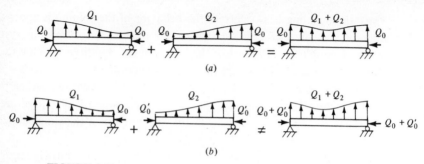

FIGURE 9.12

(a) Loadings for which the total deformation of the beam column can be obtained by means of superposition; (b) loadings for which superposition is not valid.

in which C_1, C_2, C_3, and C_4 are arbitrary constants of integration and u_p represents any particular solution to Eq. (9.5.2). The constants of integration are determined from the beam column's four end conditions. The form of the particular solution u_p depends upon the form of the nonhomogeneous term in Eq. (9.5.2); that is, it depends upon the distribution of the lateral load $Q(x_1)$.

Owing to the linearity of the governing differential equation, the complete solution (9.5.3) for u_2 is proportional to the amplitude of the lateral load $Q(x_1)$. The linearity also implies that the deflection produced by several lateral loads applied simultaneously is equal to the sum of the deflections produced by each load individually. For example, the displacement curve for the beam column shown on the right side of Fig. 9.12a may be found by adding the displacements caused by the two sets of loadings to the left. The same linear differential equation governs the three systems. It is noted, however, that superposition is not valid if the axial force Q_0 is different for each system. For instance, the deflections of the two bars on the left of Fig. 9.12b cannot be added to obtain the deflection of the bar to the right. Each displacement curve represents the solution to a different equation (the value of β^2 is different in each case), and consequently superposition is invalid.

The deformation and stability characteristics for beam columns subject to various loading conditions are investigated in the following examples. Based upon the approximate (small curvature) theory, it is found that instability occurs when the axial force reaches the Euler load for the column, regardless of the magnitude of the lateral load. The deflected shape of the member, on the other hand, depends upon the magnitudes of both the transverse and the axial applied loads.

Example 9.4   Beam column with a distributed lateral load   As a first example of the application of the beam column equation, let us consider a

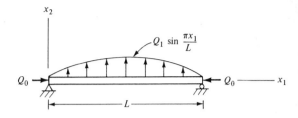

FIGURE 9.13
Beam column subject to a sinusoidal transverse load.

simply supported member carrying a sinusoidal transverse load, as shown in Fig. 9.13. To find the deflected shape of the bar we must solve the differential equation

$$u_2{}^{iv} + \beta^2 u_2'' = \frac{Q_1}{EI} \sin \frac{\pi x_1}{L} \qquad (9.5.4)$$

subject to the boundary conditions

$$u_2(0) = u_2(L) = u_2''(0) = u_2''(L) = 0 \qquad (9.5.5)$$

A particular solution to the differential equation (9.5.4) can be obtained using the method of *undetermined coefficients*. Since the nonhomogeneous term in Eq. (9.5.4) contains the harmonic function $\sin(\pi x_1/L)$, we assume a particular solution of the form

$$u_p = A \sin \frac{\pi x_1}{L} + B \cos \frac{\pi x_1}{L} \qquad (9.5.6)$$

The undetermined constants A and B are found by substituting the assumed solution into Eq. (9.5.4), and then equating to zero the coefficients of the sine and cosine functions. Solving the resulting two equations for A and B, and recalling that $\beta^2 = Q_0/EI$, we obtain

$$A = \frac{Q_1 L^2/\pi^2}{\pi^2 EI/L^2 - Q_0} \qquad B = 0 \qquad (9.5.7)$$

Thus the complete solution for u_2 is

$$u_2 = C_1 \cos \beta x_1 + C_2 \sin \beta x_1 + C_3 x_1 + C_4 + \frac{Q_1 L^2/\pi^2}{\pi^2 EI/L^2 - Q_0} \sin \frac{\pi x_1}{L} \qquad (9.5.8)$$

Application of the boundary conditions (9.5.5) leads to the result that $C_1 = C_2 = C_3 = C_4 = 0$. Hence the deflection curve is given by

$$u_2 = \frac{Q_1 L^2/\pi^2}{\pi^2 EI/L^2 - Q_0} \sin \frac{\pi x_1}{L} \qquad (9.5.9)$$

To assess the stability of the beam column, we now consider the system's total potential energy. From Eq. (9.3.10) we have

$$\Pi = \int_0^L \left[\frac{EI}{2} (u_2'')^2 - \frac{Q_0}{2} (u_2')^2 - Qu_2 \right] dx_1 \qquad (9.5.10)$$

Since the bar's deflected shape is of the form

$$u_2 = a \sin \frac{\pi x_1}{L} \qquad (9.5.11)$$

the potential energy expression becomes

$$\Pi = \int_0^L \left[\frac{EI}{2} \left(-a \frac{\pi^2}{L^2} \sin \frac{\pi x_1}{L} \right)^2 - \frac{Q_0}{2} \left(a \frac{\pi}{L} \cos \frac{\pi x_1}{L} \right)^2 \right.$$

$$\left. - Q_1 \sin \frac{\pi x_1}{L} \left(a \sin \frac{\pi x_1}{L} \right) \right] dx_1$$

$$= \frac{\pi^2}{4L} \left(\frac{\pi^2 EI}{L^2} - Q_0 \right) a^2 - \frac{Q_1 La}{2} \qquad (9.5.12)$$

The requirement for equilibrium is

$$\delta \Pi = \frac{d\Pi}{da} \delta a = \left[\frac{\pi^2}{2L} \left(\frac{\pi^2 EI}{L^2} - Q_0 \right) a - \frac{Q_1 L}{2} \right] \delta a = 0 \qquad (9.5.13)$$

Therefore

$$a = \frac{Q_1 L^2 / \pi^2}{\pi^2 EI / L^2 - Q_0} \qquad (9.5.14)$$

and

$$u_2 = \frac{Q_1 L^2 / \pi^2}{\pi^2 EI / L^2 - Q_0} \sin \frac{\pi x_1}{L} \qquad (9.5.15)$$

This verifies our previous solution (9.5.9). The type of equilibrium associated with this deflected shape depends upon the sign of the second variation of Π:

$$\delta^2 \Pi = \frac{\pi^2}{2L} \left(\frac{\pi^2 EI}{L^2} - Q_0 \right) (\delta a)^2 \qquad (9.5.16)$$

The beam column is in stable equilibrium ($\delta^2 \Pi > 0$) so long as $Q_0 < \pi^2 EI / L^2$. It becomes unstable when Q_0 reaches the critical value

$$(Q_0)_{\text{critical}} = \frac{\pi^2 EI}{L^2} \qquad (9.5.17)$$

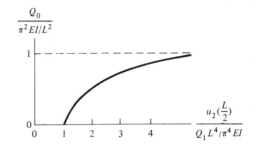

FIGURE 9.14
Axial force-versus-transverse displacement curve for the beam column.

which we recognize as the Euler load for the column. Thus, according to the small curvature theory, the lateral loading does not affect the bar's stability. However, the presence of even a very small transverse load will result in considerable lateral deflection when Q_0 is close to the critical value (9.5.17). The dependence of the midpoint displacement $u_2(L/2)$ upon Q_0 is shown in Fig. 9.14. Note that $u_2(L/2)$ is equal to $Q_1 L^4/\pi^4 EI$ when $Q_0 = 0$; it becomes infinite as Q_0 approaches the Euler load. The axial force-versus-displacement curve is clearly nonlinear. On the other hand, it can been seen from Eq. (9.5.15) that the deflection is proportional to the amplitude of the lateral load Q_1. The principle of superposition could therefore be applied in order to obtain the response of the bar to several lateral loads, providing that the axial force Q_0 remained constant as each of the lateral loads was applied. ////

Example 9.5 Beam column with a concentrated lateral load We shall now compute the deflection curve for a simply supported beam column which carries a concentrated transverse load Q_1 (Fig. 9.15). Due to the symmetry of the applied forces, the deflection curve is symmetrical about the beam's center line, and it is only necessary to consider one-half of the bar, say $0 < x_1 < L/2$. Because the lateral load $Q(x_1)$ is zero over this portion of the member, the differential equation is homogeneous; that is,

$$u_2{}^{iv} + \beta^2 u_2'' = 0 \qquad 0 < x_1 < \frac{L}{2} \qquad (9.5.18)$$

The general solution to this equation is

$$u_2 = C_1 \cos \beta x_1 + C_2 \sin \beta x_1 + C_3 x_1 + C_4 \qquad (9.5.19)$$

where the constants of integration C_1, C_2, C_3, and C_4 must be evaluated from the conditions at the ends of the element. At $x_1 = 0$, the displacement and the bending

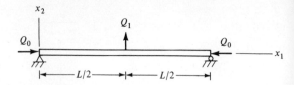

FIGURE 9.15
Beam column subject to a concentrated transverse load.

moment are zero; and at $x_1 = L/2$, the slope is zero, and the shear force is equal to $Q_1/2$. Therefore

$$u_2(0) = 0$$

$$M(0) = EIu_2''(0) = 0$$

$$u'\left(\frac{L}{2}\right) = 0 \qquad (9.5.20)$$

$$V\left(\frac{L}{2}\right) = -EIu_2'''\left(\frac{L}{2}\right) - Q_0 u_2'\left(\frac{L}{2}\right) = \frac{Q_1}{2}$$

These four conditions are fulfilled if

$$C_1 = C_4 = 0$$

$$C_2 = \frac{Q_1}{2\beta^3 EI \cos(\beta L/2)} \qquad (9.5.21)$$

$$C_3 = \frac{-Q_1}{2\beta^2 EI}$$

Substitution of these values into Eq. (9.5.19) yields the deflection curve

$$u_2 = \frac{Q_1}{2\beta^3 EI \cos(\beta L/2)} \sin \beta x_1 - \frac{Q_1}{2\beta^2 EI} x_1 \qquad (9.5.22)$$

or since $\beta^2 = Q_0/EI$

$$u_2 = \frac{Q_1}{2Q_0 \beta} \left[\frac{\sin \beta x_1}{\cos(\beta L/2)} - \beta x_1 \right] \qquad (9.5.23)$$

The maximum deflection occurs at the center of the beam column. Substituting $x_1 = L/2$ into Eq. (9.5.23) gives

$$u_2\left(\frac{L}{2}\right) = \frac{Q_1}{2Q_0 \beta} \left(\tan \frac{\beta L}{2} - \frac{\beta L}{2} \right) \qquad (9.5.24)$$

Recognizing that $\tan(\beta L/2)$ becomes infinite as β approaches π/L, we note that the center displacement $u_2(L/2)$ is infinite when Q_0 reaches the critical value

$$(Q_0)_{\text{critical}} = \frac{\pi^2 EI}{L^2} \qquad (9.5.25)$$

Here, as in the previous example, we conclude that when the axial compressive force approaches the bar's Euler load, even a small lateral force may produce extremely large displacements. In interpreting our results, we must keep in mind the fact that the theory which we have been using is based on the assumption of small displacements. Thus Eq. (9.5.24) ceases to be valid when the axial load is close to the critical value (9.5.25). ////

9.6 RAYLEIGH-RITZ METHOD

Obtaining an exact solution to the beam column equation (9.3.14) may be difficult or even impossible if either the transverse load $Q(x_1)$ or the bending stiffness $EI(x_1)$ is a complicated function of x_1. The Rayleigh-Ritz method introduced in Sec. 5.10 provides a convenient means for computing approximate values of the displacements and buckling loads in such situations.

In review, the Rayleigh-Ritz method reduces the continuous structure to a system having n degrees of freedom. The deflection curve $u_2(x_1)$ is approximated by a series of functions containing n independent coefficients a_i ($i = 1, 2, \ldots, n$). Each function must satisfy the beam column's kinematic boundary conditions, but need not satisfy the static end conditions. If the structure is in equilibrium, the variation of the total potential energy of the system is zero, in which case

$$\frac{\partial \Pi}{\partial a_i} = 0 \qquad i = 1, 2, \ldots, n \qquad (9.6.1)$$

These n equations represent a set of simultaneous algebraic equations in the unknown coefficients a_i. By solving this set for the values of a_i, one obtains the approximate deflected shape of the structure. In the event that the Eqs. (9.6.1) are homogeneous, equating the determinant of the coefficients to zero yields approximate values for the buckling loads. Since the idealized n-degree-of-freedom system is effectively stiffer than the actual structure (see the discussion in Sec. 5.10), the approximate displacements are generally smaller, and the critical axial load larger than the corresponding exact values. Hence the Rayleigh-Ritz method gives an upper bound for the bar's buckling load.

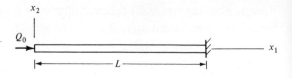

FIGURE 9.16
Fixed-free column.

Example 9.6   Rayleigh-Ritz method for a fixed-free column   To demonstrate the Rayleigh-Ritz procedure for calculating the buckling load of an elastic column, we consider the member shown in Fig. 9.16. At the left end of the column the shear force and bending moment are zero, while the displacement and slope vanish at the right end. The boundary conditions are therefore

$$V(0) = M(0) = 0 \qquad (9.6.2)$$

$$u_2(L) = u_2'(L) = 0 \qquad (9.6.3)$$

A deflection curve must be chosen which satisfies the kinematic boundary conditions (9.6.3); it need not necessarily satisfy the static conditions (9.6.2). Consider the polynomial representation

$$u_2 = a_0 + a_1 \frac{x_1}{L} + a_2 \left(\frac{x_1}{L}\right)^2 + a_3 \left(\frac{x_1}{L}\right)^3 \qquad (9.6.4)$$

Applying the kinematic conditions (9.6.3) to this displacement curve gives

$$u_2(L) = a_0 + a_1 + a_2 + a_3 = 0$$

$$u_2'(L) = \frac{a_1}{L} + \frac{2a_2}{L} + \frac{3a_3}{L} = 0 \qquad (9.6.5)$$

from which it is seen that the coefficients a_0, a_1, a_2, and a_3 cannot all be chosen independently. Two of the coefficients can be eliminated from the assumed solution by making use of Eqs. (9.6.5). Solving for a_0 and a_1, for example, gives

$$a_0 = a_2 + 2a_3$$

$$a_1 = -2a_2 - 3a_3 \qquad (9.6.6)$$

Substituting these relations into Eq. (9.6.4) yields

$$u_2 = a_2 \left(\frac{x_1}{L} - 1\right)^2 + a_3 \left[\left(\frac{x_1}{L}\right)^3 - 3\frac{x_1}{L} + 2\right] \qquad (9.6.7)$$

where a_2 and a_3 are now arbitrary and independent constants.

Assuming that the bar has a uniform cross section, the potential energy Π is

$$\Pi = \int_0^L \left[\frac{EI}{2} (u_2'')^2 - \frac{Q_0}{2} (u_2')^2 \right] dx_1$$

$$= \int_0^L \left\{ \frac{EI}{2} \left(\frac{2a_2}{L^2} + \frac{6a_3 x_1}{L^3} \right)^2 - \frac{Q_0}{2} \left[\frac{2a_2}{L} \left(\frac{x_1}{L} - 1 \right) + \frac{3a_3}{L} \left(\frac{x_1^2}{L^2} - 1 \right) \right]^2 \right\} dx_1$$

$$= \frac{2}{L^3} \left(EI - \frac{1}{3} Q_0 L^2 \right) a_2{}^2 + \frac{1}{L^3} \left(6EI - \frac{5}{2} Q_0 L^2 \right) a_2 a_3 + \frac{6}{L^3} \left(EI - \frac{2}{5} Q_0 L^2 \right) a_3{}^2$$

$$(9.6.8)$$

For equilibrium

$$\frac{\partial \Pi}{\partial a_2} = \frac{4}{L^3} \left(EI - \frac{1}{3} Q_0 L^2 \right) a_2 + \frac{1}{L^3} \left(6EI - \frac{5}{2} Q_0 L^2 \right) a_3 = 0$$

$$\frac{\partial \Pi}{\partial a_3} = \frac{1}{L^3} \left(6EI - \frac{5}{2} Q_0 L^2 \right) a_2 + \frac{12}{L^3} \left(EI - \frac{2}{5} Q_0 L^2 \right) a_3 = 0 \qquad (9.6.9)$$

A nontrivial solution to this eigenvalue problem exists only if the determinant of the coefficients multiplying a_2 and a_3 vanishes. Expansion of the determinant gives the characteristic equation

$$3Q_0{}^2 - 104 \frac{EI}{L^2} Q_0 + 240 \left(\frac{EI}{L^2} \right)^2 = 0 \qquad (9.6.10)$$

the roots or eigenvalues of which are

$$Q_0 = 2.49 \frac{EI}{L^2} \qquad \text{and} \qquad 32.2 \frac{EI}{L^2} \qquad (9.6.11)$$

The corresponding eigenvectors, found by substituting these values of Q_0 into either one of the Eqs. (9.6.9), are

$$\frac{a_3}{a_2} = 3.19 \qquad \left(Q_0 = 2.49 \frac{EI}{L^2} \right)$$

$$\frac{a_3}{a_2} = -0.523 \qquad \left(Q_0 = 32.2 \frac{EI}{L^2} \right) \qquad (9.6.12)$$

By introducing these ratios into Eq. (9.6.7), one obtains the deflection curves shown in Fig. 9.17. Note that the amplitudes a_2 of the displacements in the buckled configurations are indeterminate; hence these shapes represent positions of neutral equilibrium. The column becomes unstable when the magnitude of

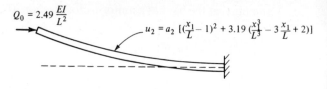

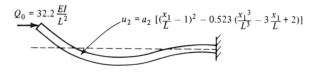

FIGURE 9.17
Approximate buckling modes for a fixed-free column.

the axial force reaches the smaller of the two values. The critical load for the member is therefore

$$(Q_0)_{\text{critical}} = 2.49 \frac{EI}{L^2} \qquad (9.6.13)$$

This value is approximately 1 percent larger than the Euler load for a fixed-free column (see Prob. 9.6). However, the second buckling load ($Q_0 = 32.2 \, EI/L^2$) is approximately $1\frac{1}{2}$ times as large as the corresponding exact value. Obtaining an accurate solution for the second and higher buckling modes is generally difficult; it requires choosing a deflected shape having a large number of independent functions and undetermined coefficients. ////

Example 9.7   Rayleigh-Ritz method for a beam column of variable cross section   Consider a beam column which is supported and loaded as shown in Fig. 9.18. The bending stiffness $EI(x_1)$ is symmetrical with respect to the bar's middle cross section and varies according to

$$EI = EI_0 \frac{2x_1}{L} \qquad 0 < x_1 < \frac{L}{2} \qquad (9.6.14)$$

A deflection curve which satisfies the kinematic boundary conditions $u(0) = u(L) = 0$ is

$$u_2 = a_1 \sin \frac{\pi x_1}{L} \qquad (9.6.15)$$

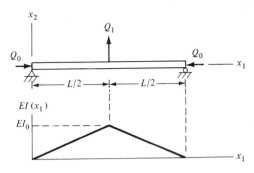

FIGURE 9.18
Beam column of variable cross section.

Owing to the symmetry of the structure, the total potential energy may be written as

$$\Pi = 2 \int_0^{L/2} \left[\frac{EI}{2} (u_2'')^2 - \frac{Q_0}{2} (u_2')^2 \right] dx_1 - Q_1 u_2 \left(\frac{L}{2} \right)$$

$$= 2 \int_0^{L/2} \left[\frac{EI_0 x_1}{L} \left(-a_1 \frac{\pi^2}{L^2} \sin \frac{\pi x_1}{L} \right)^2 - \frac{Q_0}{2} \left(a_1 \frac{\pi}{L} \cos \frac{\pi x_1}{L} \right)^2 \right] dx_1 - Q_1 a_1$$

$$= \frac{\pi^2}{4L} \left[2 \left(1 + \frac{\pi^2}{4} \right) \frac{EI_0}{L^2} - Q_0 \right] a_1^2 - Q_1 a_1 \qquad (9.6.16)$$

The requirement for equilibrium is

$$\frac{\partial \Pi}{\partial a_1} = \frac{\pi^2}{2L} \left[2 \left(1 + \frac{\pi^2}{4} \right) \frac{EI_0}{L^2} - Q_0 \right] a_1 - Q_1 = 0 \qquad (9.6.17)$$

Solving this equation for a_1 and substituting its value into Eq. (9.6.15) yields the approximate deflection curve

$$u_2 = \frac{(2L/\pi^2) Q_1}{[2(1 + \pi^2/4) EI_0/L^2 - Q_0]} \sin \frac{\pi x_1}{L} \qquad (9.6.18)$$

From this expression we see that the lateral displacements become infinite when the axial force Q_0 reaches the critical value

$$(Q_0)_{\text{critical}} = 2 \left(1 + \frac{\pi^2}{4} \right) \frac{EI_0}{L^2} = 6.94 \frac{EI_0}{L^2} \qquad (9.6.19)$$

It should be kept in mind that the Rayleigh-Ritz procedure gives a buckling load which is too large; hence buckling actually occurs before the axial load reaches this critical value. [The exact result is $(Q_0)_{\text{critical}} = 5.78\, EI_0/L^2$.]   ////

PROBLEMS

9.1   Find the critical load for a rigid bar supported and loaded as shown. Assume the rotation of the bar remains small (infinitesimal).

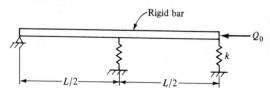

PROBLEM 9.1

9.2   Three rigid bars are connected and supported as shown. Compute the critical loads for the system, and investigate the corresponding modes of buckling. Assume small rotations.

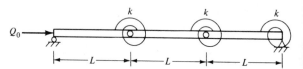

PROBLEM 9.2

9.3   Derive the differential equation (9.3.1) for a beam column by considering the equilibrium of a differential element of the beam. Also show that the shear force at an arbitrary cross section is given by

$$V = -\frac{d}{dx_1}\left(EI\frac{d^2u_2}{dx_1{}^2} + Q_0 u_2\right)$$

9.4   The differential equation of equilibrium for an initially bent column is

$$EIu_2{}^{iv} + Q_0 u_2'' = -Q_0 u_0''$$

where $u_0(x_1)$ is the initial deflected shape and $u_2(x_1)$ is the additional deflection produced by the axial force Q_0. Derive this differential equation by using the pmpe. Then compute the critical load for a pin-ended column having the initial shape $u_0 = a \sin(\pi x_1/L)$, where a is a constant.

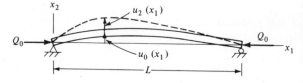

PROBLEM 9.4

9.5   Use the pmpe to derive the differential equation of equilibrium and the natural boundary conditions for a column which rests on an elastic foundation of stiffness k

[that is, $-ku_2(x_1)$ represents the restoring force per unit length exerted on the beam by the foundation]. Then compute the critical load for the pin-ended column shown.

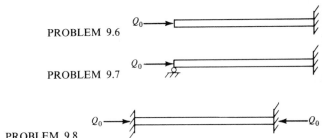

PROBLEM 9.5

9.6–9.8 Compute the buckling load for a column of uniform flexural rigidity EI which is supported as shown.

PROBLEM 9.6

PROBLEM 9.7

PROBLEM 9.8

9.9 The beam column shown is subject to the temperature change $\Theta = \Theta_0 \sin(\pi x_1/L)$. At what value of Θ_0 will instability occur?

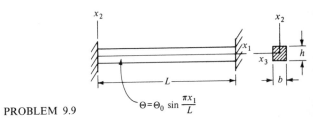

PROBLEM 9.9

$\Theta = \Theta_0 \sin \dfrac{\pi x_1}{L}$

9.10 The free end of a cantilevered beam column is subject to a compressive force Q_0 and a couple M_0. Compute the maximum lateral displacement and the maximum bending moment in the beam. Sketch the bending moment diagram $M(x_1)$ for the cases

(a) $\quad Q_0 = \dfrac{\pi^2 EI}{4L^2}$

(b) $\quad Q_0 = \dfrac{\pi^2 EI}{16L^2}$

(c) $\quad Q_0 = 0$

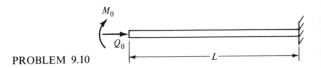

PROBLEM 9.10

9.11　By solving the appropriate differential equations, compute the maximum displacement and the maximum bending moment in the beam column shown if

(a)　$Q_0 = \dfrac{\pi^2 EI}{4L^2}$　　(compression)

(b)　$Q_0 = 0$

(c)　$Q_0 = -\dfrac{\pi^2 EI}{4L^2}$　　(tension)

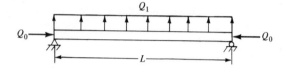

PROBLEM 9.11

9.12–9.13　Solve Probs. 9.7 to 9.8 using the Rayleigh-Ritz method. Compare your result with the exact solution.

9.14–9.17　Use the Rayleigh-Ritz method to find an approximate value of the buckling load for the column shown.

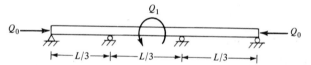

PROBLEM 9.14

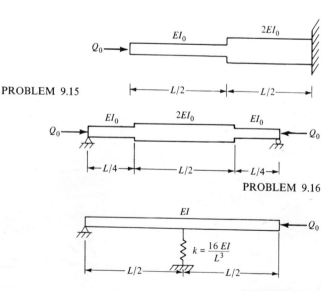

PROBLEM 9.15

PROBLEM 9.16

PROBLEM 9.17

9.18 Use the Rayleigh-Ritz method to compute the buckling load of a pin-ended column which rests on an elastic foundation of stiffness k. Compare your answer with the exact solution (see Prob. 9.5) for the case $k < 4\pi^4 EI/L^4$.

9.19 Use the Rayleigh-Ritz method to determine the maximum transverse displacement in the beam column shown. Assume $Q_0 < 4\pi^2 EI/L^2$.

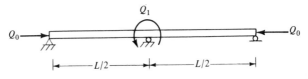

$$|\!\longleftarrow\!\!-\!\!-\!\!-L/2\!\!-\!\!-\!\!-\!\!\longrightarrow\!\!|\!\longleftarrow\!\!-\!\!-\!\!-L/2\!\!-\!\!-\!\!-\!\!\longrightarrow\!\!|$$

PROBLEM 9.19

9.20 Solve Prob. 9.11 using the Rayleigh-Ritz method, and compare your results with the exact solutions.

REFERENCES

For a comprehensive treatment of the buckling of bars (elastic and inelastic, for large and small deformations), frames, plates, and shells:

9.1 BLEICH, F.: "Buckling Strength of Metal Structures," McGraw-Hill, New York, 1952.

9.2 TIMOSHENKO, S. P., and J. M. GERE: "Theory of Elastic Stability," 2d ed., McGraw-Hill, New York, 1961.

For a study of the stability of elastic structures, with particular emphasis on energy methods:

9.3 HOFF, N. J.: "The Analysis of Structures," Wiley, New York, 1956.

9.4 LANGHAAR, H. L.: "Energy Methods in Applied Mechanics," Wiley, New York, 1962.

9.5 ODEN, J. T.: "Mechanics of Elastic Structures," McGraw-Hill, New York, 1967.

9.6 RUBINSTEIN, M. F.: "Structural Systems—Statics, Dynamics and Stability," Prentice-Hall, Englewood Cliffs, N.J., 1970.

For a presentation of the theory of dynamic stability of elastic bodies, with applications to both linear and nonlinear systems:

9.7 BOLOTIN, V. V.: "The Dynamic Stability of Elastic Systems," (in Russian); English translation by V. I. Weingarten et al., Holden-Day, San Francisco, 1964.

9.8 ZIEGLER, H.: "Principles of Structural Stability," Blaisdell, Waltham, Mass., 1968.

PART THREE

Dynamic Behavior of Structures

10

STRUCTURES WITH ONE DEGREE OF FREEDOM

10.1 INTRODUCTION

Part II of this book has been concerned with the behavior of structures under static loads. It was assumed, in effect, that the loads were applied so slowly that all elements of the structure remained in equilibrium at every instant of time. However, engineering structures are often subjected to suddenly applied forces or to forces whose amplitudes vary continuously with time. Examples of such situations include: the response of bridges and roads to moving vehicles, the action of gust loads and landing impact upon aircraft, and the vibrations of structures produced by oscillating machinery. Under these types of loading conditions, the elements of the structure are set in motion; it is therefore necessary to apply the principles of dynamics rather than those of statics to determine the response. We shall see that the maximum deflections, strains, stresses, and various other response quantities are generally more severe when loads of a given amplitude are applied dynamically rather than statically.

Part III of the text contains three chapters. The present one is concerned with the dynamic behavior of a *single-degree-of-freedom* structure; that is, a structure whose position at any instant of time is defined completely by a single coordinate. *Multi-degree-of-freedom* structures, or systems for which more than

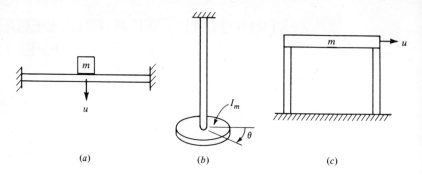

FIGURE 10.1
Structures which can be idealized as single-degree-of-freedom systems.

one coordinate is needed in order to describe the motion, are treated in Chap. 11. The dynamics of *continuous* or *distributed mass* systems is the subject of Chap. 12.

While every structure has, in reality, a continuous distribution of mass, it is possible to approximate some structures by systems having one or more discrete mass points. Such an approximation reduces the number of degrees of freedom of the structure from infinity to a finite value. Consider for example an elastic beam which carries a rigid body of mass m, as shown in Fig. 10.1a. In order to describe this system's configuration at a particular instant of time, it is necessary to specify the displacement at every cross section of the beam, i.e., at an infinite number of points. However, if the mass of the beam is small compared with the attached mass m, the structure may be idealized as a single rigid mass resting on a flexible support. The displacement u shown in the figure then suffices to describe the system's configuration.

As a second example consider the torsional deformation of an elastic shaft which supports a rigid disk, as illustrated in Fig. 10.1b. Assuming that the polar mass moment of inertia I_m of the disk is much greater than that of the shaft, the angle of twist θ defines the position of this system at every instant of time.

Likewise, the horizontal motion of the frame in Fig. 10.1c is characterized by the single displacement u, providing that the mass m of the rigid girder is large in proportion to the masses of the flexible vertical columns.

The dynamic behavior of structures of the type just described will be investigated in the following articles. While the amount of useful information which can be obtained when a continuous structure is approximated by a single-degree-of-freedom system is limited, it is often possible to obtain a reasonably accurate estimate of the structure's maximum displacement and also its period of vibration.

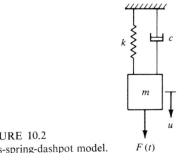

FIGURE 10.2
Mass-spring-dashpot model.

10.2 EQUATION OF MOTION

A convenient model for use in studying the dynamic response of a single-degree-of-freedom structure is the mass-spring-dashpot system shown in Fig. 10.2. Here m denotes the structure's mass (or the moment of inertia of the mass if the model is used to represent a structure undergoing a rotational motion, as in Fig. 10.1b). Applied to m is a time-dependent, generalized external force $F(t)$ which produces a generalized displacement u at time t.

The structure's elastic restoring action is represented by the spring element in the model. Assuming linearly elastic material behavior, the force tending to restore the structure to its original position is $-ku$, where k denotes the structure's stiffness coefficient; that is, k is the generalized force required to produce a generalized displacement of unit magnitude (see Chap. 5).

The effects of damping are represented by the model's dashpot. We shall assume that the resistance to motion arising from all sources (e.g., internal molecular friction, joint-interface slip, air resistance, etc.) is of the *viscous* type, that is, proportional to velocity.[1] Accordingly, the damping force is expressed as $-c\,du/dt = -c\dot{u}$, where c is referred to as the *coefficient of viscous damping*.

Application of Newton's second law of motion to the mass m in Fig. 10.2 yields

$$m\ddot{u} = -ku - c\dot{u} + F(t) \qquad (10.2.1)$$

[1]Since Coulomb's initial efforts approximately 200 years ago, there has been considerable research on damping in structures. Various models have been proposed for representing the phenomenon, each model being appropriate for a limited class of materials and structures. A linear viscous idealization, in which the damping force is proportional to velocity, has been found to yield results which are generally acceptable for engineering purposes. When damping is not of the viscous type, it is sometimes possible to approximate the actual dissipation by an equivalent viscous damping (see L. S. Jacobsen, Steady Forced Vibration as Influenced by Damping, *Trans. ASME*, vol. 52, 1930). For these reasons, and also because viscous damping is easy to deal with analytically, we shall restrict our attention to this type.

which after a rearrangement of terms gives

$$\ddot{u} + 2n\dot{u} + p^2 u = \frac{F(t)}{m} \qquad (10.2.2)$$

where

$$2n = \frac{c}{m} \qquad p^2 = \frac{k}{m} \qquad (10.2.3)$$

The quantities n and p are called the *damping factor* and the *circular frequency*, respectively. Equation (10.2.2) is the equation of motion for the mass-spring-dashpot model. It must be understood, however, that every linearly elastic, viscously damped, single-degree-of-freedom structure has an equation of motion of exactly this form. By examining the solutions to Eq. (10.2.2) for various initial disturbances and forcing functions $F(t)$, we are in fact studying the response of a large class of structures. Moreover, we shall make use of these solutions when we study the behavior of more complicated multi-degree-of-freedom and continuous systems.

10.3   FREE VIBRATION

Let us first consider the type of motion which can occur in the absence of any externally applied forces. The equation of motion is then

$$\ddot{u} + 2n\dot{u} + p^2 u = 0 \qquad (10.3.1)$$

To solve this homogeneous, linear differential equation we let

$$u = Ce^{\lambda t} \qquad (10.3.2)$$

in which case the characteristic equation is

$$\lambda^2 + 2n\lambda + p^2 = 0 \qquad (10.3.3)$$

The roots of this quadratic expression are

$$\lambda_{1,2} = -n \pm \sqrt{n^2 - p^2} \qquad (10.3.4)$$

Most structures are relatively "lightly" damped; that is, the damping factor n is much smaller than the circular frequency p. (A typical value of the ratio n/p for a steel structure is 0.02, and for a concrete structure 0.10.) It is then convenient to write Eq. (10.3.4) as

$$\lambda_{1,2} = -n \pm i\bar{p} \qquad (10.3.5)$$

where

$$\bar{p} = \sqrt{p^2 - n^2} \qquad (10.3.6)$$

is a positive, real constant. Substituting the two roots (10.3.5) into the assumed solution (10.3.2) gives

$$u = C_1 e^{(-n+i\bar{p})t} + C_2 e^{(-n-i\bar{p})t}$$
$$= e^{-nt}(C_1 e^{i\bar{p}t} + C_2 e^{-i\bar{p}t})$$
$$= e^{-nt}(A \cos \bar{p}t + B \sin \bar{p}t) \qquad (10.3.7)$$

In obtaining Eq. (10.3.7) use has been made of the trigonometric identity $e^{i\phi} = \cos \phi + i \sin \phi$; also the arbitrary constants C_1 and C_2 have been replaced by two new constants, A and B. These constants are determined by applying the prescribed initial conditions. Letting u_0 and $\dot{u}_0$ represent the displacement and velocity at time $t = 0$, it is found that

$$u = e^{-nt}\left[u_0 \cos \bar{p}t + \left(\frac{\dot{u}_0 + nu_0}{\bar{p}}\right) \sin \bar{p}t\right] \qquad (10.3.8)$$

In the case of zero damping ($n = 0$), Eq. (10.3.8) reduces to

$$u = u_0 \cos pt + \frac{\dot{u}_0}{p} \sin pt \qquad (10.3.9)$$

Both the damped (10.3.8) and undamped (10.3.9) free-vibration solutions are shown in Fig. 10.3. Note that the effect of damping is to decrease the amplitude of the response and to increase the period of the vibrations. When damping is present, the *period* $\bar{\tau}$ and the *frequency* $\bar{f}$ (defined as the reciprocal of the period) are

$$\bar{\tau} = \frac{2\pi}{\bar{p}} = \frac{2\pi}{\sqrt{p^2 - n^2}} \qquad \bar{f} = \frac{\bar{p}}{2\pi} = \frac{\sqrt{p^2 - n^2}}{2\pi} \qquad (10.3.10)$$

In the absence of damping, these quantities are given by

$$\tau = \frac{2\pi}{p} \qquad f = \frac{p}{2\pi} \qquad (10.3.11)$$

If the period ($\bar{\tau}$ or τ) is measured in seconds, the frequency ($\bar{f}$ or f) gives the number of cycles occurring in one second (cps). The circular frequency ($\bar{p}$ or p), which is sometimes also referred to as simply the frequency, has the dimensions of radians per second.

Example 10.1 Free vibration of a mass on a beam An elastic, cantilever beam having a flexural rigidity EI supports a body of mass m as shown in Fig. 10.4. The mass of the beam is small compared to m, so that the structure may be treated as a single-degree-of-freedom system. Damping is assumed to be of the viscous type, and the ratio of the damping factor to the circular frequency is

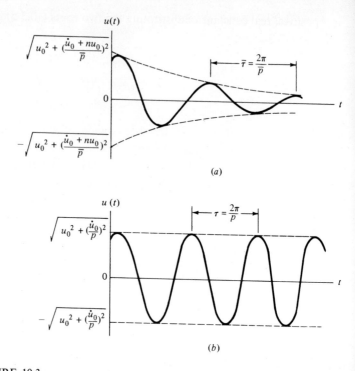

(a)

(b)

FIGURE 10.3
(a) Free-vibration response of a damped system; (b) free-vibration response of an undamped system.

$n/p = 0.05$. We wish to determine the response of the system when the mass m is given an initial displacement u_0 and then released.

Using one of the energy methods given in Chap. 5, the stiffness coefficient, or the ratio of the magnitude of a force at the tip of the beam to the corresponding displacement, can be shown to be $k = 3EI/L^3$. If u is measured from the position of static equilibrium (so that the elastic restoring force associated with the static deflection cancels the force of gravity mg), then the resultant force on m is the sum of the restoring force $-ku$ and the damping force $-c\dot{u}$. As before, the equation of motion may be written in the standard form

$$\ddot{u} + 2n\dot{u} + p^2 u = 0 \qquad (10.3.12)$$

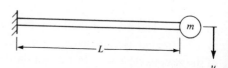

FIGURE 10.4
Cantilever beam with an attached mass.

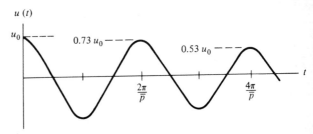

FIGURE 10.5
Response to an initial displacement.

where for this system $n = 0.05p$ and

$$p = \sqrt{\frac{k}{m}} = \sqrt{\frac{3EI}{mL^3}}$$

$$\bar{p} = \sqrt{p^2 - n^2} = \sqrt{p^2 - 0.0025p^2} = 0.999\sqrt{\frac{3EI}{mL^3}} \qquad (10.3.13)$$

For the given initial conditions $u(0) = u_0$ and $\dot{u}(0) = 0$, the solution (10.3.8) gives

$$u = u_0 e^{-0.05pt}\left(\cos \bar{p}t + \frac{0.05p}{\bar{p}}\sin \bar{p}t\right) \qquad (10.3.14)$$

The response (10.3.14) is shown in Fig. 10.5. Although the damping has a negligible effect upon the period of the vibration ($\bar{\tau} = 2\pi/\bar{p} \cong 2\pi/p$), it causes significant attenuation of the motion. After just two cycles of oscillation, the peak displacement decays to approximately one-half of its initial value u_0.   ////

10.4   FORCED VIBRATION

In contrast to the free vibration which results when a structure is given an initial displacement or velocity, a *forced vibration* is defined as the oscillatory motion produced by a time-dependent, externally applied load. In this section we shall study the forced vibrations generated by several different load-time functions.

The standard form of the equation of motion for a linearly elastic, viscously damped structure having one degree of freedom is

$$\ddot{u} + 2n\dot{u} + p^2 u = \frac{F(t)}{m} \qquad (10.4.1)$$

The general solution to this nonhomogeneous, linear differential equation is given by the solution to the associated homogeneous equation [the free-vibration solution (10.3.7)], plus any particular solution to Eq. (10.4.1); that is,

$$u = e^{-nt}(A \cos \bar{p}t + B \sin \bar{p}t) + u_p \qquad (10.4.2)$$

Depending upon the form of the load-time function $F(t)$, the particular solution u_p can often be found using the method of undetermined coefficients.[1]

Several illustrative problems are presented below in which the effect of viscous damping has been included. We shall see that when a structure is subjected to a periodic or pulsating load, damping has a significant influence upon the system's long-time response. On the other hand if a structure is loaded suddenly, and if only the initial or short-time response is of interest, then the effect of damping may be relatively unimportant.

Example 10.2   Response to a harmonic force   Consider a harmonic force of amplitude F_0 and circular frequency ω. Substituting

$$F(t) = F_0 \sin \omega t \qquad (10.4.3)$$

into the general equation of motion (10.4.1) gives

$$\ddot{u} + 2n\dot{u} + p^2 u = \frac{F_0}{m} \sin \omega t \qquad (10.4.4)$$

A particular solution to this equation may be obtained by assuming a trial solution of the form $u_p = A \cos \omega t + B \sin \omega t$, or what is equivalent

$$u_p = C \sin(\omega t + \delta) \qquad (10.4.5)$$

Substituting Eq. (10.4.5) and its time derivatives into (10.4.4) and solving for the undetermined constants C and δ yields

$$C = \frac{F_0/m}{\sqrt{(p^2 - \omega^2)^2 + 4n^2\omega^2}} \qquad \delta = \tan^{-1}\left(\frac{-2n\omega}{p^2 - \omega^2}\right) \qquad (10.4.6)$$

The complete solution is therefore

$$u = e^{-nt}(A \cos \bar{p}t + B \sin \bar{p}t) + \frac{F_0/m}{\sqrt{(p^2 - \omega^2)^2 + 4n^2\omega^2}} \sin(\omega t + \delta) \qquad (10.4.7)$$

---

[1]The method of undetermined coefficients is applicable when the forcing function is a linear combination of any of the following functions: e^{kt}, $\sin kt$, $\cos kt$, and t^n (where k is a constant and n is a positive integer).

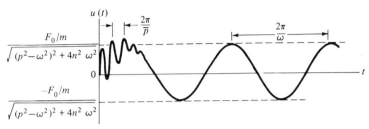

FIGURE 10.6
Response of a damped system to a harmonic force.

in which the arbitrary constants A and B must be determined from the initial conditions. The response (10.4.7) consists of a superposition of two types of motion. The complementary solution, or the first term on the right-hand side of Eq. (10.4.7), represents a damped vibration of period $2\pi/\bar{p}$. Since the amplitude of these oscillations eventually approaches zero, this part of the motion is called the *transient response*. The particular solution, or the second term in Eq. (10.4.7), represents a constant amplitude vibration having the same period as the forcing function. This motion continues after the transients die out, and is therefore referred to as the *steady-state response*. Whereas the transient behavior depends upon the initial conditions, the steady-state motion is independent of the initial conditions and depends only upon the forcing function and the parameters of the structure. Figure 10.6 shows a typical response for the case $\bar{p} > \omega$.

It is seen from Eq. (10.4.7) that extremely large displacements can occur when the circular frequency ω of the forcing function is close to the structure's undamped circular frequency p. This condition is known as *resonance*. Theoretically the displacements would become infinite at resonance if the system were undamped ($n = 0$). The effect of damping is to decrease the amplitude of the steady-state vibrations.

In designing a structure to withstand periodic loads, it is desirable to select a stiffness k and a mass distribution m such that resonant vibrations are avoided. If this is impossible (as, for example, when the frequency of the forcing function is unknown), then it is important to ensure that the structure has sufficient damping to prevent the build-up of excessively large deformations. In problems of this nature, damping plays an important role, and its presence cannot be ignored. ////

Example 10.3   Response to a step force   Consider now the response to the rectangular-step forcing function shown in Fig. 10.7. Substituting $F(t) = F_0$ into the equation of motion (10.4.1) gives

$$\ddot{u} + 2n\dot{u} + p^2 u = \frac{F_0}{m} \qquad (10.4.8)$$

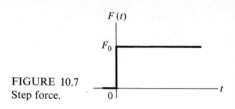

FIGURE 10.7
Step force.

A particular solution in this case is

$$u_p = \frac{F_0}{mp^2} = \frac{F_0}{k} \quad (10.4.9)$$

and the complete solution is therefore

$$u = e^{-nt}(A \cos \bar{p}t + B \sin \bar{p}t) + \frac{F_0}{k} \quad (10.4.10)$$

Assuming that the structure is originally at rest, the initial conditions $u(0) = \dot{u}(0) = 0$ can be used to eliminate the arbitrary constants A and B from Eq. (10.4.10). This gives

$$u = \frac{F_0}{k}\left[1 - e^{-nt}\left(\cos \bar{p}t + \frac{n}{\bar{p}}\sin \bar{p}t\right)\right] \quad (10.4.11)$$

In the case of zero damping, the response becomes

$$u = \frac{F_0}{k}(1 - \cos pt) \quad (10.4.12)$$

The damped (10.4.11) and undamped (10.4.12) solutions are shown in Fig. 10.8. It is seen that in the damped case, the maximum displacement is somewhat less than twice the displacement which would occur if the load were applied statically. When damping is ignored, the maximum deflection is exactly twice the static value.

Note, if one is interested only in the maximum response of the structure, the undamped system can be used to obtain an upper bound.   ////

In the preceding examples the forced vibration solutions were constructed by combining a particular solution to the equation of motion with the complementary, or free-vibration solution. While the method of undetermined coefficients offers one means of finding a particular solution, it is applicable only for certain types of excitations. We shall now develop a general solution which is applicable in the case of an arbitrary forcing function.

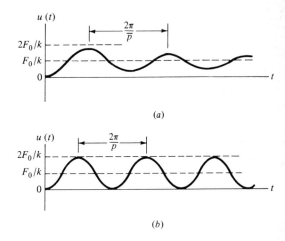

FIGURE 10.8
(a) Response of a damped system to a step force; (b) response of an undamped system to a step force.

Suppose that the structure in question is excited by the force $F(t)$ shown in Fig. 10.9. The function $F(t)$ may be represented by a sequence of impulses of infinitesimal duration. Let us begin by determining the response to the elemental impulse $F(\xi)\,d\xi$ which occurs at time ξ. Assuming that the structure is at rest prior to time ξ, application of the impulse-momentum form of the equation of motion for mass m gives

$$F(\xi)\,d\xi = m\dot{u}(\xi) \quad (10.4.13)$$

Immediately after the impulse occurs, the velocity and displacement of m have the values

$$\dot{u}(\xi) = \frac{F(\xi)\,d\xi}{m} \qquad u(\xi) = 0 \quad (10.4.14)$$

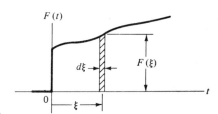

FIGURE 10.9
Arbitrary forcing function.

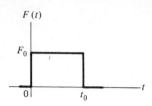

FIGURE 10.10
Rectangular-pulse forcing function.

For time $t > \xi$ the mass performs free vibration described by the equation

$$u = e^{-nt}(A \cos \bar{p}t + B \sin \bar{p}t) \quad (10.4.15)$$

Using the conditions (10.4.14) to eliminate the constants of integration from Eq. (10.4.15), one obtains

$$u = \frac{F(\xi)\, d\xi}{m\bar{p}} e^{-n(t-\xi)} \sin \bar{p}(t - \xi) \quad (10.4.16)$$

Equation (10.4.16) gives the displacement at time t caused by the elemental impulse at time ξ. The displacement due to the infinity of impulses occurring during the time interval 0 to t is, by virtue of the principle of superposition,

$$u = \frac{1}{m\bar{p}} \int_0^t F(\xi)e^{-n(t-\xi)} \sin \bar{p}(t - \xi)\, d\xi \quad (10.4.17)$$

This relation is valid only in the case of zero initial conditions. In order to account for the arbitrary conditions $u(0) = u_0$ and $\dot{u}(0) = \dot{u}_0$, we must superimpose the free-vibration solution (10.3.8) upon (10.4.17). This gives

$$u = e^{-nt}\left(u_0 \cos \bar{p}t + \frac{\dot{u}_0 + nu_0}{\bar{p}} \sin \bar{p}t\right)$$

$$+ \frac{1}{m\bar{p}} \int_0^t F(\xi)e^{-n(t-\xi)} \sin \bar{p}(t - \xi)\, d\xi \quad (10.4.18)$$

The integral portion of expression (10.4.18) is known as the *Duhamel's integral*, or the *convolution integral*.

In the case of zero damping, Duhamel's form of the general solution becomes

$$u = u_0 \cos pt + \frac{\dot{u}_0}{p} \sin pt + \frac{1}{mp} \int_0^t F(\xi)\sin p(t - \xi)\, d\xi \quad (10.4.19)$$

Example 10.4   Forced vibration of a mass on a beam   A mass m, attached to the tip of a cantilever beam of length L and bending stiffness EI, is subjected to the rectangular-pulse forcing function shown in Fig. 10.10. We wish to determine the displacement $u(t)$, assuming that the beam is initially at rest. The mass of the beam is small compared to m, in which case the structure may

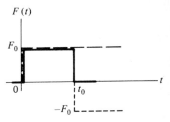

FIGURE 10.11
Representation of a rectangular-pulse forcing function as the sum of a positive and a negative step function.

be treated as a single-degree-of-freedom system. The stiffness coefficient for the system is $k = 3EI/L^3$, and the natural circular frequency is $p = \sqrt{3EI/mL^3}$. Damping will be ignored in this example.

During the time interval $0 < t < t_0$, the applied force is $F(t) = F_0$. Since the initial conditions are $u(0) = \dot{u}(0) = 0$, the convolution integral in Eq. (10.4.19) represents the complete solution to the problem. Accordingly

$$
\begin{aligned}
u &= \frac{F_0}{mp} \int_0^t \sin p(t - \xi)\, d\xi \\
&= \frac{F_0}{mp} \left[\frac{\cos p(t - \xi)}{p} \right]_{\xi=0}^{\xi=t} \\
&= \frac{F_0}{k} (1 - \cos pt) \\
&= \frac{F_0 L^3}{3EI} \left(1 - \cos \sqrt{\frac{3EI}{mL^3}}\, t \right) \qquad 0 < t < t_0 \quad (10.4.20)
\end{aligned}
$$

The result naturally agrees with the solution (10.4.12) which we obtained for a step force excitation using the method of undetermined coefficients.

In the region $t > t_0$ the force $F(t) = 0$, and the convolution integral gives

$$
\begin{aligned}
u &= \frac{F_0}{mp} \int_0^{t_0} \sin p(t - \xi)\, d\xi + \frac{1}{mp} \int_{t_0}^t 0 \\
&= \frac{F_0}{mp} \left[\frac{\cos p(t - \xi)}{p} \right]_{\xi=0}^{\xi=t_0} \\
&= \frac{F_0}{k} [\cos p(t - t_0) - \cos pt] \\
&= \frac{F_0 L^3}{3EI} \left[\cos \sqrt{\frac{3EI}{mL^3}}\, (t - t_0) - \cos \sqrt{\frac{3EI}{mL^3}}\, t \right] \qquad t > t_0
\end{aligned}
$$

$$(10.4.21)$$

Alternatively, the system's response at time $t > t_0$ can be found by superimposing the responses to the two step functions shown in Fig. 10.11. Equation

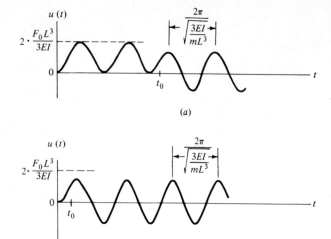

FIGURE 10.12
(a) Response for the case $t_0 > \tau$; (b) response for the case $t_0 < \tau$.

(10.4.20) gives the displacement produced by the force F_0 acting at $t = 0$; the displacement due to the step $-F_0$ at $t = t_0$ can be obtained from Eq. (10.4.20) by replacing t by $t - t_0$. Thus

$$u = \frac{F_0}{k}(1 - \cos pt) - \frac{F_0}{k}[1 - \cos p(t - t_0)]$$

$$= \frac{F_0}{k}[\cos p(t - t_0) - \cos pt]$$

$$= \frac{F_0 L^3}{3EI}\left[\cos\sqrt{\frac{3EI}{mL^3}}(t - t_0) - \cos\sqrt{\frac{3EI}{mL^3}}t\right] \qquad t > t_0 \quad (10.4.22)$$

as before.

Typical responses are shown in Fig. 10.12, corresponding to the two cases: (a) a pulse duration t_0 which exceeds the structure's period of free vibration τ; and (b) a duration t_0 which is small relative to the period τ. In the first case the maximum displacement is exactly two times the corresponding static deflection $F_0 L^3/3EI$, while in the second situation the maximum response is generally less than twice the static value. In both instances the structure performs free vibration for $t > t_0$, the amplitude of vibration depending upon the displacement and velocity of the system at time t_0. ////

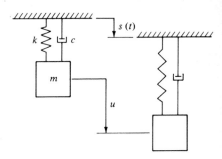

FIGURE 10.13
Mass-spring-dashpot model subject to a
foundation motion $s(t)$.

10.5 RESPONSE TO A FOUNDATION MOTION

A state of forced vibration can arise through motion of a structure's foundation rather than by the application of an external load. Common examples include the response of buildings to earthquakes and the oscillations of airplane wings resulting from flight maneuvers.

The equation of motion for the one-degree-of-freedom system shown in Fig. 10.13 is

$$m\ddot{u} = -k(u - s) - c(\dot{u} - \dot{s}) \qquad (10.5.1)$$

where $s(t)$ is the imposed foundation motion. Again letting $p^2 = k/m$ and $2n = c/m$, Eq. (10.5.1) becomes

$$\ddot{u} + 2n\dot{u} + p^2u = 2n\dot{s} + p^2s \qquad (10.5.2)$$

Alternatively, the equation of motion can be expressed in terms of the relative displacement $w = u - s$ as

$$\ddot{w} + 2n\dot{w} + p^2w = -\ddot{s} \qquad (10.5.3)$$

It is generally convenient to use Eq. (10.5.2) for problems in which the foundation's displacement is prescribed; Eq. (10.5.3) is more appropriate when the acceleration of the foundation is specified. Solutions to either equation may be found using the methods employed earlier in the forced-vibration problems.

Example 10.5 Response to an acceleration pulse Suppose that the structure considered in Example 10.4 is excited by the half-sine acceleration pulse shown in Fig. 10.14. We wish to determine the relative displacement $w(t)$ and the absolute acceleration $\ddot{u}(t)$ of mass m, assuming that the structure is initially at rest and that damping is negligible.

During the time interval $0 < t < t_0$, the equation of motion is

$$\ddot{w} + p^2w = -\ddot{s} = -a_0 \sin \frac{\pi t}{t_0} \qquad (10.5.4)$$

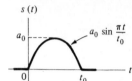

FIGURE 10.14
Half-sine acceleration pulse.

where $p^2 = 3EI/mL^3$. The method of undetermined coefficients can be used to obtain the particular solution

$$w_p = \frac{a_0}{p^2(\pi^2/p^2t_0{}^2 - 1)} \sin \frac{\pi t}{t_0} \qquad (10.5.5)$$

The complete solution is therefore

$$w = A \cos pt + B \sin pt + \frac{a_0}{p^2(\pi^2/p^2t_0{}^2 - 1)} \sin \frac{\pi t}{t_0} \qquad (10.5.6)$$

After eliminating the constants of integration A and B through application of the initial conditions $w(0) = \dot{w}(0) = 0$, one obtains

$$w = \frac{a_0}{p^2(\pi^2/p^2t_0{}^2 - 1)} \left(\sin \frac{\pi t}{t_0} - \frac{\pi}{pt_0} \sin pt \right) \qquad 0 < t < t_0 \qquad (10.5.7)$$

In order to determine the response at time $t > t_0$, the acceleration pulse is regarded as the sum of the two continuous sine curves of amplitude a_0 shown dotted in Fig. 10.15. Equation (10.5.7) gives the response associated with the curve beginning at $t = 0$; the response corresponding to the curve starting at

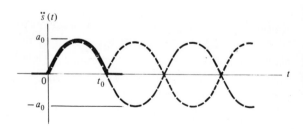

FIGURE 10.15
Representation of a half-sine pulse as the sum of two continuous sine curves.

$t = t_0$ is given by Eq. (10.5.7) with t replaced by $t - t_0$. Application of the principle of superposition then yields

$$w = \frac{a_0}{p^2(\pi^2/p^2 t_0^2 - 1)} \left[\sin \frac{\pi t}{t_0} - \frac{\pi}{p t_0} \sin pt + \sin \frac{\pi(t - t_0)}{t_0} - \frac{\pi}{p t_0} \sin p(t - t_0) \right]$$

$$= \frac{-a_0}{p^2(\pi^2/p^2 t_0^2 - 1)} \frac{\pi}{p t_0} [\sin pt + \sin p(t - t_0)] \qquad t > t_0 \qquad (10.5.8)$$

From the equation of motion (10.5.4) it is seen that the absolute acceleration of m is given by

$$\ddot{u} = \ddot{w} + \ddot{s} = -p^2 w \qquad (10.5.9)$$

Using Eqs. (10.5.7) and (10.5.8) we then obtain

$$\ddot{u} = \begin{cases} \dfrac{-a_0}{\pi^2/p^2 t_0^2 - 1} \left(\sin \dfrac{\pi t}{t_0} - \dfrac{\pi}{p t_0} \sin pt \right) & 0 < t < t_0 \\[3mm] \dfrac{a_0}{\pi^2/p^2 t_0^2 - 1} \dfrac{\pi}{p t_0} [\sin pt + \sin p(t - t_0)] & t > t_0 \end{cases} \qquad (10.5.10)$$

////

10.6   NUMERICAL INTEGRATION

Our study of the dynamic behavior of structures has been—and for the most part will continue to be—confined to problems involving linearly elastic, viscously damped or undamped systems. For the simple types of excitations considered thus far, it has been possible to obtain closed-form solutions to the equations of motion.

In practice, however, it may be necessary to investigate the dynamic response of a structure characterized by inelastic material behavior and nonviscous damping. The governing equations are then nonlinear, and generally the integration must be performed numerically. Numerical integration may also be required if the excitations are of such an irregular form that they cannot be represented in terms of simple mathematical functions.

Various step-by-step integration techniques exist, most of which are well suited for automatic computation. In fact nearly every modern computer facility has its own special subroutines for solving nonlinear differential equations. For this reason we shall not concern ourselves with the particulars of any specific integration scheme. References are listed at the end of the chapter which describe some of the more commonly used procedures.

While the response of structures governed by nonlinear equations will not be dealt with in any depth, a few problems requiring the use of a computer have been included (Probs. 10.13 and 10.14).

10.7 SUMMARY

Since the response of a single-degree-of-freedom system forms the basis for much of the theory presented in Chaps. 11 and 12, the following summary will be helpful.

RESPONSE OF A LINEARLY ELASTIC, VISCOUSLY DAMPED, SINGLE-DEGREE-OF-FREEDOM SYSTEM

Equation of motion

$$\ddot{u} + 2n\dot{u} + p^2 u = \frac{F(t)}{m} \tag{10.7.1}$$

$$2n = \frac{c}{m} \qquad p^2 = \frac{k}{m} \tag{10.7.2}$$

General solution

$$u = e^{-nt}(A \cos \bar{p}t + B \sin \bar{p}t) + u_p \tag{10.7.3}$$

$$\bar{p} = \sqrt{p^2 - n^2} \tag{10.7.4}$$

$$u_p = \text{any particular solution} \tag{10.7.5}$$

$$A = u(0) - u_p(0) \qquad B = \frac{\dot{u}(0) - \dot{u}_p(0)}{\bar{p}} + \frac{n[u(0) - u_p(0)]}{\bar{p}} \tag{10.7.6}$$

Alternate form of the general solution

$$u = e^{-nt}(a \cos \bar{p}t + b \sin \bar{p}t) + \frac{1}{m\bar{p}} \int_0^t F(\xi)e^{-n(t-\xi)} \sin \bar{p}(t - \xi) \, d\xi \tag{10.7.7}$$

$$a = u(0) \qquad b = \frac{\dot{u}(0)}{\bar{p}} + \frac{nu(0)}{\bar{p}} \tag{10.7.8}$$

PROBLEMS

10.1–10.2 The idealized one-degree-of-freedom system shown has a coefficient of viscous damping $c = 80$ lb-sec/in., mass $m = 10$ lb-sec^2/in., length $L = 100$ in., moment of inertia $I = 40$ in.^{4}, and Young's modulus $E = 30 \times 10^6$ lb$_f$/in.^{2}. Compute: (a) the frequency $\bar{f}$ and period $\bar{\tau}$ of free vibration; (b) the response $u(t)$ to an initial displacement $u(0) = 0.1$ in., assuming $\dot{u}(0) = 0$; and (c) the maximum displacement and the maximum bending moment caused by an initial velocity $\dot{u}(0) = 10$ in./sec, when $u(0) = 0$.

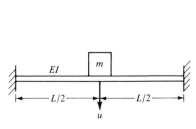

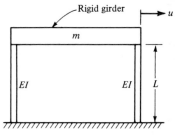

PROBLEM 10.1   PROBLEM 10.2

10.3   A mass-spring-dashpot system is given an initial displacement $u(0) = u_0$ and released. Determine the motion of the system in the case of (*a*) "heavy" damping ($n^2 > p^2$), and (*b*) "critical" damping ($n^2 = p^2$).

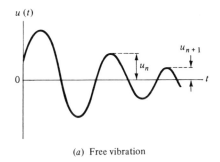

(*a*) Free vibration

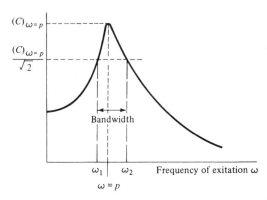

(*b*) Resonance curve

PROBLEM 10.4

10.4   One measure of the amount of damping in a structure is the *logarithmic decrement* Δ, defined as the natural logarithm of the ratio of any two successive peaks of free vibration. That is, $\Delta = \ln(u_n/u_{n+1})$ where u_n and u_{n+1} are the amplitudes shown in figure (a).

Another damping measure is the *bandwidth of the resonance curve*; the resonance curve is a plot of the amplitude of a steady-state vibration versus the circular frequency ω of the forcing function. As shown in figure (b), the bandwidth $\omega_2 - \omega_1$ represents the width of the curve where the amplitude is $1/\sqrt{2}$ times the resonant amplitude.

Show that for the case of very small damping $n^2 \ll p^2$: (a) $\Delta = 2\pi n/p$ and (b) $\omega_2 - \omega_1 = 2n = p\Delta/\pi$.

10.5 Find the response $u(t)$ of a linearly elastic, undamped, single-degree-of-freedom system to the forcing function shown. Assume that the system is at rest when the force is applied. Make a rough sketch of the response corresponding to several different values of the ratio t_0/τ.

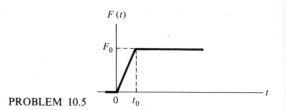

PROBLEM 10.5

10.6 Using the convolution integral, compute the response of a viscously damped system to the forcing functions: (a) $F(t) = F_0$ (a step function), and (b) $F(t) = F_0 e^{-t/t_0}$, where F_0 and t_0 are constants. Assume that the system is at rest initially.

10.7 A simply supported beam carries a motor of total mass m, as shown. During operation of the motor, a small unbalanced mass m_1 of eccentricity e rotates at an angular velocity ω. Assuming that the system is subject to viscous damping, investigate the steady-state vertical vibrations which result.

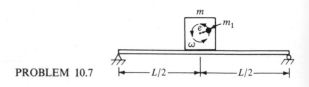

PROBLEM 10.7

10.8–10.9 The structure shown supports a mass m which is large compared with the mass of the structure. Assuming that damping is negligible, compute: (a) the system's natural frequency f and period τ; (b) the response to an initial displacement $u(0) = u_0$

with $\dot{u}(0) = 0$; and (c) the response to the forcing function $F(t)$ given in Prob. 10.5, for the case in which the system is initially at rest. *Note:* The stiffness coefficient for the structure can be obtained from the solution to Prob. 5.25 or 6.10.

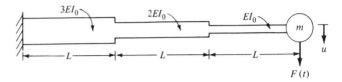

PROBLEM 10.8

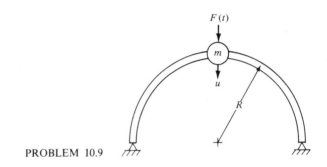

PROBLEM 10.9

10.10   A circular shaft of diameter D is made from a linearly elastic material of shear modulus G. The shaft carries a disk which has a mass moment of inertia I_m about the axis of the shaft. The system is at rest when a step torque of magnitude $\mathcal{T}_0$ is applied, as shown. Assuming that the shaft's mass moment of inertia is very small compared to I_m, and that damping is negligible, determine the angle of twist $\theta(t)$. What is the maximum shear stress developed in the shaft?

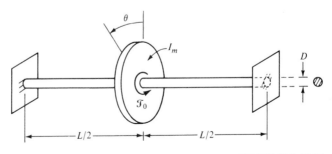

PROBLEM 10.10

10.11   A mass m is attached to the midpoint of a simply supported beam, as shown. The beam has a rectangular cross section, and is made from a linearly elastic material of Young's modulus E. Assuming that the mass of the beam is small compared to m, that damping is negligible, and that the system is initially at rest, derive expressions for the relative displacement $w(t)$ corresponding to the foundation disturbances: (a) $s = v_0 t$, and (b) $\ddot{s} = a_0$, where v_0 and a_0 are constants. What is the maximum bending stress developed in each case?

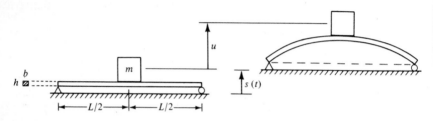

PROBLEM 10.11

10.12   Show that a complete solution to the problem of a linearly elastic, viscously damped, single-degree-of-freedom structure, subject to an arbitrary foundation acceleration $\ddot{s}(t)$ is

$$w = e^{-nt}\left[w(0)\cos \bar{p}t + \frac{\dot{w}(0) + nw(0)}{\bar{p}}\sin \bar{p}t \right]$$
$$- \frac{1}{\bar{p}} \int_0^t \ddot{s}(\xi) e^{-n(t-\xi)}\sin \bar{p}(t - \xi)\, d\xi$$

Problems 10.13 and 10.14 can be solved most efficiently using numerical integration and a digital computer. Most computer facilities can provide a subroutine program suitable for this purpose.

10.13   A structure having a nonlinear force-versus-displacement behavior is represented by a mass-spring model, in which the spring force is of the form $F_{\text{Spring}} = k(1 + a^2u^2)u$ where k and a^2 are positive constants. Since the force increases more rapidly than a linear function of the displacement, the spring is said to be a "stiffening" spring. The structure is at rest initially and is subject to a step forcing function $F(t) = F_0$.

(a) Show that the equation of motion can be written in the nondimensional form

$$\frac{d^2u^*}{dt^{*2}} + (1 + A^2u^{*2})u^* = 1$$

where $\qquad u^* = \dfrac{u}{F_0/k} \qquad t^* = pt \qquad A = a\dfrac{F_0}{k} \qquad p = \sqrt{\dfrac{k}{m}}$

(b) Compute and plot the response $u*(t*)$ which occurs during the first several cycles of oscillation when the nonlinearity in the spring is given by : (a) $A = 0.1$, (b) $A = 1.0$, and (c) $A = 10.0$. Compare your results with the response for a linear spring $(A = 0)$.

10.14   A structure has an elastic-plastic type of material behavior, characterized by the restoring force-versus-displacement curve shown. The system's weight $W = 100,000$ lb is initially at rest and is subject to a step force $F(t) = F_0$. Compute the maximum displacement occurring during the first cycle of vibration if (a) $F_0 = 20,000$ lb and (b) $F_0 = 30,000$ lb.

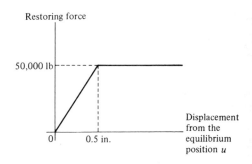

PROBLEM 10.14

10.15   Make a study of the use of analytical methods for finding approximate solutions to nonlinear differential equations. In particular, study the *perturbation method,* the *method of the variation of parameters,* and *averaging methods based on residuals* (see Ref. 10.5). Note that all these methods are limited in their application to equations in which the degree of nonlinearity is relatively small. Use one or more of these techniques to investigate the free vibration of the structure of Prob. 10.13. Assume that the motion is initiated by a displacement $u(0) = u_0$, and that the nonlinearity is given by $A = 0.1$.

REFERENCES

For additional examples of the vibration of single-degree-of-freedom systems, see any standard textbook on mechanical vibrations. References which are concerned exclusively with the dynamic behavior of engineering structures include:

10.1   NORRIS, C. H. et al.: "Structural Design for Dynamic Loads," McGraw-Hill, New York, 1959.

10.2   ROGERS, G. L.: "Dynamics of Framed Structures," Wiley, New York, 1959.

10.3   BIGGS, J. M.: "Introduction to Structural Dynamics," McGraw-Hill, New York, 1964.

For a description of various analytical, numerical, and graphical methods for analyzing systems in which the restoring forces are not proportional to the displacements and in which the dissipative forces are not of the viscous type:

10.4 TIMOSHENKO, S. P., and D. H. YOUNG: "Vibration Problems in Engineering," 3d ed., chap. II, Van Nostrand, Princeton, N.J., 1955.

10.5 CUNNINGHAM, W. J.: "Introduction to Nonlinear Analysis," McGraw-Hill, New York, 1958.

10.6 JACOBSEN, L. S., and R. S. AYRE: "Engineering Vibrations; With Applications to Structures and Machinery," chap. 6, McGraw-Hill, New York, 1958.

For a discussion of several of the more common numerical integration techniques and the application of digital computers for solving nonlinear differential equations:

10.7 SOUTHWORTH, R. W., and S. L. DELEEUW: "Digital Computation and Numerical Methods," McGraw-Hill, New York, 1965.

STRUCTURES WITH MANY DEGREES OF FREEDOM

11.1 INTRODUCTION

As pointed out in Chap. 10, every real structure has a continuous distribution of mass, and hence, an infinite number of degrees of freedom. In other words, it takes an infinite number of coordinates to specify the configuration of an actual structure undergoing vibration. We saw in Chap. 10, however, that depending upon the mass distribution and the constraints of the structure, the dynamic response can sometimes be determined with sufficient accuracy by treating the structure as a single-degree-of-freedom system. Such an idealization leads to considerable simplifications in the analysis. However, if a more detailed or more accurate description of the dynamic behavior is required, or if the structure has a complicated mass distribution or is not constrained to vibrate in just one direction, then it becomes necessary to consider a more sophisticated model, i.e., one having more than a single degree of freedom.

Let us consider, for example, a beam which carries two rigid bodies of mass m_1 and m_2, as shown in Fig. 11.1a. If the beam's mass is small compared with m_1 and m_2, and if each rigid body is assumed to move in the vertical direction only, then the displacements u_1 and u_2 suffice to describe the system's configuration at every instant of time. The idealized structure therefore has two degrees of freedom.

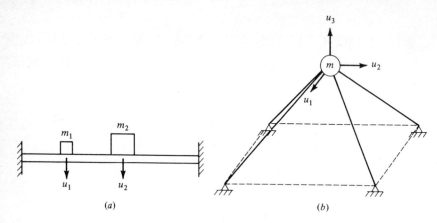

FIGURE 11.1
Structures which can be idealized as discrete-mass systems.

As another example, suppose we wish to compute the dynamic response of the mass m shown in Fig. 11.1b. We shall assume that the mass of the truss which supports m is negligible compared with m. Although the idealized system then consists of a single mass, the mass is free to move in three directions. The coordinates u_1, u_2, and u_3 can be used to describe the motion of this three-degree-of-freedom system.

Now consider the airplane wing shown in Fig. 11.2a. Unlike the previous examples in which the masses of the structures themselves were negligible, the mass in this case is distributed continuously along the wing's span. Nevertheless it is still possible to approximate the wing by a system having a finite number of degrees of freedom. For instance, if one is only interested in the vertical bending modes of vibration, then the model shown in Fig. 11.2b can be used. Here the wing has been divided into n sections, and the mass within each section is assumed to be concentrated one-half at each end of that section. The beam elements joining the masses are taken to be massless, and are assumed to have the same bending stiffness as the corresponding sections of the actual wing. If it is further assumed that the masses move only in the vertical direction, the system has n degrees of freedom, u_i $(i = 1, 2, \ldots, n)$. Generally speaking, the larger n is, the greater the accuracy of the computed response will be.

Depending upon the geometry, stiffness, and mass distribution of the actual wing, a purely bending mode of vibration may not be possible. For example, vertical translations of the elements of the beam may be accompanied by either horizontal or torsional deformations. In order to account for such coupled modes of vibration, a three-dimensional model of the type shown in Fig. 11.2c. may be used. Here the continuous mass distribution is again approximated

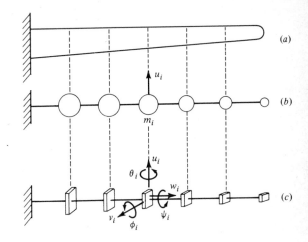

FIGURE 11.2
(a) Airplane wing; (b) discrete-mass model suitable for studying the lateral vibration of the wing; (c) discrete-mass model suitable for studying three-dimensional motions of the wing.

by discrete masses, and in addition the mass moment of inertia distributions are replaced by lumped-mass moments of inertia. Since three translations (u_i, v_i, w_i) and three rotations $(\theta_i, \phi_i, \psi_i)$ are required in order to specify the position of each rigid body in the model, the system has a total of $6n$ degrees of freedom.

The dynamic behavior of multi-degree-of-freedom structures of the type described above will be investigated in the following sections of this chapter.

11.2 EQUATIONS OF MOTION— LAGRANGE'S EQUATIONS

As a first step in finding the response of a discrete-mass system, the differential equations which govern the motion of the masses must be derived. This may be accomplished by a direct application of Newton's laws of motion. An alternative approach, which is generally far more convenient when complicated structures are involved, is known as Lagrange's method.[1] As we shall see, the lagrangian approach is closely related to the energy methods presented in Chap. 5.

Consider an n-degree-of-freedom structure consisting of rigid masses joined by flexible structural elements. Such a system is shown in Fig. 11.3, in which the

[1] This method was introduced by J. L. Lagrange in his treatise " Mécanique Analytique " in 1788.

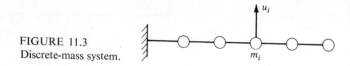

FIGURE 11.3
Discrete-mass system.

displacements of the masses m_i are denoted by $u_i (i = 1, 2, \ldots, n)$. If F_i represents the resultant force acting on a particular mass m_i in the direction of u_i, then Newton's second law of motion states that[1]

$$F_i = m_i \ddot{u}_i \quad (11.2.1)$$

Multiplying this equation by δu_i, where δu_i denotes a virtual displacement (see Sec. 5.4) of mass m_i, we obtain

$$F_i \, \delta u_i = m_i \ddot{u}_i \, \delta u_i \quad (11.2.2)$$

Note that $F_i \, \delta u_i$ represents the virtual work resulting from a variation in just one of the coordinates, namely u_i. Summing the work associated with all n coordinates yields the total virtual work[2]

$$\delta W = \sum_{i=1}^{n} F_i \, \delta u_i = \sum_{i=1}^{n} m_i \ddot{u}_i \, \delta u_i \quad (11.2.3)$$

In order to arrive at a system of equations which will be valid for an arbitrary set of independent generalized coordinates $q_1, q_2, \ldots, q_n$, we express the displacements u_i in terms of the coordinates q_j by the functional relations

$$u_i = u_i(q_1, q_2, \ldots, q_n) \qquad i = 1, 2, \ldots, n \quad (11.2.4)$$

The virtual displacements δu_i may then be written as

$$\delta u_i = \sum_{j=1}^{n} \frac{\partial u_i}{\partial q_j} \delta q_j \quad (11.2.5)$$

so that Eq. (11.2.3) becomes

$$\delta W = \sum_{j=1}^{n} \sum_{i=1}^{n} F_i \frac{\partial u_i}{\partial q_j} \delta q_j = \sum_{j=1}^{n} \sum_{i=1}^{n} m_i \ddot{u}_i \frac{\partial u_i}{\partial q_j} \delta q_j \quad (11.2.6)$$

Defining the generalized forces Q_j by

$$Q_j = \sum_{i=1}^{n} F_i \frac{\partial u_i}{\partial q_j} \quad (11.2.7)$$

---

[1] No summation is implied by the repeated indices in Eqs. (11.2.1) and (11.2.2).
[2] The expression (11.2.3) for the virtual work δW resembles Eq. (5.4.9) which defines the virtual work δW_E done by the external forces applied to the structure. In Eq. (11.2.3), however, the forces F_i are the resultants of internal elastic forces, external applied loads, and damping forces.

we are led to the virtual work expression

$$\delta W = \sum_{j=1}^{n} Q_j\, \delta q_j = \sum_{j=1}^{n} \sum_{i=1}^{n} m_i \ddot{u}_i \frac{\partial u_i}{\partial q_j}\, \delta q_j \qquad (11.2.8)$$

Since the virtual quantities δq_j are arbitrary, Eq. (11.2.8) requires that

$$Q_j = \sum_{i=1}^{n} m_i \ddot{u}_i \frac{\partial u_i}{\partial q_j} \qquad (11.2.9)$$

From Eq. (11.2.4) it is noted that

$$\dot{u}_i = \sum_{j=1}^{n} \frac{\partial u_i}{\partial q_j}\, \dot{q}_j \qquad (11.2.10)$$

and therefore

$$\frac{\partial \dot{u}_i}{\partial \dot{q}_j} = \frac{\partial u_i}{\partial q_j} \qquad \text{and} \qquad \frac{d}{dt}\left(\frac{\partial \dot{u}_i}{\partial \dot{q}_j}\right) = \frac{\partial \dot{u}_i}{\partial q_j} \qquad (11.2.11)$$

Substituting Eqs. (11.2.11) into (11.2.9) yields

$$Q_j = \sum_{i=1}^{n} m_i \ddot{u}_i \frac{\partial \dot{u}_i}{\partial \dot{q}_j}$$

$$= \frac{d}{dt} \sum_{i=1}^{n} m_i \dot{u}_i \frac{\partial \dot{u}_i}{\partial \dot{q}_j} - \sum_{i=1}^{n} m_i \dot{u}_i \frac{\partial \dot{u}_i}{\partial q_j} \qquad (11.2.12)$$

Finally, the generalized forces Q_j may be expressed as

$$Q_j = \frac{d}{dt}\left(\frac{\partial T}{\partial \dot{q}_j}\right) - \frac{\partial T}{\partial q_j} \qquad j = 1, 2, \ldots, n \qquad (11.2.13)$$

where the kinetic energy T of the structure is defined by the relation

$$T = \tfrac{1}{2} \sum_{i=1}^{n} m_i \dot{u}_i^2 \qquad (11.2.14)$$

The n equations (11.2.13) are called *Lagrange's equations of motion.*

The generalized forces Q_j will, in general, consist of both conservative and nonconservative forces. We may therefore write

$$Q_j = (Q_j)_C + (Q_j)_N \qquad (11.2.15)$$

where $(Q_j)_C$ and $(Q_j)_N$ denote, respectively, the conservative and nonconservative components of Q_j. The conservative components can, by definition,[1] be expressed in terms of a potential energy V as

$$(Q_j)_C = -\frac{\partial V}{\partial q_j} \qquad (11.2.16)$$

---

[1] A force Q_j is defined as conservative if: (1) it is a function of position only, and (2) is derivable from a scalar potential as indicated in Eq. (11.2.16).

After substituting Eqs. (11.2.15) and (11.2.16) into (11.2.13), we arrive at the following alternate form of Lagrange's equations:

$$\frac{d}{dt}\left(\frac{\partial T}{\partial \dot{q}_j}\right) - \frac{\partial T}{\partial q_j} + \frac{\partial V}{\partial q_j} = (Q_j)_N \qquad j = 1, 2, \ldots, n \quad (11.2.17)$$

In applying Lagrange's equations (11.2.13) or (11.2.17) to a discrete-mass structure, the damping forces and, in most cases, the applied loads are nonconservative and must be represented by the forces $(Q_j)_N$. The internal forces, on the other hand, depend only on the elastic deformation, and are derivable from the strain energy U of the structure. Recalling Castigliano's first theorem (5.9.6), we have

$$(Q_j)_C = -\frac{\partial U}{\partial q_j} \quad (11.2.18)$$

where the minus sign accounts for the fact that the resultant internal forces exerted on the masses are equal in magnitude but opposite in direction to the internal forces which act on the adjoining elastic elements. Hence, for these forces the potential energy V may be replaced by the strain energy, and Eq. (11.2.17) may be written as

$$\frac{d}{dt}\left(\frac{\partial T}{\partial \dot{q}_j}\right) - \frac{\partial T}{\partial q_j} + \frac{\partial U}{\partial q_j} = (Q_j)_N \qquad j = 1, 2, \ldots, n \quad (11.2.19)$$

It should be noted that while Lagrange's equations were derived here for a system of masses (particles), they are applicable to rigid bodies as well. This follows from the fact that a rigid body can be treated as a collection of particles, and must therefore obey the same general equations of motion. When applying Lagrange's equations to a rigid body, one must remember to include the body's rotational kinetic energy in the expression for T.

Example 11.1 Lagrange's equations for a two-degree-of-freedom system

A cantilever beam of length L and bending stiffness EI supports a rigid body at its free end, as shown in Fig. 11.4a. The mass and mass moment of inertia of the rigid body about its center of mass (point c) are m and I_m, respectively. The mass of the beam is assumed to be small compared with m. A force $F(t)$ and a couple $M(t)$ are applied at point c. Let us now use Lagrange's method in order to derive the equations of motion for the rigid body.

Noting that the structure has two degrees of freedom, we select the coordinates q_1 and q_2 shown in the figure as the independent generalized coordinates. To obtain the corresponding generalized forces Q_1 and Q_2 we compute the virtual

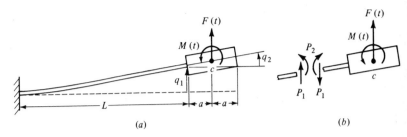

FIGURE 11.4
(a) Cantilever beam with an attached rigid body; (b) free-body diagram for the rigid body.

work done during virtual displacements δq_1 and δq_2. From Fig. 11.4b it is seen that

$$\delta W = [F(t) - P_1]\,\delta q_1 + [aF(t) + M(t) - P_2]\,\delta q_2 \quad (11.2.20)$$

where P_1 and P_2 are the internal elastic forces acting on the rigid body. According to Eq. (11.2.8)

$$\delta W = Q_1\,\delta q_1 + Q_2\,\delta q_2 \quad (11.2.21)$$

and thus

$$Q_1 = F(t) - P_1$$
$$Q_2 = aF(t) + M(t) - P_2 \quad (11.2.22)$$

Since the applied forces $F(t)$ and $M(t)$ are nonconservative, and the internal forces P_1 and P_2 are conservative, we let

$$(Q_1)_N = F(t) \qquad (Q_1)_C = -P_1 = -\frac{\partial U}{\partial q_1}$$

$$(11.2.23)$$

$$(Q_2)_N = aF(t) + M(t) \qquad (Q_2)_C = -P_2 = -\frac{\partial U}{\partial q_2}$$

where the strain energy U for the structure is (see Sec. 7.2)

$$U = \frac{1}{2}\left(\frac{12EI}{L^3}q_1{}^2 - \frac{12EI}{L^2}q_1 q_2 + \frac{4EI}{L}q_2{}^2\right) \quad (11.2.24)$$

The kinetic energy of the rigid body, assuming small deformations, may be expressed in terms of q_1 and q_2 as

$$T = \tfrac{1}{2}m(\dot{q}_1 + a\dot{q}_2)^2 + \tfrac{1}{2}I_m\dot{q}_2{}^2 \quad (11.2.25)$$

Finally, substituting the relations (11.2.23) to (11.2.25) into either Eq. (11.2.13) or (11.2.19) gives the two equations of motion

$$m\ddot{q}_1 + ma\ddot{q}_2 + \frac{12EI}{L^3}q_1 - \frac{6EI}{L^2}q_2 = F(t)$$

$$ma\ddot{q}_1 + (ma^2 + I_m)\ddot{q}_2 - \frac{6EI}{L^2}q_1 + \frac{4EI}{L}q_2 = aF(t) + M(t)$$

(11.2.26)

////

11.3　MATRIX FORMULATION

As seen in Chap. 7, it is convenient to formulate the analysis of a complicated structure in matrix notation. With the aid of Lagrange's equations we shall now develop a matrix equation which governs the vibration of a damped, n-degree-of-freedom, discrete-mass structure.

The expression for the kinetic energy is

$$T = \tfrac{1}{2} \sum_{i=1}^{n} m_i \dot{u}_i{}^2 \qquad (11.3.1)$$

in which $\dot{u}_i$ is the velocity of the mass m_i. This equation may be written in matrix notation as

$$T = \tfrac{1}{2}\{\dot{u}\}^T[m]\{\dot{u}\} \qquad (11.3.2)$$

where $\{\dot{u}\}$ is a column matrix of the velocities $\dot{u}_1, \dot{u}_2, \ldots, \dot{u}_n$, and $\{\dot{u}\}^T$ is the transpose of $\{\dot{u}\}$. The matrix $[m]$, known as the *inertia matrix*, is a diagonal square matrix of order n having m_i as its diagonal elements.

From Eq. (11.3.1) it is seen that

$$\frac{\partial T}{\partial \dot{u}_i} = m_i \dot{u}_i \qquad \text{and} \qquad \frac{\partial T}{\partial u_i} = 0 \qquad i = 1, 2, \ldots, n \qquad (11.3.3)$$

or in matrix notation

$$\left\{\frac{\partial T}{\partial \dot{u}}\right\} = [m]\{\dot{u}\} \qquad \text{and} \qquad \left\{\frac{\partial T}{\partial u}\right\} = \{0\} \qquad (11.3.4)$$

The resultant force F_i acting on the mass m_i is assumed to consist of three different types of forces: internal elastic forces, damping forces, and applied loads. As discussed previously, the internal elastic forces are conservative and have as their potential the strain energy U. Assuming that U is the only source of potential energy, then [see Eq. (7.2.9)]

$$V = U = \tfrac{1}{2}\{u\}^T[k]\{u\} \qquad (11.3.5)$$

where $[k]$ is the structure's stiffness matrix. Since $[k]$ is symmetric, it follows from Eq. (11.3.5) that

$$\left\{\frac{\partial V}{\partial u}\right\} = [k]\{u\} \qquad (11.3.6)$$

in which $\{\partial V/\partial u\}$ is a column matrix of the elements $\partial V/\partial u_1, \partial V/\partial u_2, \ldots, \partial V/\partial u_n$.

We shall denote the damping force acting on m_i by $(F_i)_d$ and the applied force by $(F_i)_a$. Together these forces represent the resultant nonconservative force acting on m_i; that is, $(F_i)_N = (F_i)_d + (F_i)_a$, or in matrix notation

$$\{F\}_N = \{F\}_d + \{F\}_a \qquad (11.3.7)$$

The n Lagrange's equations (11.2.17) for the coordinates u_i and forces F_i may be expressed in matrix form as

$$\frac{d}{dt}\left\{\frac{\partial T}{\partial \dot{u}}\right\} - \left\{\frac{\partial T}{\partial u}\right\} + \left\{\frac{\partial V}{\partial u}\right\} = \{F\}_N \qquad (11.3.8)$$

Substituting Eqs. (11.3.4), (11.3.6), and (11.3.7) into (11.3.8) gives

$$[m]\{\ddot{u}\} + [k]\{u\} = \{F\}_d + \{F\}_a \qquad (11.3.9)$$

If we assume that the damping forces are of a viscous type, then we may express $\{F\}_d$ as

$$\{F\}_d = -[d]\{\dot{u}\} \qquad (11.3.10)$$

where $[d]$ is called the *damping matrix*. In this case Eq. (11.3.9) gives

$$[m]\{\ddot{u}\} + [d]\{\dot{u}\} + [k]\{u\} = \{F\}_a \qquad (11.3.11)$$

Equation (11.3.11) is the matrix equation of motion for a discrete-mass structure. If a set of generalized coordinates q_j (different from the absolute displacements u_i) are used in order to specify the structure's configuration, the equations of motion will be of the same form as (11.3.11); that is, (see Prob. 11.3)

$$[m]\{\ddot{q}\} + [d]\{\dot{q}\} + [k]\{q\} = \{Q\}_a \qquad (11.3.12)$$

The elements of the matrices $[m]$, $[d]$, $[k]$ for a structure will, of course, be different when referred to different sets of coordinates. These three matrices will always be symmetric; they may or may not be diagonal.

Example 11.2 Matrix equation of motion for a two-degree-of-freedom system

Let us now derive the matrix equation governing the motion of the structure considered in Example 11.1. As before, the displacements q_1 and q_2

in Fig. 11.4 will be used as the generalized coordinates. The kinetic energy expression (11.2.25) may be written in matrix notation as

$$T = \tfrac{1}{2}\{\dot{q}\}^T[m]\{\dot{q}\} \tag{11.3.13}$$

where
$$\{\dot{q}\} = \begin{bmatrix} \dot{q}_1 \\ \dot{q}_2 \end{bmatrix} \quad \text{and} \quad [m] = \begin{bmatrix} m & ma \\ ma & ma^2 + I_m \end{bmatrix} \tag{11.3.14}$$

The strain energy expression (11.2.24) is now written as

$$U = \tfrac{1}{2}\{q\}^T[k]\{q\} \tag{11.3.15}$$

where the stiffness matrix $[k]$ for the cantilever beam is

$$[k] = \begin{bmatrix} \dfrac{12EI}{L^3} & \dfrac{-6EI}{L^2} \\ \dfrac{-6EI}{L^2} & \dfrac{4EI}{L} \end{bmatrix} \tag{11.3.16}$$

The applied generalized forces $(Q_j)_a$ can be found in the same way as before, i.e., by considering the work done by these forces during the virtual displacements δq_1 and δq_2. Writing the results in matrix form gives

$$\{Q\}_a = \begin{bmatrix} F(t) \\ aF(t) + M(t) \end{bmatrix} \tag{11.3.17}$$

Substituting the above expressions for $[m]$, $[k]$, and $\{Q\}_a$ into Eq. (11.3.12) gives the matrix equation of motion

$$\begin{bmatrix} m & ma \\ ma & ma^2 + I_m \end{bmatrix} \begin{bmatrix} \ddot{q}_1 \\ \ddot{q}_2 \end{bmatrix} + \begin{bmatrix} \dfrac{12EI}{L^3} & \dfrac{-6EI}{L^2} \\ \dfrac{-6EI}{L^2} & \dfrac{4EI}{L} \end{bmatrix} \begin{bmatrix} q_1 \\ q_2 \end{bmatrix} = \begin{bmatrix} F(t) \\ aF(t) + M(t) \end{bmatrix} \tag{11.3.18}$$

which agrees with Eq. (11.2.26).   ////

11.4   FREE, UNDAMPED VIBRATION

Referred to a system of generalized coordinates q_j, the matrix equation of motion for the free vibration of a discrete-mass structure which has negligible damping is

$$[m]\{\ddot{q}\} + [k]\{q\} = \{0\} \tag{11.4.1}$$

Premultiplying this equation by the structure's flexibility matrix $[c]$, which is equal to the inverse of the stiffness matrix $[k]$, gives

$$[c][m]\{\ddot{q}\} + [I]\{q\} = \{0\} \qquad (11.4.2)$$

or

$$[D]\{\ddot{q}\} + \{q\} = \{0\} \qquad (11.4.3)$$

where $[I]$ represents the unit diagonal matrix; we have also introduced the *dynamical matrix* $[D]$, defined as

$$[D] = [c][m] \qquad (11.4.4)$$

As a trial solution to the set of homogeneous equations (11.4.3) it will be assumed that each mass vibrates at the same circular frequency p, but has a different amplitude of displacement. Letting $\{\phi\}$ be a column matrix of these amplitudes, our trial solution is

$$\{q\} = \{\phi\}\sin(pt + \delta) \qquad (11.4.5)$$

For such a motion, the generalized accelerations are

$$\{\ddot{q}\} = -p^2\{\phi\}\sin(pt + \delta) \qquad (11.4.6)$$

and Eq. (11.4.3) then gives

$$-p^2[D]\{\phi\} + \{\phi\} = \{0\} \qquad (11.4.7)$$

This equation may be rewritten as

$$([D] - \lambda[I])\{\phi\} = \{0\} \qquad (11.4.8)$$

where

$$\lambda = \frac{1}{p^2} \qquad (11.4.9)$$

Equation (11.4.8) represents a system of n homogeneous, linear algebraic equations in the amplitudes $\{\phi\}$. Equations of this form are referred to as eigenvalue problems (see Appendix B). For a nontrivial solution, the determinant of the coefficient matrix must be zero. Thus we have

$$|[D] - \lambda[I]| = \begin{vmatrix} D_{11} - \lambda & D_{12} & \cdots & D_{1n} \\ D_{21} & D_{22} - \lambda & \cdots & D_{2n} \\ \cdots & \cdots & \cdots & \cdots \\ D_{n1} & D_{n2} & \cdots & D_{nn} - \lambda \end{vmatrix} = 0 \qquad (11.4.10)$$

Expansion of this determinant gives an nth order polynomial in λ. The n roots $\lambda_1, \lambda_2, \ldots, \lambda_n$ of this polynomial equation represent the system's eigenvalues; each eigenvalue is related to a circular frequency p by relation (11.4.9) An n-degree-of-freedom system therefore has n natural frequencies $p_1, p_2, \ldots, p_n$. These frequencies are generally arranged in ascending order with p_1 being the lowest or *fundamental* frequency. Corresponding to each distinct frequency p_j (or eigenvalue λ_j), there will be a set of amplitudes or eigenvectors $\{\phi\}_j$. These amplitudes are obtained by substituting the value of the eigenvalue λ_j into Eq. (11.4.8). Because Eq. (11.4.8) represents a system of homogeneous equations, only relative values or ratios of the amplitudes may be found. One of the amplitudes may be taken equal to unity, and the remaining values then determined; the resulting vector $\{\phi\}_j$ is referred to as the *j*th *normal mode* of vibration.

Hence we have shown that a possible motion of the structure is a harmonic vibration at a circular frequency p_j, with displacements having the relative amplitudes $\{\phi\}_j$; or

$$\{q\} = \{\phi\}_j \sin(p_j t + \delta_j) \quad (11.4.11)$$

Multiplying each of the n such solutions by an arbitrary constant C_j and summing, gives the general solution to Eq. (11.4.1); namely,

$$\{q\} = \{\phi\}_1 C_1 \sin(p_1 t + \delta_1) + \{\phi\}_2 C_2 \sin(p_2 t + \delta_2)$$
$$+ \cdots + \{\phi\}_n C_n \sin(p_n t + \delta_n) \quad (11.4.12)$$

Equation (11.4.12) may be written more compactly as

$$\{q\} = [\phi]\{C \sin(pt + \delta)\} \quad (11.4.13)$$

where the *j*th column of the so-called "modal matrix" $[\phi]$ is the normal mode $\{\phi\}_j$. In other words, the element ϕ_{ij} of $[\phi]$ is the relative amplitude of the displacement q_i corresponding to the *j*th mode of vibration. The column matrix $\{C \sin(pt + \delta)\}$ in Eq. (11.4.13) is given by

$$\{C \sin(pt + \delta)\} = \begin{bmatrix} C_1 \sin(p_1 t + \delta_1) \\ C_2 \sin(p_2 t + \delta_2) \\ \vdots \\ C_n \sin(p_n t + \delta_n) \end{bmatrix} \quad (11.4.14)$$

The general solution (11.4.13) contains $2n$ arbitrary constants $C_1, C_2, \ldots, C_n$, $\delta_1, \delta_2, \ldots, \delta_n$, which must be determined from the prescribed initial conditions. An alternate form of the solution, which is often more useful in initial condition problems, is

$$\{q\} = [\phi]\{A \cos pt + B \sin pt\} \quad (11.4.15)$$

where the column matrix $\{A \cos pt + B \sin pt\}$ has the form

$$\{A \cos pt + B \sin pt\} = \begin{bmatrix} A_1 \cos p_1 t + B_1 \sin p_1 t \\ A_2 \cos p_2 t + B_2 \sin p_2 t \\ \vdots \\ A_n \cos p_n t + B_n \sin p_n t \end{bmatrix} \quad (11.4.16)$$

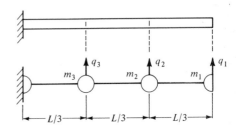

FIGURE 11.5
Uniform cantilever beam idealized as a
three-degree-of-freedom system.

The $2n$ arbitrary constants $A_1, A_2, \ldots, A_n, B_1, B_2, \ldots, B_n$ are related to $C_1,$ $C_2, \ldots, C_n, \delta_1, \delta_2, \ldots, \delta_n$ by

$$\begin{aligned} A_j &= C_j \sin \delta_j \\ B_j &= C_j \cos \delta_j \end{aligned} \qquad j = 1, 2, \ldots, n \quad (11.4.17)$$

In order to determine these constants, the generalized displacements $\{q\}$ and velocities $\{\dot{q}\}$ must be known at some instant of time. For instance, let us assume that at time zero the displacements and velocities have the values $\{q_0\}$ and $\{\dot{q}_0\}$, respectively. From the general solution (11.4.15) we obtain

$$\begin{aligned} \{q_0\} &= [\phi]\{A\} \\ \{\dot{q}_0\} &= [\phi]\{pB\} \end{aligned} \qquad (11.4.18)$$

where

$$\{A\} = \begin{bmatrix} A_1 \\ A_2 \\ \vdots \\ A_n \end{bmatrix} \quad \text{and} \quad \{pB\} = \begin{bmatrix} p_1 B_1 \\ p_2 B_2 \\ \vdots \\ p_n B_n \end{bmatrix} \qquad (11.4.19)$$

Solving Eqs. (11.4.18) for the arbitrary constants gives

$$\begin{aligned} \{A\} &= [\phi]^{-1}\{q_0\} \\ \{pB\} &= [\phi]^{-1}\{\dot{q}_0\} \end{aligned} \qquad (11.4.20)$$

An example which demonstrates the computation of the natural frequencies and the modal matrix for a particular discrete-mass structure is given below.

Example 11.3   Free vibration of a three-mass cantilevered beam As discussed earlier, the dynamic response of a continuous structure can often be determined with sufficient accuracy by first idealizing the structure as a discrete-mass system. With this in mind, we shall represent a uniform cantilever beam by the three-mass model shown in Figure 11.5. The free vibration of the model will be examined in this example; the results will be compared with the characteristics of the actual distributed-mass structure in Chap. 12.

The beam has been divided into three sections of length $L/3$, as shown. In this problem it will be assumed that the mass within each section $m/3$ is proportioned one-half to the right end of the section and one-half to the left end. Based on this criterion, the lumped masses are $m_1 = m/6, m_2 = m_3 = m/3$. The concentration of mass at the fixed end of the beam has no influence upon the structure's vibration and is therefore ignored.

Since the actual beam's bending stiffness EI is constant over the length L, it is assumed that the elastic elements which interconnect the discrete masses also have a uniform stiffness EI.

The motion of the model can be described by the displacements q_1, q_2, and q_3 shown, in which case the kinetic energy is

$$T = \tfrac{1}{2}m_1 \dot{q}_1{}^2 + \tfrac{1}{2}m_2 \dot{q}_2{}^2 + \tfrac{1}{2}m_3 \dot{q}_3{}^2 \quad (11.4.21)$$

or in matrix notation

$$T = \tfrac{1}{2}\{\dot{q}\}^T [m]\{\dot{q}\} \quad (11.4.22)$$

where the inertia matrix is

$$[m] = \begin{bmatrix} 0.1667 & 0 & 0 \\ 0 & 0.3333 & 0 \\ 0 & 0 & 0.3333 \end{bmatrix} m \quad (11.4.23)$$

Using any one of the methods described in Chap. 5, the flexibility matrix for the structure is found to be

$$[c] = \begin{bmatrix} 0.3333 & 0.1728 & 0.0494 \\ 0.1728 & 0.0988 & 0.0309 \\ 0.0494 & 0.0309 & 0.0124 \end{bmatrix} \frac{L^3}{EI} \quad (11.4.24)$$

Premultiplying $[m]$ by $[c]$ gives the dynamical matrix

$$[D] = [c][m] = \begin{bmatrix} 0.0556 & 0.0576 & 0.0165 \\ 0.0288 & 0.0329 & 0.0103 \\ 0.0082 & 0.0103 & 0.0041 \end{bmatrix} \frac{mL^3}{EI} \quad (11.4.25)$$

Following the procedure outlined earlier, the eigenvalues λ of the matrix $[D]$ are found by solving the following determinant equation

$$|[D] - \lambda[I]| = \begin{vmatrix} 0.0556 - \lambda \dfrac{EI}{mL^3} & 0.0576 & 0.0165 \\[2mm] 0.0288 & 0.0329 - \lambda \dfrac{EI}{mL^3} & 0.0103 \\[2mm] 0.0082 & 0.0103 & 0.0041 - \lambda \dfrac{EI}{mL^3} \end{vmatrix} \frac{mL^3}{EI} = 0$$

$$(11.4.26)$$

The three roots which are obtained upon expansion of this determinant are

$$\lambda_1 = 0.0893 \frac{mL^3}{EI} \qquad \lambda_2 = 0.00280 \frac{mL^3}{EI} \qquad \lambda_3 = 0.000452 \frac{mL^3}{EI} \qquad (11.4.27)$$

Since $\lambda_j = 1/p_j^2$, the natural circular frequencies are given by

$$p_1 = 3.36 \sqrt{\frac{EI}{mL^3}} \qquad p_2 = 18.9 \sqrt{\frac{EI}{mL^3}} \qquad p_3 = 47.2 \sqrt{\frac{EI}{mL^3}} \qquad (11.4.28)$$

These values are 5, 14, and 24 percent smaller, respectively, than the lowest three frequencies of a distributed-mass cantilever beam.

Substitution of each eigenvalue λ_j into Eq. (11.4.8) yields the corresponding eigenvector, or mode shape, $\{\phi\}_j$. For example, for the first mode we write

$$\frac{mL^3}{EI} \begin{bmatrix} 0.0556 - 0.0893 & 0.0576 & 0.0165 \\ 0.0288 & 0.0329 - 0.0893 & 0.0103 \\ 0.0082 & 0.0103 & 0.0041 - 0.0893 \end{bmatrix} \begin{bmatrix} \phi_{11} \\ \phi_{21} \\ \phi_{31} \end{bmatrix} = \begin{bmatrix} 0 \\ 0 \\ 0 \end{bmatrix}$$

$$(11.4.29)$$

Arbitrarily choosing $\phi_{11} = 1.0$, any two of the equations represented by Eq. (11.4.29) are solved to obtain $\phi_{21} = 0.540$ and $\phi_{31} = 0.162$. The eigenvectors for the second and third modes are obtained in a similar fashion; the resulting modal matrix is

$$[\phi] = \begin{bmatrix} 1.0 & 1.0 & 1.0 \\ 0.540 & -0.707 & -1.592 \\ 0.162 & -0.731 & 2.224 \end{bmatrix} \qquad (11.4.30)$$

The three mode shapes are shown in Fig. 11.6.

Now consider the response of the beam to an initially displaced configuration. Assume, for example, that at time zero the displacements and velocities have the following values:

$$\{q_0\} = \begin{bmatrix} 3.0 \\ 2.0 \\ 1.0 \end{bmatrix} q_0 \qquad \{\dot{q}_0\} = \begin{bmatrix} 0 \\ 0 \\ 0 \end{bmatrix} \qquad (11.4.31)$$

The six arbitrary constants in the general solution (11.4.15) can be found by inverting the modal matrix and using Eqs. (11.4.20). This gives

$$\{A\} = [\phi]^{-1}\{q_0\} = \begin{bmatrix} 0.611 & 0.660 & 0.198 \\ 0.326 & -0.461 & -0.477 \\ 0.063 & -0.200 & 0.279 \end{bmatrix} \begin{bmatrix} 3.0 \\ 2.0 \\ 1.0 \end{bmatrix} q_0 = \begin{bmatrix} 3.353 \\ -0.421 \\ 0.068 \end{bmatrix} q_0$$

$$(11.4.32)$$

$$\{pB\} = [\phi]^{-1}\{\dot{q}_0\} = \{0\}$$

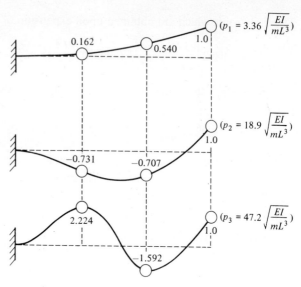

FIGURE 11.6
Normal modes of vibration for the idealized beam.

Finally, substitution of Eqs. (11.4.32) into (11.4.15) yields the displacements

$$\{q\} = [\phi]\{A \cos pt + B \sin pt\}$$

$$= \begin{bmatrix} 1.0 & 1.0 & 1.0 \\ 0.540 & -0.707 & -1.592 \\ 0.162 & -0.731 & 2.224 \end{bmatrix} \begin{bmatrix} 3.353 & \cos p_1 t \\ -0.421 & \cos p_2 t \\ 0.068 & \cos p_3 t \end{bmatrix} q_0 \qquad (11.4.33)$$

or $\quad q_1 = 3.353 q_0 \cos 3.36 \sqrt{\dfrac{EI}{mL^3}} t - 0.421 q_0 \cos 18.9 \sqrt{\dfrac{EI}{mL^3}} t$

$$+ 0.068 q_0 \cos 47.2 \sqrt{\frac{EI}{mL^3}} t \qquad (11.4.34)$$

with similar expressions for the displacements q_2 and q_3. An upper bound to the absolute value of the displacement of any mass can be obtained by disregarding the phase relationships between the modes. For example, if we assume that at some instant of time, $\cos p_1 t = -\cos p_2 t = \cos p_3 t = 1$, then the maximum displacement of mass m_1 is found to be

$$|q_1|_{\max} = (3.353 + 0.421 + 0.068) q_0 = 3.84 q_0 \qquad (11.4.35)$$

////

11.5   ORTHOGONALITY RELATIONS

The eigenvectors or relative amplitudes $\{\phi\}_j$ obtained from the free-vibration problem satisfy certain relationships known as the *orthogonality conditions*. These conditions can be used to advantage in computing the response of structures subject to initial disturbances or applied loads. A derivation of the orthogonality relations follows.

A solution to the equations of motion (11.4.1) for the free vibration of an undamped, discrete-mass system was found to be

$$\{q\} = \{\phi\}_j \sin(p_j t + \delta_j) \qquad (11.5.1)$$

This equation represents motion in the jth normal mode. Substituting Eq. (11.5.1) into (11.4.1) gives

$$p_j^2 [m]\{\phi\}_j = [k]\{\phi\}_j \qquad (11.5.2)$$

Similarly, for vibration in the ith mode,

$$p_i^2 [m]\{\phi\}_i = [k]\{\phi\}_i \qquad (11.5.3)$$

Premultiplying Eq. (11.5.2) by the transpose of $\{\phi\}_i$, and premultiplying Eq. (11.5.3) by the transpose of $\{\phi\}_j$ gives

$$p_j^2 \{\phi\}_i^T [m]\{\phi\}_j = \{\phi\}_i^T [k]\{\phi\}_j \qquad (11.5.4)$$

and

$$p_i^2 \{\phi\}_j^T [m]\{\phi\}_i = \{\phi\}_j^T [k]\{\phi\}_i \qquad (11.5.5)$$

The transpose of Eq. (11.5.5) is

$$p_i^2 \{\phi\}_i^T [m]\{\phi\}_j = \{\phi\}_i^T [k]\{\phi\}_j \qquad (11.5.6)$$

where use has been made of the reversal law of transposition (see Appendix B) and of the fact that the inertia and stiffness matrices are symmetric. Subtracting Eq. (11.5.6) from (11.5.4) yields

$$(p_j^2 - p_i^2)\{\phi\}_i^T [m]\{\phi\}_j = 0 \qquad (11.5.7)$$

Hence for distinct natural frequencies $p_i^2 \neq p_j^2$

$$\{\phi\}_i^T [m]\{\phi\}_j = 0 \qquad (11.5.8)$$

Then from Eq. (11.5.4) we see that

$$\{\phi\}_i^T [k]\{\phi\}_j = 0 \qquad (11.5.9)$$

If we let i and j refer to the same mode, so that $p_i^2 = p_j^2$, then the quantity $\{\phi\}_i^T [m]\{\phi\}_j$ in Eq. (11.5.7) is, in general, a nonzero constant. We denote this constant by $\mathcal{M}_j$; that is,

$$\{\phi\}_j^T [m]\{\phi\}_j = \mathcal{M}_j \qquad (11.5.10)$$

where $\mathcal{M}_j$ is called the *generalized mass* for the jth mode. The values of the n generalized masses are, in general, different. Since each eigenvector represents a set of relative amplitudes, each component of the eigenvector can be multiplied by a constant value without changing the form of the general solution. Hence the eigenvectors can always be adjusted so as to make the generalized mass for each mode have the same value. This process is known as *normalization*.

Substitution of Eq. (11.5.10) into (11.5.4) gives, for the case $i = j$,

$$\{\phi\}_j^T[k]\{\phi\}_j = \mathcal{M}_j p_j^2 \qquad (11.5.11)$$

Summarizing the above results,

$$\{\phi\}_i^T[m]\{\phi\}_j = \begin{cases} 0 & i \neq j \\ \mathcal{M}_j & i = j \end{cases}$$

$$\{\phi\}_i^T[k]\{\phi\}_j = \begin{cases} 0 & i \neq j \\ \mathcal{M}_j p_j^2 & i = j \end{cases} \qquad (11.5.12)$$

These expressions are the orthogonality relations which the eigenvectors satisfy. Since $\{\phi\}_j$ represents the jth column of the modal matrix $[\phi]$, the orthogonality conditions may also be written as

$$[\phi]^T[m][\phi] = [\mathcal{M}]$$
$$[\phi]^T[k][\phi] = [\mathcal{K}] \qquad (11.5.13)$$

where the *generalized inertia matrix* $[\mathcal{M}]$ and the *generalized stiffness matrix* $[\mathcal{K}]$ are diagonal matrices defined, respectively, as

$$[\mathcal{M}] = \begin{bmatrix} \mathcal{M}_1 & 0 & \cdots & 0 \\ 0 & \mathcal{M}_2 & \cdots & 0 \\ \cdots & \cdots & \cdots & \cdots \\ 0 & 0 & \cdots & \mathcal{M}_n \end{bmatrix} \qquad [\mathcal{K}] = \begin{bmatrix} \mathcal{M}_1 p_1^2 & 0 & \cdots & 0 \\ 0 & \mathcal{M}_2 p_2^2 & \cdots & 0 \\ \cdots & \cdots & \cdots & \cdots \\ 0 & 0 & \cdots & \mathcal{M}_n p_n^2 \end{bmatrix}$$

$$(11.5.14)$$

11.6 NORMAL COORDINATES

The matrix equation (11.4.1) governing the vibration of an undamped structure represents, in general, a system of coupled differential equations. That is, the motion of one of the masses cannot be obtained without considering the motion of the other masses. In fact it is only when the structure's inertia matrix $[m]$ and stiffness matrix $[k]$ are diagonal that the equations become uncoupled. Since great simplifications arise in this case, we now seek a set of coordinates which will lead to this uncoupling.

Consider the system of *normal coordinates* $r_1, r_2, \ldots, r_n$, or $\{r\}$, defined by the relation

$$\{q\} = [\phi]\{r\} \qquad (11.6.1)$$

where $\{q\}$ represents the set of generalized coordinates for an n-degree-of-freedom system, and $[\phi]$ is the modal matrix. We will now show that the transformation of coordinates (11.6.1) reduces the expressions for the kinetic and potential energies to sums of squares of the coordinates. In terms of the coordinates $\{q\}$, the kinetic energy may be expressed as

$$T = \tfrac{1}{2}\{\dot{q}\}^T[m]\{\dot{q}\} \qquad (11.6.2)$$

Substitution of Eq. (11.6.1) into (11.6.2) gives

$$T = \tfrac{1}{2}\{\dot{r}\}^T[\phi]^T[m][\phi]\{\dot{r}\} \qquad (11.6.3)$$

By virtue of the orthogonality conditions (11.5.13), Eq. (11.6.3) becomes

$$T = \tfrac{1}{2}\{\dot{r}\}^T[\mathcal{M}]\{\dot{r}\} \qquad (11.6.4)$$

Thus when normal coordinates are used, the inertia matrix becomes the generalized inertia matrix $[\mathcal{M}]$ defined by Eq. (11.5.14). We recall that this matrix is diagonal.

Likewise, the system's strain energy becomes

$$U = \tfrac{1}{2}\{q\}^T[k]\{q\}$$
$$= \tfrac{1}{2}\{r\}^T[\phi]^T[k][\phi]\{r\} \qquad (11.6.5)$$

Again using the orthogonality relations (11.5.13), Eq. (11.6.5) yields

$$U = \tfrac{1}{2}\{r\}^T[\mathcal{K}]\{r\} \qquad (11.6.6)$$

in which the generalized stiffness matrix $[\mathcal{K}]$ is a diagonal matrix.

As we shall see in the following section, the normal coordinates $\{r\}$ are very useful in the study of the forced vibration of a structure.

11.7   FORCED VIBRATION

The response of an n-degree-of-freedom structure to time-varying applied forces $\{Q\}_a$ is now examined. Damping will be neglected for the present. In this case the motion is governed by the matrix equation (11.3.12)

$$[m]\{\ddot{q}\} + [k]\{q\} = \{Q\}_a \qquad (11.7.1)$$

The general solution of Eq. (11.7.1) for the coordinates $\{q\}$ may be written as the sum of the solutions to the associated homogeneous equations plus particular solutions to the nonhomogeneous equations. Since the homogeneous equations are the same as those governing the structure's free vibration, their solution is given by Eq. (11.4.15). Particular solutions to the equations of motion can be found in much the same way as for single-degree-of-freedom systems. However, the computations are rather tedious due to the existence of coupling between the coordinates. The problem of finding particular solutions can be simplified greatly if use is made of the normal coordinates $\{r\}$ introduced in the previous section. Substituting

$$\{q\} = [\phi]\{r\} \qquad (11.7.2)$$

into Eq. (11.7.1) gives

$$[m][\phi]\{\ddot{r}\} + [k][\phi]\{r\} = \{Q\}_a \qquad (11.7.3)$$

Premultiplying each term of this equation by the transpose of the modal matrix yields

$$[\phi]^T[m][\phi]\{\ddot{r}\} + [\phi]^T[k][\phi]\{r\} = [\phi]^T\{Q\}_a \qquad (11.7.4)$$

Using the orthogonality relations, namely,

$$[\phi]^T[m][\phi] = [\mathcal{M}]$$
$$[\phi]^T[k][\phi] = [\mathcal{K}] \qquad (11.7.5)$$

Eq. (11.7.4) becomes

$$[\mathcal{M}]\{\ddot{r}\} + [\mathcal{K}]\{r\} = \{R\}_a \qquad (11.7.6)$$

where the generalized applied forces $\{R\}_a$ associated with the normal coordinates $\{r\}$ are defined by

$$\{R\}_a = [\phi]^T\{Q\}_a \qquad (11.7.7)$$

Recalling that the generalized inertia matrix $[\mathcal{M}]$ and the generalized stiffness matrix $[\mathcal{K}]$ are diagonal, we see that the equations of motion (11.7.6) are uncoupled. These equations could also have been obtained by substituting the kinetic and potential energy expressions (11.6.4) and (11.6.6) into Lagrange's equations.

The jth equation of (11.7.6) states that

$$\mathcal{M}_j\ddot{r}_j + \mathcal{M}_j p_j{}^2 r_j = (R_j)_a \qquad (11.7.8)$$

or

$$\ddot{r}_j + p_j{}^2 r_j = \frac{(R_j)_a}{\mathcal{M}_j} \qquad (11.7.9)$$

where the force $(R_j)_a$ is given by

$$(R_j)_a = \phi_{1j}(Q_1)_a + \phi_{2j}(Q_2)_a + \cdots + \phi_{nj}(Q_n)_a = \{\phi\}_j{}^T\{Q\}_a \qquad (11.7.10)$$

We recognize Eq. (11.7.9) as being the equation of motion for a single-degree-of-freedom system whose mass is $\mathcal{M}_j$ and whose natural circular frequency is p_j. By analogy with the results for the one-degree-of-freedom case, the solution to Eq. (11.7.9) is

$$r_j = A_j \cos p_j t + B_j \sin p_j t + (r_j)_p \quad (11.7.11)$$

where A_j and B_j are arbitrary constants and $(r_j)_p$ is any particular solution. The expression (11.7.11) represents that portion of the structure's response which is associated with the jth normal mode of vibration. To find the total response, we superimpose these modal solutions. That is, by substituting the solution (11.7.11) for the normal coordinate r_j into equation (11.7.2), we obtain the following general solution for the displacements:

$$\{q\} = [\phi]\{A \cos pt + B \sin pt + r_p\} \quad (11.7.12)$$

in which

$$\{A \cos pt + B \sin pt + r_p\} = \begin{bmatrix} A_1 \cos p_1 t + B_1 \sin p_1 t + (r_1)_p \\ A_2 \cos p_2 t + B_2 \sin p_2 t + (r_2)_p \\ \vdots \\ A_n \cos p_n t + B_n \sin p_n t + (r_n)_p \end{bmatrix} \quad (11.7.13)$$

The $2n$ arbitrary constants $A_1, A_2, \ldots, A_n, B_1, B_2, \ldots, B_n$ are determined by applying the appropriate initial conditions. If the initial displacements and velocities are $\{q_0\}$ and $\{\dot{q}_0\}$, respectively, then

$$\{q_0\} = [\phi]\{A + r_p(0)\}$$
$$\{\dot{q}_0\} = [\phi]\{pB + \dot{r}_p(0)\} \quad (11.7.14)$$

Solving for the arbitrary constants gives

$$\{A\} = [\phi]^{-1}\{q_0\} - \{r_p(0)\}$$
$$\{pB\} = [\phi]^{-1}\{\dot{q}_0\} - \{\dot{r}_p(0)\} \quad (11.7.15)$$

Rather than inverting the modal matrix to obtain $[\phi]^{-1}$, it is generally simpler to make use of the relation

$$[\phi]^{-1} = [\mathcal{M}]^{-1}[\phi]^T[m] \quad (11.7.16)$$

which follows from the orthogonality conditions (Prob. 11.12). Note that since $[\mathcal{M}]$ is a diagonal matrix, $[\mathcal{M}]^{-1}$ is also diagonal.

An alternate form of the general solution for r_j is [see Eq. (10.7.7)]

$$r_j = a_j \cos p_j t + b_j \sin p_j t + D_j(t) \quad (11.7.17)$$

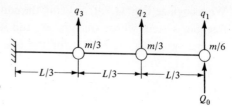

FIGURE 11.7
Three-mass cantilevered beam subject to
a step force Q_0.

where $D_j(t)$ denotes Duhamel's integral

$$D_j(t) = \frac{1}{\mathcal{M}_j p_j} \int_0^t [R_j(\xi)]_a \sin p_j(t - \xi) \, d\xi \quad (11.7.18)$$

The displacements $\{q\}$, found by substituting Eq. (11.7.17) into (11.7.2), are then given by

$$\{q\} = [\phi]\{a \cos pt + b \sin pt + D(t)\} \quad (11.7.19)$$

We recall that in deriving Duhamel's integral we assumed zero initial conditions; that is, $D(0) = \dot{D}(0) = 0$. Hence for the prescribed initial displacements and velocities $\{q_0\}$ and $\{\dot{q}_0\}$, we have

$$\{q_0\} = [\phi]\{a\}$$
$$\{\dot{q}_0\} = [\phi]\{pb\} \quad (11.7.20)$$

Inverting these expressions, we find that the $2n$ arbitrary constants $a_1, a_2, \ldots, a_n$, $b_1, b_2, \ldots, b_n$ are given by

$$\{a\} = [\phi]^{-1}\{q_0\} = [\mathcal{M}]^{-1}[\phi]^T[m]\{q_0\}$$
$$\{pb\} = [\phi]^{-1}\{\dot{q}_0\} = [\mathcal{M}]^{-1}[\phi]^T[m]\{\dot{q}_0\} \quad (11.7.21)$$

Example 11.4   Forced vibration of a three-mass cantilevered beam   We shall now examine the forced vibration of the three-mass cantilevered beam considered in Example 11.3. It will be assumed that the structure is initially at rest and that a force of constant magnitude Q_0 is suddenly applied to the tip of the beam, as indicated in Fig. 11.7.

Following the procedure described above, we introduce normal coordinates $\{r\}$ defined by

$$\{q\} = [\phi]\{r\} \quad (11.7.22)$$

The equations of motion, written in terms of the normal coordinates, are

$$[\mathcal{M}]\{\ddot{r}\} + [\mathcal{K}]\{r\} = \{R\}_a \quad (11.7.23)$$

From the free-vibration solution of the beam (Example 11.3), we have

$$p_1 = 3.36 \sqrt{\frac{EI}{mL^3}} \qquad p_2 = 18.9 \sqrt{\frac{EI}{mL^3}} \qquad p_3 = 47.2 \sqrt{\frac{EI}{mL^3}} \quad (11.7.24)$$

and

$$[\phi] = \begin{bmatrix} 1.0 & 1.0 & 1.0 \\ 0.540 & -0.707 & -1.592 \\ 0.162 & -0.731 & 2.224 \end{bmatrix} \quad (11.7.25)$$

The generalized inertia matrix $[\mathcal{M}]$ becomes

$$[\mathcal{M}] = [\phi]^T[m][\phi] = \begin{bmatrix} 0.273 & 0 & 0 \\ 0 & 0.511 & 0 \\ 0 & 0 & 2.660 \end{bmatrix} m \quad (11.7.26)$$

and the generalized applied force matrix $\{R\}_a$ is

$$\{R\}_a = [\phi]^T\{Q\}_a = [\phi]^T \begin{bmatrix} Q_0 \\ 0 \\ 0 \end{bmatrix} = \begin{bmatrix} \phi_{11} Q_0 \\ \phi_{12} Q_0 \\ \phi_{13} Q_0 \end{bmatrix} \quad (11.7.27)$$

Therefore the jth equation of (11.7.23) may be written as

$$\ddot{r}_j + p_j^2 r_j = \frac{(R_j)_a}{\mathcal{M}_j} = \frac{\phi_{1j} Q_0}{\mathcal{M}_j} \quad (11.7.28)$$

The solution to this equation is

$$r_j = A_j \cos p_j t + B_j \sin p_j t + \frac{\phi_{1j} Q_0}{\mathcal{M}_j p_j^2} \quad (11.7.29)$$

For the case of the zero initial conditions $\{q_0\} = \{\dot{q}_0\} = \{0\}$, the arbitrary constants are found from Eq. (11.7.15) to be

$$A_j = -\frac{\phi_{1j} Q_0}{\mathcal{M}_j p_j^2} \qquad B_j = 0 \quad (11.7.30)$$

Substituting Eqs. (11.7.29) and (11.7.30) into (11.7.22) yields the generalized displacements $\{q\}$. The displacement at the tip of the beam is, for example,

$$q_1 = \phi_{11} \frac{\phi_{11} Q_0}{\mathcal{M}_1 p_1^2} (1 - \cos p_1 t) + \phi_{12} \frac{\phi_{12} Q_0}{\mathcal{M}_2 p_2^2} (1 - \cos p_2 t)$$

$$+ \phi_{13} \frac{\phi_{13} Q_0}{\mathcal{M}_3 p_3^2} (1 - \cos p_3 t) \quad (11.7.31)$$

or, in terms of the given numerical data

$$q_1 = [0.328(1 - \cos p_1 t) + 0.005(1 - \cos p_2 t) + 0.0002(1 - \cos p_3 t)] \frac{Q_0 L^3}{EI}$$

$$\quad (11.7.32)$$

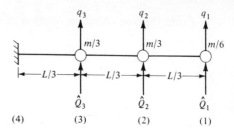

FIGURE 11.8
Three-mass cantilevered beam subject to
applied forces $\hat{Q}_i = (Q_i)_a - m_i\ddot{q}_i$.

Similar expressions are obtained for the displacements q_2 and q_3. From Eq.
(11.7.32) it is clear that although the response represents a superposition of all
three normal modes, the contribution of the fundamental mode is much larger
than that of the second, which in turn is larger than that of the third.

By disregarding the phase relationships between the different modes, an
upper bound to the displacement q_1 is found to be

$$|q_1|_{max} = [0.328(2.0) + 0.005(2.0) + 0.0002(2.0)] \frac{Q_0 L^3}{EI}$$

$$= 0.667 \frac{Q_0 L^3}{EI} \tag{11.7.33}$$

It is noted that the maximum displacement q_1 produced by the suddenly applied
force is twice as large as the displacement produced by a statically applied force
of the same magnitude.

Now suppose that we wish to calculate the bending moment distribution in
the vibrating beam. Our approach will be to treat the "inertia forces" as additional
applied loads and to make use of the general methods of statics (D'Alembert's
principle). Thus each mass m_i is subject to an inertia force $-m_i \ddot{q}_i$ as well as to
an applied load $(Q_i)_a$. The resultant forces applied to the beam, say $\hat{Q}_i$ (Fig. 11.8),
can therefore be expressed as

$$\{\hat{Q}\} = \{Q\}_a - [m]\{\ddot{q}\} \tag{11.7.34}$$

The bending moments at the cross sections (2), (3), and (4), denoted by
M_2, M_3, and M_4 respectively, are

$$M_2 = \frac{L}{3} \hat{Q}_1$$

$$M_3 = \frac{2L}{3} \hat{Q}_1 + \frac{L}{3} \hat{Q}_2 \tag{11.7.35}$$

$$M_4 = L\hat{Q}_1 + \frac{2L}{3} \hat{Q}_2 + \frac{L}{3} \hat{Q}_3$$

These equations may be written in matrix form as

$$\{M\} = [L]\{\hat{Q}\} \quad (11.7.36)$$

where the matrix $[L]$ is

$$[L] = \begin{bmatrix} 0.3333 & 0 & 0 \\ 0.6667 & 0.3333 & 0 \\ 1.0 & 0.6667 & 0.3333 \end{bmatrix} L \quad (11.7.37)$$

Note that a typical element L_{ij} of the matrix $[L]$ represents the bending moment at section $i + 1$ due to a unit generalized force applied at section j; for example, L_{32} represents the bending moment M_4 resulting from the force $\hat{Q}_2 = 1.0$.

The bending moments $\{M\}$ can be obtained by substituting Eq. (11.7.34) into (11.7.36). However, in order to avoid the task of differentiating the displacements $\{q\}$ in order to find $\{\ddot{q}\}$, we shall proceed as follows. The forces $\{\hat{Q}\}$ may be written in terms of the displacements $\{q\}$ by introducing Eq. (11.7.34) into the matrix equation of motion (11.7.1). This yields

$$\{\hat{Q}\} = [k]\{q\} \quad (11.7.38)$$

Therefore

$$\{M\} = [L][k]\{q\}$$
$$= [L][k][\phi]\{r\} \quad (11.7.39)$$

Equation (11.7.39) is applicable to any framed structure. For the beam considered here it is found that

$$M_4 = [1.113(1 - \cos p_1 t) - 0.138(1 - \cos p_2 t)$$
$$+ 0.025(1 - \cos p_3 t)]Q_0 L \quad (11.7.40)$$

with similar expressions for M_2 and M_3. An upper bound to the amplitude of M_4 is

$$|M_4|_{max} = [1.113(2.0) - 0.138(0) + 0.025(2.0)]Q_0 L$$
$$= 2.28 Q_0 L \quad (11.7.41)$$

Thus, the maximum bending moment produced by the suddenly applied force is 2.28 times greater than the moment produced by a statically applied force of the same magnitude.   ////

In conclusion it is noted that although the basic matrix equations developed above are applicable to any discrete-mass structure, the calculations generally become very laborious for systems which have a large number of degrees of freedom. Digital computers offer a convenient, if not indispensable tool in such situations.

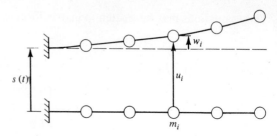

FIGURE 11.9
Discrete-mass beam subject to a foundation motion $s(t)$.

It should also be emphasized that when normal coordinates are used, it is rarely necessary to consider more than the lowest few natural modes of vibration. As demonstrated in the preceding example, the higher modes generally have little effect upon the system's overall response. Herein lies the main advantage of using normal coordinates.

11.8   RESPONSE TO A FOUNDATION MOTION

The vibration resulting from the motion of a structure's foundation will now be investigated. Let us again consider an undamped, n-degree-of-freedom structure, such as the beam shown in Fig. 11.9. Due to the foundation motion $s(t)$, each mass m_i is assumed to undergo an absolute displacement u_i. The displacement of m_i relative to the foundation is denoted by w_i.† Thus $w_i = u_i - s(t)$, or in matrix notation

$$\{w\} = \{u\} - \{s\} \quad (11.8.1)$$

where, for the system shown, each element of the column matrix $\{s\}$ has the same value, namely, $s(t)$. Since the kinetic energy is defined in terms of absolute velocities, we have

$$T = \tfrac{1}{2}\{\dot{u}\}^T[m]\{\dot{u}\} \quad (11.8.2)$$

in which $[m]$ is the structure's inertia matrix.

Substitution of Eq. (11.8.1) into (11.8.2) gives

$$T = \tfrac{1}{2}\{\dot{w}\}^T[m]\{\dot{w}\} + \{\dot{w}\}^T[m]\{\dot{s}\} + \tfrac{1}{2}\{\dot{s}\}^T[m]\{\dot{s}\} \quad (11.8.3)$$

---

†Since it is important to differentiate between the absolute and relative coordinates in this problem, the symbols u_i and w_i (rather than q_i) will be used for the generalized displacements.

For later use in Lagrange's equations we compute the following derivatives of T

$$\left(\frac{\partial T}{\partial \dot{w}}\right) = [m]\{\dot{w}\} + [m]\{\dot{s}\}$$

$$\left(\frac{\partial T}{\partial w}\right) = \{0\} \tag{11.8.4}$$

Assuming that the structure's strain energy U is the only source of potential energy, we have

$$V = U = \tfrac{1}{2}\{w\}^T[k]\{w\} \tag{11.8.5}$$

where $[k]$ is the structure's stiffness matrix. Thus

$$\left(\frac{\partial V}{\partial w}\right) = [k]\{w\} \tag{11.8.6}$$

Substituting Eqs. (11.8.4) and (11.8.6) into Lagrange's equation (11.3.8) yields

$$[m]\{\ddot{w}\} + [k]\{w\} = -[m]\{\ddot{s}\} \tag{11.8.7}$$

Equation (11.8.7) is identical in form to the matrix equation of motion for forced vibrations, except that inertia forces $-[m]\{\ddot{s}\}$ have replaced the applied forces. This result could have been anticipated on the basis of D'Alembert's principle.

The determination of the general solution to (11.8.7) is facilitated by introducing normal coordinates $\{r\}$, defined in terms of the relative displacements $\{w\}$ by

$$\{w\} = [\phi]\{r\} \tag{11.8.8}$$

Here $[\phi]$ is the eigenvector matrix of the free-vibration solution. Substituting Eq. (11.8.8) into (11.8.7) and making use of the orthogonality conditions (11.5.13), it can be shown that

$$[\mathcal{M}]\{\ddot{r}\} + [\mathcal{K}]\{r\} = -[\phi]^T[m]\{\ddot{s}\} \tag{11.8.9}$$

where $[\mathcal{M}]$ and $[\mathcal{K}]$ have the same definitions as before.

It is also possible to formulate the problem in terms of absolute rather than relative displacements. Substitution of the relations (11.8.1) into Eq. (11.8.7) yields

$$[m]\{\ddot{u}\} + [k]\{u\} = [k]\{s\} \tag{11.8.10}$$

The simultaneous equations represented by (11.8.10) can be reduced to a system of uncoupled equations by means of the coordinate transformation

$$\{u\} = [\phi]\{r\} \tag{11.8.11}$$

In this case we obtain the equations

$$[\mathcal{M}]\{\ddot{r}\} + [\mathcal{K}]\{r\} = [\phi]^T[k]\{s\} \tag{11.8.12}$$

It is generally convenient to use Eq. (11.8.9) for problems in which the acceleration of the foundation is known; Eq. (11.8.12) is more appropriate when the foundation's displacement is prescribed. Both formulations are of the same general form as the forced-vibration problem (11.7.6), and the solutions given in Sec. 11.7 are therefore applicable.

11.9 RESPONSE OF A STRUCTURE WITH RIGID-BODY DEGREES OF FREEDOM

So far we have considered the response of structures which are supported in a manner that precludes rigid-body motion. In practice, however, there are numerous situations in which structures undergo both rigid-body and vibratory motions. Airplanes and submarines, to cite two examples, move in space and also suffer deformations as a result of external forces. Other structures, such as land vehicles, cranes, helicopter blades, rotating shafts, etc., are partially constrained, such that their rigid-body displacements are confined to a plane or to a single direction. Depending upon the nature of the constraints, a system may have from zero to six unrestrained or rigid-body degrees of freedom.

If a structure has at least one rigid-body degree of freedom, the stiffness matrix $[k]$ is singular. This can be seen by recalling that the structure's strain energy may be expressed in the quadratic form $U = \frac{1}{2}\{q\}^T[k]\{q\}$. Since the strain energy is necessarily zero when $\{q\}$ represents a rigid-body motion, it follows that $[k]$ is singular,[1] and the flexibility matrix $[c] = [k]^{-1}$ does not exist. It is then impossible to construct the dynamical matrix $[D] = [c][m]$, and it becomes necessary to modify our analysis as follows. The matrix equation of motion (11.4.1) for free vibration has the form

$$[m]\{\ddot{q}\} + [k]\{q\} = \{0\} \qquad (11.9.1)$$

Premultiplying this equation $[m]^{-1}$ gives

$$[I]\{\ddot{q}\} + [m]^{-1}[k]\{q\} = \{0\} \qquad (11.9.2)$$

or

$$\{\ddot{q}\} + [E]\{q\} = \{0\} \qquad (11.9.3)$$

in which the matrix

$$[E] = [m]^{-1}[k] \qquad (11.9.4)$$

---

[1]When a quadratic form is positive semidefinite (i.e., it is nonnegative for all values of the variables, and vanishes for certain nonzero values of the variables), the associated matrix of coefficients ($[k]$ in this case) is singular. (See F. B. Hildebrand, *Methods of Applied Mathematics*, 2d ed., Prentice-Hall, Englewood Cliffs, N.J., 1965.)

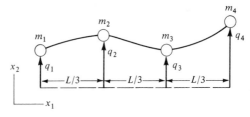

FIGURE 11.10
Unconstrained four-mass beam.

is called the *inverse dynamical matrix*. For a solution we take

$$\{q\} = \{\phi\}\sin(pt + \delta) \qquad (11.9.5)$$

in which case Eq. (11.9.3) gives

$$-p^2\{\phi\} + [E]\{\phi\} = \{0\} \qquad (11.9.6)$$

or

$$([E] - p^2[I])\{\phi\} = \{0\} \qquad (11.9.7)$$

The system's natural circular frequencies p_j and the mode shapes $\{\phi\}_j$ are found by solving the eigenvalue problem (11.9.7). Corresponding to each rigid-body degree of freedom of the structure there will be a zero eigenvalue. From this point on, however, the general methods of analysis for fully constrained structures are applicable.

It might be observed at this point that while one must use the inverse dynamical matrix for finding the natural frequencies and mode shapes of an unconstrained system, either $[E]$ or the dynamical matrix $[D]$ may be used in the case of a fully constrained structure. It is generally more convenient to use the dynamical matrix formulation for problems in which the structure's flexibility matrix is known, since $[D]$ can then be constructed without performing a matrix inversion. On the other hand if the flexibility matrix is not known but the stiffness matrix is given, then the computations may be simpler if the inverse formulation is employed.

Example 11.5   Response of an unconstrained four-mass beam   We wish
to investigate the free and forced flexural vibrations of the discrete-mass structure shown in Fig. 11.10. The rigid masses m_i have the values $m_1 = m_4 = m/6$, $m_2 = m_3 = m/3$; the masses are joined by beam elements of flexural rigidity EI. This system approximates a free-free continuous beam of mass m, length L, and uniform stiffness EI.

Assuming that each mass moves in the transverse x_2 direction only, the lumped-mass structure has four degrees of freedom, $q_1, q_2, q_3,$ and q_4. However, because the system is capable of two independent rigid-body motions (a translation in the x_2 direction, and a rotation about an axis perpendicular to the $x_1 x_2$ plane), there will be just two modes of deformation.

In order to determine the inertia matrix $[m]$ for the structure, we express the kinetic energy in terms of the coordinates q_j; this gives

$$T = \tfrac{1}{2} m_1 \dot{q}_1^2 + \tfrac{1}{2} m_2 \dot{q}_2^2 + \tfrac{1}{2} m_3 \dot{q}_3^2 + \tfrac{1}{2} m_4 \dot{q}_4^2 \qquad (11.9.8)$$

By writing Eq. (11.9.8) in the matrix form $T = \tfrac{1}{2}\{\dot{q}\}^T [m]\{\dot{q}\}$, it becomes evident that

$$[m] = \begin{bmatrix} 0.1667 & 0 & 0 & 0 \\ 0 & 0.3333 & 0 & 0 \\ 0 & 0 & 0.3333 & 0 \\ 0 & 0 & 0 & 0.1667 \end{bmatrix} m \qquad (11.9.9)$$

The structure's stiffness matrix $[k]$ may be computed using the matrix displacement method described in Sec. 7.3. It is found that

$$[k] = \begin{bmatrix} 43.2 & -97.2 & 64.8 & -10.8 \\ -97.2 & 259.2 & -226.8 & 64.8 \\ 64.8 & -226.8 & 259.2 & -97.2 \\ -10.8 & 64.8 & -97.2 & 43.2 \end{bmatrix} \frac{EI}{L^3} \qquad (11.9.10)$$

Premultiplying $[k]$ by the inverse of $[m]$ gives the inverse dynamical matrix

$$[E] = [m]^{-1}[k] = \begin{bmatrix} 259.2 & -583.2 & 388.8 & -64.8 \\ -291.6 & 777.6 & -680.4 & 194.4 \\ 194.4 & -680.4 & 777.6 & -291.6 \\ -64.8 & 388.8 & -583.2 & 259.2 \end{bmatrix} \frac{EI}{mL^3} \qquad (11.9.11)$$

The natural circular frequencies, which are given by the square roots of the eigenvalues of $[E]$ are found to be[1]

$$p_1 = p_2 = 0 \qquad p_3 = 17.1 \sqrt{\frac{EI}{mL^3}} \qquad p_4 = 42.2 \sqrt{\frac{EI}{mL^3}} \qquad (11.9.12)$$

and the corresponding modal matrix is

$$[\phi] = \begin{bmatrix} 1.0 & 1.0 & 1.0 & 1.0 \\ 1.0 & 0.333 & -0.500 & -1.500 \\ 1.0 & -0.333 & -0.500 & 1.500 \\ 1.0 & -1.0 & 1.0 & -1.0 \end{bmatrix} \qquad (11.9.13)$$

[1]The computed values of the frequencies p_3 and p_4 are 24 percent and 32 percent smaller, respectively, than the corresponding natural frequencies of a free-free distributed-mass beam. Obtaining a more accurate approximation would require increasing the number of masses in the discrete model.

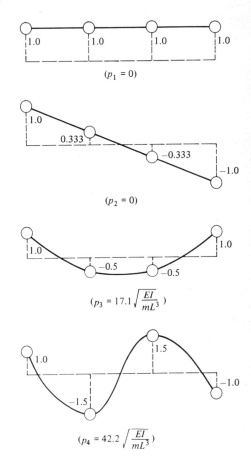

FIGURE 11.11
Normal modes of vibration for the
unconstrained beam.

The four mode shapes are shown in Fig. 11.11. Note that the first and second modes represent rigid-body motions, while the third and fourth modes involve deformation.

Now suppose that the structure is initially at rest and is subject to two step forces of amplitude $Q_0/2$, as shown in Fig. 11.12. The generalized inertia matrix $[\mathcal{M}]$, computed according to Eq. (11.7.5), is

$$[\mathcal{M}] = [\phi]^T[m][\phi] = \begin{bmatrix} 1.0 & 0 & 0 & 0 \\ 0 & 0.407 & 0 & 0 \\ 0 & 0 & 0.500 & 0 \\ 0 & 0 & 0 & 1.833 \end{bmatrix} m \qquad (11.9.14)$$

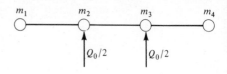

FIGURE 11.12
Unconstrained four-mass beam subject
to step forces $Q_0/2$.

and the generalized force matrix $\{R\}_a$ is, from Eq. (11.7.7),

$$\{R\}_a = [\phi]^T\{Q\}_a = \begin{bmatrix} Q_0 \\ 0 \\ -\dfrac{Q_0}{2} \\ 0 \end{bmatrix} \qquad (11.9.15)$$

Therefore the equations of motion (11.7.9) for the normal coordinates become

$$\ddot{r}_1 = \frac{Q_0}{m}$$

$$\ddot{r}_2 = 0$$

$$\ddot{r}_3 + p_3{}^2 r_3 = -\frac{Q_0}{m} \qquad (11.9.16)$$

$$\ddot{r}_4 + p_4{}^2 r_4 = 0$$

Solving Eqs. (11.9.16), for the case of zero initial conditions (where it is noted that since $\{q(0)\} = \{\dot{q}(0)\} = \{0\}$, therefore $\{r(0)\} = \{\dot{r}(0)\} = \{0\}$), we obtain

$$r_1 = \frac{Q_0}{m}\frac{t^2}{2}$$

$$r_2 = 0$$

$$r_3 = \frac{-Q_0}{mp_3{}^2}(1 - \cos p_3 t) \qquad (11.9.17)$$

$$r_4 = 0$$

The result that $r_2 = r_4 = 0$ follows from the fact that the second and fourth modes are antisymmetric with respect to the beam's center line and are therefore not excited by the symmetrical loading shown in Fig. 11.12. In fact, we could have neglected these modes at the outset. Substituting the relations (11.9.17) into Eq. (11.7.2) gives the generalized displacements

$$q_1 = q_4 = \frac{Q_0}{m} \frac{t^2}{2} - \frac{Q_0}{mp_3{}^2} (1 - \cos p_3 t)$$

$$q_2 = q_3 = \frac{Q_0}{m} \frac{t^2}{2} + \frac{Q_0}{2mp_3{}^2} (1 - \cos p_3 t)$$

(11.9.18)

in which $p_3 = 17.1\sqrt{EI/mL^3}$.   ////

11.10   DAMPED VIBRATION

At this point we return to the problem of the vibration of a discrete-mass structure in which the effects of damping are not negligible. For the case of viscous damping the matrix equation of motion (11.3.12) is

$$[m]\{\ddot{q}\} + [d]\{\dot{q}\} + [k]\{q\} = \{Q\}_a \quad (11.10.1)$$

The general solution to Eq. (11.10.1) can be obtained using standard mathematical techniques (e.g., the Laplace transform). However, for a general type of viscous damping (an arbitrary matrix $[d]$), the calculations are extremely laborious for structures having more than two or three degrees of freedom. For this reason we shall only consider special types of viscous damping, in particular, ones for which the equations of motion can be uncoupled.

We begin by assuming that the free, undamped vibration problem has been solved, so that the natural frequencies p_j ($j = 1, 2, \ldots, n$) and the modal matrix $[\phi]$ are known. As in the undamped problem we define a set of normal coordinates $\{r\}$ by the relation

$$\{q\} = [\phi]\{r\} \quad (11.10.2)$$

Substituting Eq. (11.10.2) into (11.10.1), and making use of the orthogonality relations (11.5.13), we obtain

$$[\mathcal{M}]\{\ddot{r}\} + [\mathcal{D}]\{\dot{r}\} + [\mathcal{K}]\{r\} = \{R\}_a \quad (11.10.3)$$

where $[\mathcal{M}]$, $[\mathcal{K}]$, $\{R\}_a$ have the same definitions as in the undamped case, and

$$[\mathcal{D}] = [\phi]^T [d][\phi] \quad (11.10.4)$$

We recall that $[\mathcal{M}]$ and $[\mathcal{K}]$ are diagonal matrices. However, the matrix $[\mathcal{D}]$ is in general nondiagonal, and consequently the equations of motion (11.10.3) are still coupled. The jth equation, for example, is

$$\mathcal{M}_j \ddot{r}_j + \sum_{k=1}^{n} \mathcal{D}_{jk} \dot{r}_k + \mathcal{M}_j p_j{}^2 r_j = (R_j)_a \quad (11.10.5)$$

There are two special cases for which the "velocity coupling" in Eq. (11.10.5) disappears. First, suppose that the damping matrix $[d]$ is proportional to the inertia matrix $[m]$; that is,

$$[d] = 2\zeta[m] \quad (11.10.6)$$

where ζ is a constant. Then, by virtue of Eq. (11.10.4) and the orthogonality relations (11.5.13), we obtain

$$[\mathscr{D}] = 2\zeta[\phi]^T[m][\phi] = 2\zeta[\mathscr{M}] \quad (11.10.7)$$

in which case Eq. (11.10.5) yields

$$\ddot{r}_j + 2\zeta\dot{r}_j + p_j{}^2 r_j = \frac{(R_j)_a}{\mathscr{M}_j} \quad (11.10.8)$$

We note that Eq. (11.10.8) represents the equation of motion of a damped one-degree-of-freedom system whose mass is $\mathscr{M}_j$ and whose undamped natural frequency is p_j. The solution to Eq. (11.10.8) describes that portion of the structure's response which is associated with the jth mode of vibration. By analogy with the result for a one-degree-of-freedom system, the general solution to Eq. (11.10.8) is

$$r_j = e^{-\zeta t}(A_j \cos \bar{p}_j t + B_j \sin \bar{p}_j t) + (r_j)_p \quad (11.10.9)$$

where A_j and B_j are arbitrary constants of integration, $(r_j)_p$ is any particular solution, and $\bar{p}_j = \sqrt{p_j{}^2 - \zeta^2}$. The generalized displacements $\{q\}$ are found by substituting Eq. (11.10.9) into (11.10.2). It is clear from Eq. (11.10.9) that for this type of damping, the amplitude of the vibration decays at the same rate $(-\zeta)$ for each normal mode.

As a second special case, suppose that the damping matrix $[d]$ is proportional to the structure's stiffness matrix $[k]$; that is,

$$[d] = 2\gamma[k] \quad (11.10.10)$$

where γ is a constant. Equation (11.10.4) then gives

$$[\mathscr{D}] = 2\gamma[\phi]^T[k][\phi] = 2\gamma[\mathscr{K}] \quad (11.10.11)$$

The equation of motion (11.10.5) becomes

$$\ddot{r}_j + 2\gamma p_j{}^2 \dot{r}_j + p_j{}^2 r_j = \frac{(R_j)_a}{\mathscr{M}_j} \quad (11.10.12)$$

the solution to which is

$$r_j = e^{-\gamma p_j{}^2 t}(A_j \cos \bar{p}_j t + B_j \sin \bar{p}_j t) + (r_j)_p \quad (11.10.13)$$

where $\bar{p}_j = p_j\sqrt{1 - \gamma^2 p_j^2}$ is the damped natural frequency. In this case the rate of decay of the transient solution is proportional to the square of the natural frequency, and consequently the higher modes suffer greater damping than the lower ones.

It can also be shown (Prob. 11.19) that the equations of motion (11.10.3) decouple if the damping matrix $[d]$ is equal to a linear combination of the inertia and stiffness matrices; that is,

$$[d] = 2\zeta[m] + 2\gamma[k] \quad (11.10.14)$$

It should be mentioned that the specification of damping in an actual structure is generally subject to much uncertainty. It is often based upon insufficient or unreliable experimental data. For this reason there is generally no point in considering a more complicated form of damping than that given by Eq. (11.10.14).

PROBLEMS

11.1–11.2 A uniform rigid bar of mass m and length L is supported and loaded as shown. Use Lagrange's equation to derive the equation (or equations) of motion describing small oscillations of the bar about its equilibrium position.

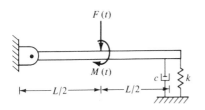

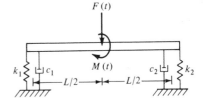

PROBLEM 11.1   PROBLEM 11.2

11.3 Consider a set of n independent generalized coordinates $\{q\}$ which are related to the n absolute displacements $\{u\}$ of a discrete-mass structure by the linear transformation $\{u\} = [A]\{q\}$. Show that the equations of motion may be written in the matrix form [see Eq. (11.3.12)]

$$[m]\{\ddot{q}\} + [d]\{\dot{q}\} + [k]\{q\} = \{Q\}_a$$

where $[m]$, $[d]$, and $[k]$ are symmetric matrices, providing these same matrices are symmetric for the coordinates $\{u\}$.

11.4 A uniform, elastic, simply supported beam supports two masses m_1 and $m_2 = 2m_1$ which are subject to external forces $F_1(t)$ and $F_2(t)$, as shown. Assuming that the mass of the beam is small compared with m_1 and m_2, and that damping is negligible, develop a matrix equation of motion for the system.

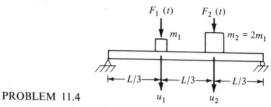

PROBLEM 11.4

11.5 A two-story framed building is represented by the discrete-mass system shown. Develop a matrix equation of motion for the system's undamped, forced vibrations.

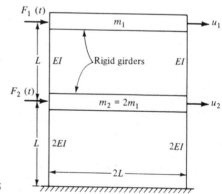

PROBLEM 11.5

11.6 Three rigid disks of axial mass moments of inertia, Im_1, Im_2, and Im_3, respectively, are attached to a massless elastic shaft which is supported by frictionless bearings. Each disk is subject to an applied torque, as shown. Derive a matrix equation of motion for this torsional system.

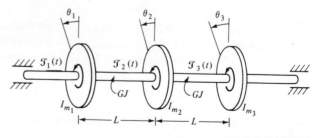

PROBLEM 11.6

11.7   A cantilever beam consists of three uniform sections, each section having the mass per unit length μ_i and bending stiffness EI_i shown. Represent the continuous beam by a three-mass model and develop a matrix equation of motion for the model.

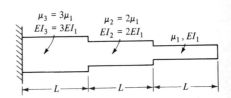

$$\mu_3 = 3\mu_1 \qquad \mu_2 = 2\mu_1$$
$$EI_3 = 3EI_1 \qquad EI_2 = 2EI_1 \qquad \mu_1, EI_1$$

PROBLEM 11.7

11.8–11.9   For the structure of Prob. 11.4–11.5 compute: (*a*) the natural frequencies and mode shapes and (*b*) the free vibration response to the initial displacements $u_1(0) = 2u_2(0) = u_0$. Take $\dot{u}_1(0) = \dot{u}_2(0) = 0$.

11.10–11.11   Show that the eigenvectors for the structure of Prob. 11.4–11.5 are orthogonal with respect to the inertia and stiffness matrices.

11.12   Verify Eq. (11.7.16).

11.13   The lumped-mass beam shown is subject to an applied load $F(t) = F_0 e^{-t/t_0}$, where F_0 and t_0 are constants. Make use of the free-vibration solution obtained in Example 11.3 to determine an upper bound for the displacement at the free end of the beam for the case in which $t_0 = \tau_1/2\pi$ (where τ_1 is the beam's fundamental period). Assume that the structure is at rest initially.

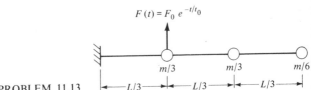

$$F(t) = F_0 e^{-t/t_0}$$

PROBLEM 11.13

11.14–11.15   The structure of Prob. 11.4–11.5 is subject to the applied forces $F_1(t) = 0.5F_2(t) = F(t)$, where $F(t)$ is a step force of amplitude F_0. Compute the maximum displacements of the masses and the maximum bending moment in the structure, assuming that the system is at rest initially. Compare your results with the corresponding static values.

11.16   The beam in Prob. 11.4 is subject to a vertical foundation motion given by $\ddot{s} = a_0$, where a_0 is a constant. Determine the displacement of each mass relative to the foundation. Also compute the maximum bending moment developed in the beam. Assume the system is at rest initially. (See figure on following page.)

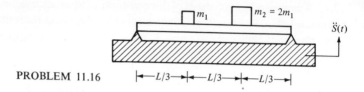

PROBLEM 11.16 $\longleftarrow L/3 \longrightarrow\!\!\longleftarrow L/3 \longrightarrow\!\!\longleftarrow L/3 \longrightarrow$

11.17   Solve Prob. 11.16 for the case of the foundation motion $s = v_0 t$ where v_0 is a constant.

11.18   Investigate the motion of the torsional system of Prob. 11.6, for the case in which a step torque is applied to the first disk; that is $\mathcal{T}_1(t) = \mathcal{T}_0$ (constant), $\mathcal{T}_2(t) = \mathcal{T}_3(t) = 0$. Assume that the system is at rest initially, and that $I_{m_1} = 2I_{m_2} = 2I_{m_3}$.

11.19   Show that the equations of motion for a discrete-mass system decouple when the viscous damping matrix $[d]$ is given by Eq. (11.10.14).

11.20   Investigate the response of the system of Prob. 11.2 to a step force $F(t) = F_0$. Assume that the system is at rest initially, and that $M(t) = 0$. Let $k_1 = 2k_2$, $c_1 = 2c_2$.

Problems 11.21 to 11.25 require a determination of the eigenvalues and eigenvectors of a matrix of order at least 4. The calculations are sufficiently involved that a digital computer is essential. Most computer centers will be able to furnish a subroutine program for this purpose. (The *Scientific Subroutine Program* NROOT is available in most libraries and is a particularly appropriate subroutine for these problems.)

11.21   An airplane wing is idealized as two rigid bodies connected by massless, flexible bars, as shown. Each body has a mass m and a mass moment of inertia $I_m = mL^2/100$ about the mass center. The mass centers are located at a distance $e = L/20$ behind the wing's *elastic axis* (the elastic axis is defined as the locus of points along which a transverse load will produce bending without torsion). The wing is assumed to have a constant bending stiffness EI and a torsional stiffness $GJ = EI/4$ about the elastic axis. Determine the natural frequencies and mode shapes for vertical bending and twisting of the beam. *Hint:* Use the generalized coordinates

$$\{q\} = \begin{bmatrix} u_1 \\ u_2 \\ e\theta_1 \\ e\theta_2 \end{bmatrix}$$

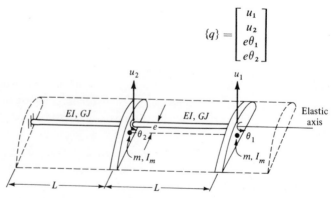

PROBLEM 11.21

11.22   A cantilever beam of negligible mass carries two rigid bodies as shown. Each body has a mass m and a mass moment of inertia (about the body's centroidal axis perpendicular to the plane of motion) $I_m = mL^2$.

(a) Compute the natural frequencies and mode shapes; sketch the mode shapes.

(b) Determine the displacement response to the following initial conditions:

$$u_1(0) = u_2(0) = 0 \qquad \theta_1(0) = -\theta_2(0) = \theta_0$$
$$\dot{u}_1(0) = \dot{u}_2(0) = \dot{\theta}_1(0) = \dot{\theta}_2(0) = 0$$

(c) Determine the displacement response to the suddenly applied couple M_0 (step function) shown, assuming the structure is at rest initially.

Hint: Use the generalized coordinates

$$\{q\} = \begin{bmatrix} u_1 \\ u_2 \\ L\theta_1 \\ L\theta_2 \end{bmatrix}$$

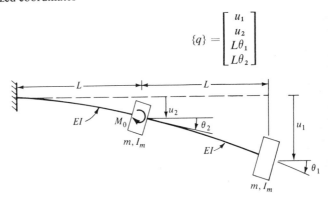

PROBLEM 11.22

11.23   A railroad bridge is idealized as the lumped-mass system shown.

(a) Compute the natural frequencies and mode shapes for the system.

(b) Determine the displacements of the masses produced by the forces:

$$F_1(t) = F_3(t) = F_5(t) = F_7(t) = F_0 \qquad \text{(step functions)}$$
$$F_2(t) = F_4(t) = F_6(t) = F_8(t) = 0$$

Assume that the structure is at rest initially.

Note: The stiffness and flexibility matrices for the truss were computed in Prob. 7.7.

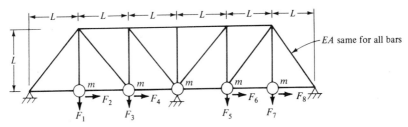

PROBLEM 11.23

11.24 A uniform, elastic beam resting on flexible supports is idealized as the lumped-mass system shown.

(*a*) Compute the natural frequencies and mode shapes for the system.

(*b*) Determine the relative displacements $\{w\}$ and the absolute accelerations $\{\ddot{u}\}$ produced by the foundation acceleration $\ddot{s}(t) = a_0$, where a_0 is a constant. Assume that the system is at rest initially.

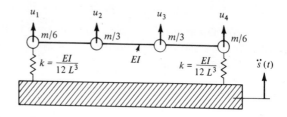

PROBLEM 11.24

11.25 An airplane is idealized as the discrete-mass system shown.

(*a*) Compute the natural frequencies and mode shapes for the vertical motion of the masses.

(*b*) Determine the displacements due to the load $F(t)$ shown.

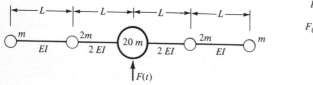
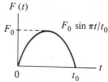

PROBLEM 11.25

REFERENCES

For additional reading on the dynamics of discrete-mass structures:

11.1 SCANLAN, R. H., and R. ROSENBAUM: "Aircraft Vibration and Flutter," Macmillan, New York, 1951.

11.2 NORRIS, C. H., et al.: "Structural Design for Dynamic Loads," McGraw-Hill, New York, 1959.

11.3 ROGERS, G. L.: "Dynamics of Framed Structures," Wiley, New York, 1959.

11.4 HURTY, W. C., and M. F. RUBINSTEIN: "Dynamics of Structures," Prentice-Hall, Englewood Cliffs, N.J., 1964.

11.5 ANDERSON, R. A.: "Fundamentals of Vibrations," Macmillan, New York, 1967.

12

CONTINUOUS STRUCTURES

12.1 INTRODUCTION

The dynamic behavior of structures having continuously distributed mass and stiffness will be studied in this chapter. In particular, the longitudinal and lateral vibrations of linearly elastic bars will be considered. The general methods of analysis to be used in these problems are applicable to other types of continuous structures, including rings, plates, and shells. Such structures can be approximated by equivalent lumped-mass systems, following the procedure outlined in Chap. 11. However, it is sometimes simpler to solve the equations of motion for the actual structure than to reduce the system to an idealized model. Moreover, the results which are obtained will be exact within the framework of the theory (the classical beam theory in the present case).

Since a continuous structure has an infinite number of degrees of freedom, it will also possess an infinite number of normal modes of vibration. Although an arbitrary initial disturbance or an applied load may excite all the modes, we shall see that it is generally the few lowest modes which are of greatest importance; the amplitudes associated with the higher modes of vibration are usually extremely small. By introducing normal coordinates, it will be possible to compute the effect of each mode separately. The normal mode solutions may then be super-imposed to obtain the structure's total response.

Before proceeding to the vibration problem, we shall first develop a general technique for finding the equations of motion and the appropriate boundary conditions for a continuous structure.

12.2   EQUATIONS OF MOTION—HAMILTON'S PRINCIPLE

As was seen in the previous chapter, Lagrange's equations provide a convenient way of establishing the equations of motion for multi-degree-of-freedom systems. An analogous approach for continuous structures involves a well-known energy theorem called Hamilton's principle.[1] This principle represents a generalization of the principle of minimum potential energy (Sec. 5.7) to include dynamic effects. The development of Hamilton's principle is given below.

Consider a structure subject to prescribed surface forces T_i and body forces f_i. The differential equations of motion can be obtained from the equations of equilibrium (2.4.17) by invoking D'Alembert's principle, that is, by adding the forces of inertia to the components of the body forces. Letting ρ denote the mass density of the material, and taking u_i as the components of the displacement vector, the inertia forces acting on an infinitesimal element of volume of the structure are $-\rho\, \partial^2 u_i/\partial t^2$. Hence the equations of motion for the element are

$$\sigma_{ij,j} + f_i = \rho \frac{\partial^2 u_i}{\partial t^2} \qquad (12.2.1)$$

Now consider the virtual work δW_E done by the applied forces T_i and f_i during a virtual distortion δu_i. As before, we require that δu_i be kinematically admissible or consistent with the prescribed constraint conditions. Now, however, the deformation varies with time t as well as position x_j. In this case we shall further restrict the virtual distortion by demanding that the variations δu_i be zero at all points in the body at two arbitrary instants of time, t_1 and t_2. Thus, we require that

$$\delta u_i(x_j,t_1) = \delta u_i(x_j,t_2) = 0 \qquad (12.2.2)$$

In other words, we assume that the displacement field is known at the instants t_1 and t_2; the problem is to find the motion of the body during the interval $t_1 < t < t_2$.

---

[1]This principle, derived by Sir W. R. Hamilton (1805–1865), is oftentimes regarded as the fundamental law of dynamics. It is broader than the lagrangian formulation, since it is applicable to continuous as well as discrete systems.

As before [see Eq. (5.4.1)], the external virtual work is

$$\delta W_E = \int_{\mathscr{S}} T_i \, \delta u_i \, d\mathscr{S} + \int_{\mathscr{V}} f_i \, \delta u_i \, d\mathscr{V} \qquad (12.2.3)$$

It was shown previously that the surface integral in Eq. (12.2.3) can be written as

$$\int_{\mathscr{S}} T_i \, \delta u_i \, d\mathscr{S} = \int_{\mathscr{V}} (\sigma_{ij} \, \delta u_i)_{,j} \, d\mathscr{V} \qquad (12.2.4)$$

in which case the expression for virtual work becomes

$$\delta W_E = \int_{\mathscr{V}} [\sigma_{ij} \, \delta u_{i,j} + (\sigma_{ij,j} + f_i) \, \delta u_i] \, d\mathscr{V} \qquad (12.2.5)$$

Using Eq. (12.2.1) and the definition of the vertical strain tensor, namely, $\delta e_{ij} = \frac{1}{2}(\delta u_{i,j} + \delta u_{j,i})$, it is shown easily that

$$\delta W_E = \int_{\mathscr{V}} \sigma_{ij} \, \delta e_{ij} \, d\mathscr{V} + \int_{\mathscr{V}} \rho \, \frac{\partial^2 u_i}{\partial t^2} \, \delta u_i \, d\mathscr{V} \qquad (12.2.6)$$

The first volume integral in Eq. (12.2.6) is, by definition, the internal virtual work δU [see Eq. (5.4.8)]. Hence

$$\delta W_E = \delta U + \int_{\mathscr{V}} \rho \, \frac{\partial^2 u_i}{\partial t^2} \, \delta u_i \, d\mathscr{V} \qquad (12.2.7)$$

To obtain Hamilton's principle, we integrate Eq. (12.2.7) with respect to time from t_1 to t_2. This gives

$$\int_{t_1}^{t_2} (\delta W_E - \delta U) \, dt = \int_{t_1}^{t_2} \int_{\mathscr{V}} \rho \, \frac{\partial^2 u_i}{\partial t^2} \, \delta u_i \, d\mathscr{V} \, dt \qquad (12.2.8)$$

By inverting the order of the integrations on the right-hand side of Eq. (12.2.8) and then integrating by parts, one obtains

$$\int_{t_1}^{t_2} (\delta W_E - \delta U) \, dt = \int_{\mathscr{V}} \rho \, \frac{\partial u_i}{\partial t} \, \delta u_i \, d\mathscr{V} \Big|_{t_1}^{t_2} - \int_{\mathscr{V}} \int_{t_1}^{t_2} \rho \, \frac{\partial u_i}{\partial t} \, \frac{\partial \delta u_i}{\partial t} \, dt \, d\mathscr{V} \qquad (12.2.9)$$

The first term on the right side of Eq. (12.2.9) is zero by virtue of the requirements (12.2.2). Noting that $\partial(\delta u_i)/\partial t = \delta(\partial u_i/\partial t)$, Eq. (12.2.9) can be written as

$$\int_{t_1}^{t_2} (\delta W_E - \delta U) \, dt = -\int_{t_1}^{t_2} \delta \int_{\mathscr{V}} \frac{\rho}{2} \frac{\partial u_i}{\partial t} \frac{\partial u_i}{\partial t} \, d\mathscr{V} \, dt \qquad (12.2.10)$$

or

$$\int_{t_1}^{t_2} (\delta W_E - \delta U + \delta T) \, dt = 0 \qquad (12.2.11)$$

where

$$T = \int_{\mathscr{V}} \frac{\rho}{2} \frac{\partial u_i}{\partial t} \frac{\partial u_i}{\partial t} \, d\mathscr{V} \qquad (12.2.12)$$

Note that T is the product of an element of mass $\rho\, d\mathcal{V}$ times one-half the square of the velocity, integrated over the volume; hence T represents the kinetic energy of the continuous structure. Equation (12.2.11) is a general statement of Hamilton's principle.

For an elastic structure which possesses a strain energy U and is subject to external forces which are derivable from a scalar potential,[1] Eq. (12.2.11) may be written as

$$\delta \int_{t_1}^{t_2} (U - T + V_E)\, dt = 0 \tag{12.2.13}$$

where
$$\delta V_E = -\delta W_E = -\int_{\mathscr{S}} T_i\, \delta u_i\, d\mathscr{S} - \int_{\mathscr{V}} f_i\, \delta u_i\, d\mathscr{V} \tag{12.2.14}$$

In particular, if the external forces T_i and f_i are independent of the elastic displacements u_i, which is generally the case, then it is evident from Eq. (12.2.14) that

$$V_E = -\int_{\mathscr{S}} T_i u_i\, d\mathscr{S} - \int_{\mathscr{V}} f_i u_i\, d\mathscr{V} \tag{12.2.15}$$

In the event that the applied loads are discrete rather than distributed, the expression for V_E involves a summation rather than an integration. For a system of discrete generalized forces $Q_i(t)$ $(i = 1, 2, \ldots, n)$ which are independent of the corresponding displacements q_i, for example, the potential function is

$$V_E = -\sum_{i=1}^{n} Q_i q_i \tag{12.2.16}$$

Hamilton's principle, in the form of Eq. (12.2.13), may be stated as follows:

Hamilton's principle   The motion of an elastic structure during the time interval $t_1 < t < t_2$ is such that the time integral of the *total dynamic potential* $U - T + V_E$ is an extremum.

Note that when dynamic effects are neglected, $T = 0$ and Hamilton's principle reduces to the principle of minimum potential energy (5.7.11).

Example 12.1   Derivation of the partial differential equation and the boundary conditions governing the longitudinal vibration of a bar

An elastic bar of variable cross-sectional area $A(x_1)$, mass per unit length $\mu(x_1) = \rho\, d\mathcal{V}/dx_1$, and length L is subject to a distributed axial load (force per unit

---

[1]The restrictions mentioned here are the same as those introduced in deriving the principle of minimum potential energy (Sec. 5.7). Specifically, there must exist potential functions $G(u_i)$ and $g(u_i)$ such that

$$T_i = -\frac{\partial G}{\partial u_i} \quad \text{and} \quad f_i = -\frac{\partial g}{\partial u_i}$$

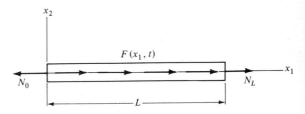

FIGURE 12.1
Bar subject to a distributed axial load.

length) $F(x_1,t)$, as shown in Fig. 12.1. It is assumed that the motion occurs only in a direction parallel to the axis of the bar, and that the cross sections of the bar remain planar. For generality, we shall not specify a particular set of boundary conditions; the displacements $u_1(0,t)$ and $u_1(L,t)$ and the end loads N_0 and N_L are presumed to be arbitrary.

Under the above assumptions, the strain energy is

$$U = \int_{\mathscr{V}} \frac{1}{2E} \sigma_{11}{}^2 \, d\mathscr{V} = \int_0^L \frac{1}{2E} \left(E \frac{\partial u_1}{\partial x_1} \right)^2 A \, dx_1 = \int_0^L \frac{EA}{2} (u_1')^2 \, dx_1 \quad (12.2.17)$$

in which the prime denotes differentiation with respect to the axial coordinate x_1. The kinetic energy of the vibrating bar is

$$T = \int_{\mathscr{V}} \frac{\rho}{2} \left(\frac{\partial u_1}{\partial t} \right)^2 d\mathscr{V} = \int_0^L \frac{\mu}{2} \dot{u}_1{}^2 \, dx_1 \quad (12.2.18)$$

where, as before, a superposed dot represents a time derivative. The potential V_E of the external loading is

$$V_E = -\int_0^L F(x_1,t) u_1 \, dx_1 + N_0 u_1(0,t) - N_L u_1(L,t) \quad (12.2.19)$$

Substitution of the above expressions for U, T, and V_E into Hamilton's principle (12.2.13) yields

$$\delta \int_{t_1}^{t_2} \left\{ \int_0^L \left[\frac{EA}{2} (u_1')^2 - \frac{\mu}{2} \dot{u}_1{}^2 - F(x_1,t) u_1 \right] dx_1 + N_0 u_1(0,t) - N_L u_1(L,t) \right\} dt = 0$$
$$(12.2.20)$$

Performing the indicated variation gives

$$\int_{t_1}^{t_2} \left\{ \int_0^L [EAu_1' \, \delta(u_1') - \mu \dot{u}_1 \, \delta(\dot{u}_1) - F(x_1,t) \, \delta(u_1)] \, dx_1 + N_0 \, \delta u_1(0,t) \right.$$
$$\left. - N_L \, \delta u_1(L,t) \right\} dt = 0 \quad (12.2.21)$$

Noting that $\delta(u_1') = (\delta u_1)'$ and $\delta(\dot{u}_1) = \partial(\delta u_1)/\partial t$, the first two terms in the integral may be integrated by parts as follows:

$$\int_{t_1}^{t_2} \left\{ \int_0^L EAu_1' \, \delta(u_1') \, dx_1 \right\} dt = \int_{t_1}^{t_2} \left\{ [EAu_1' \, \delta u_1]_0^L - \int_0^L (EAu_1')' \, \delta u_1 \, dx_1 \right\} dt \quad (12.2.22)$$

$$-\int_{t_1}^{t_2} \left\{ \int_0^L \mu \dot{u}_1 \, \delta(\dot{u}_1) \, dx_1 \right\} dt = -\int_0^L \left\{ \int_{t_1}^{t_2} \mu \dot{u}_1 \, \delta(\dot{u}_1) \, dt \right\} dx_1$$

$$= -\int_0^L \left\{ [\mu \dot{u}_1 \, \delta u_1]_{t_1}^{t_2} - \int_{t_1}^{t_2} \mu \ddot{u}_1 \, \delta u_1 \, dt \right\} dx_1 \quad (12.2.23)$$

The first term on the right-hand side of Eq. (12.2.23) vanishes because $\delta u_1(x_1,t_1) = \delta u_1(x_1,t_2) = 0$. Substituting Eqs. (12.2.22) and (12.2.23) into (12.2.21) gives

$$\int_{t_1}^{t_2} \left\{ \int_0^L [-(EAu_1')' + \mu \ddot{u}_1 - F(x_1,t) \, \delta u_1 \, dx_1 \right.$$

$$\left. + [N_0 - EAu_1']_{x_1=0} \, \delta u_1(0,t) + [-N_L + EAu_1']_{x_1=L} \, \delta u_1(L,t) \right\} dt = 0 \quad (12.2.24)$$

Since the variation δu_1 is arbitrary for $0 < x_1 < L$, Eq. (12.2.24) requires that

$$(EAu_1')' - \mu \ddot{u}_1 + F(x_1,t) = 0 \quad (12.2.25)$$

and
$$N_0 = [EAu_1']_{x_1=0} \quad \text{or} \quad \delta u_1(0,t) = 0$$
$$N_L = [EAu_1']_{x_1=L} \quad \text{or} \quad \delta u_1(L,t) = 0 \quad (12.2.26)$$

Equation (12.2.25) is the equation of motion for longitudinal vibrations, and Eqs. (12.2.26) are the corresponding boundary conditions. From the latter it is seen that at each end of the bar it is necessary to prescribe either the normal force N or the displacement u_1 (in which case the variation δu_1 vanishes). An example of the first type of boundary condition is a free end ($N = EAu_1' = 0$), whereas a clamped end ($u_1 = 0$) is an example of the second type. ////

Example 12.2 Derivation of the partial differential equation and the boundary conditions governing the lateral vibration of a beam A symmetrical elastic beam of flexural rigidity $EI(x_1)$, mass per unit length $\mu(x_1)$, and length L is acted upon by the transverse distributed force $F(x_1,t)$ shown in Fig. 12.2. So that we may obtain the system's natural boundary conditions as a consequence of Hamilton's principle, the boundary conditions are left unspecified. Thus, the transverse displacements $u_2(0,t)$ and $u_2(L,t)$, the rotations $u_2'(0,t)$ and $u_2'(L,t)$, the shear forces V_0 and V_L, and the bending moments M_0 and M_L are considered arbitrary.

The strain energy associated with bending of the beam is

$$U = \int_0^L \frac{EI}{2} (u_2'')^2 \, dx_1 \quad (12.2.27)$$

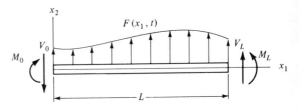

FIGURE 12.2
Beam subject to a distributed lateral load.

where a prime again denotes differentiation with respect to x_1. The kinetic energy of the beam is

$$T = \int_0^L \frac{\mu}{2} \dot{u}_2{}^2 \, dx_1 \quad (12.2.28)$$

and the potential of the external loads is

$$V_E = -\int_0^L F(x_1, t) u_2 \, dx_1 + V_0 u_2(0,t) + M_0 u_2'(0,t) - V_L u_2(L,t) - M_L u_2'(L,t)$$
$$(12.2.29)$$

Hamilton's principle then gives

$$\delta \int_{t_1}^{t_2} \left\{ \int_0^L \left[\frac{EI}{2} (u_2'')^2 - \frac{\mu}{2} \dot{u}_2{}^2 - F(x_1,t)\, u_2 \right] dx_1 \right.$$
$$\left. + V_0 u_2(0,t) + M_0 u_2'(0,t) - V_L u_2(L,t) - M_L u_2'(L,t) \right\} dt = 0 \quad (12.2.30)$$

Performing the variation as indicated leads to

$$\int_{t_1}^{t_2} \left\{ \int_0^L [EI u_2'' \, \delta(u_2'') - \mu \dot{u}_2 \, \delta(\dot{u}_2) - F(x_1,t) \, \delta u_2]\, dx_1 \right.$$
$$\left. + V_0 \, \delta u_2(0,t) + M_0 \, \delta u_2'(0,t) - V_L \, \delta u_2(L,t) - M_L \, \delta u_2'(L,t) \right\} dt = 0 \quad (12.2.31)$$

Noting that $\delta(u_2'') = (\delta u_2)''$ and $\delta(\dot{u}_2) = \partial(\delta u_2)/\partial t$, the first two terms in Eq. (12.2.31) may be integrated by parts to obtain

$$\int_{t_1}^{t_2} \left\{ \int_0^L EI u_2'' \, \delta(u_2'')\, dx_1 \right\} dt = \int_{t_1}^{t_2} \left\{ [EI u_2'' \, \delta u_2']_0^L - [(EI u_2'')' \, \delta u_2]_0^L \right.$$
$$\left. + \int_0^L (EI u_2'')'' \, \delta u_2 \, dx_1 \right\} dt \quad (12.2.32)$$

$$-\int_{t_1}^{t_2} \left\{ \int_0^L \mu \dot{u}_2 \, \delta(\dot{u}_2)\, dx_1 \right\} dt = -\int_0^L \left\{ [\mu \dot{u}_2 \, \delta u_2]_{t_1}^{t_2} - \int_{t_1}^{t_2} \mu \ddot{u}_2 \, \delta u_2 \, dt \right\} dx_1 \quad (12.2.33)$$

By the definition of the virtual distortion, $\delta u_2 = 0$ at the instants of time t_1 and t_2, and hence the first term on the right-hand side of Eq. (12.2.33) vanishes. Substitution of Eqs. (12.2.32) and (12.2.33) into (12.2.31) yields

$$\int_{t_1}^{t_2} \left\{ \int_0^L [(EIu_2'')'' + \mu\ddot{u}_2 - F(x_1,t)] \, \delta u_2 \, dx_1 \right.$$

$$+ \left[V_0 + (EIu_2'')'\right]_{x_1=0} \delta u_2(0,t) + \left[-V_L - (EIu_2'')'\right]_{x_1=L} \delta u_2(L,t)$$

$$\left. + \left[M_0 - EIu_2''\right]_{x_1=0} \delta u_2'(0,t) + \left[-M_L + EIu_2''\right]_{x_1=L} \delta u_2'(L,t) \right\} dt = 0 \quad (12.2.34)$$

Since δu_2 is arbitrary for $0 < x_1 < L$, Eq. (12.2.34) requires that

$$(EIu_2'')'' + \mu\ddot{u}_2 = F(x_1,t) \qquad (12.2.35)$$

and

$$V_0 = -\left[(EIu_2'')'\right]_{x_1=0} \quad \text{or} \quad \delta u_2(0,t) = 0$$

$$V_L = -\left[(EIu_2'')'\right]_{x_1=L} \quad \text{or} \quad \delta u_2(L,t) = 0$$

$$M_0 = \left[EIu_2''\right]_{x_1=0} \quad \text{or} \quad \delta u_2'(0,t) = 0 \qquad (12.2.36)$$

$$M_L = \left[EIu_2''\right]_{x_1=L} \quad \text{or} \quad \delta u_2'(L,t) = 0$$

Equation (12.2.35) is the equation of motion for flexural vibrations, and Eqs. (12.2.36) are the corresponding boundary conditions. It is evident from the latter equations that the shear force or the lateral displacement, and also the bending moment or the slope, must be specified at each end of the beam. ////

12.3 FREE, LONGITUDINAL VIBRATION OF A BAR

The partial differential equation governing the free longitudinal vibration of an elastic bar of uniform cross section is, according to Eq. (12.2.25)

$$EAu_1'' = \mu\ddot{u}_1 \qquad (12.3.1)$$

This equation is known as the one-dimensional wave equation and is frequently written as

$$c^2 u_1'' = \ddot{u}_1 \qquad (12.3.2)$$

in which

$$c^2 = \frac{EA}{\mu} \qquad (12.3.3)$$

The general solution to Eq. (12.3.2) can be expressed in the form[1]

$$u_1 = g(x_1 - ct) + h(x_1 + ct) \qquad (12.3.4)$$

[1] A detailed discussion of the solution (12.3.4) to the wave equation is given in Ref. 12.6.

where g and h are arbitrary functions which must be determined from prescribed initial conditions. The function $g(x_1 - ct)$ represents a wave traveling in the positive x_1 direction with speed c, while $h(x_1 + ct)$ corresponds to a wave of the same speed traveling in the negative x_1 direction. Here we shall consider a less general type of motion, namely, normal mode vibration. In particular, we shall assume that each particle in the bar vibrates harmonically with a circular frequency p, so that

$$u_1(x_1,t) = \phi(x_1) \sin(pt + \delta) \qquad (12.3.5)$$

Substitution of this trial solution into the equation of motion (12.3.2) gives

$$\phi'' + \frac{p^2}{c^2}\phi = 0 \qquad (12.3.6)$$

The problem has thus been reduced from one governed by a partial differential equation to one involving an ordinary differential equation. The general solution to Eq. (12.3.6) is

$$\phi(x_1) = C_1 \cos \frac{px_1}{c} + C_2 \sin \frac{px_1}{c} \qquad (12.3.7)$$

Introduction of the solution (12.3.7) into the boundary conditions for the bar results in an eigenvalue problem, the solution to which yields the natural circular frequencies p_j and the mode shapes or *eigenfunctions* ϕ_j $(j = 1, 2, \ldots, \infty)$. (The details of the solution are illustrated in Example 12.3, for the case of a bar clamped at both ends.) The general free-vibration solution, obtained by superposing the modal solutions, is

$$u_1(x_1,t) = \sum_{j=1}^{\infty} \phi_j(x_1)(A_j \cos p_j t + B_j \sin p_j t) \qquad (12.3.8)$$

where the constants of integration A_j and B_j are chosen to satisfy the prescribed initial conditions. In determining these constants, it is convenient to make use of certain orthogonal properties which the eigenfunctions exhibit.

To obtain the orthogonality relationships for the longitudinal vibrations, we first note that each eigenfunction ϕ_j satisfies Eq. (12.3.6); thus

$$\frac{p_j^2}{c^2}\phi_j = -\phi_j'' \qquad (12.3.9)$$

Multiplying this equation by ϕ_i and then integrating over the length of the bar, we have

$$\frac{p_j^2}{c^2} \int_0^L \phi_i \phi_j \, dx_1 = -\int_0^L \phi_i \phi_j'' \, dx_1 \qquad (12.3.10)$$

Integrating the right-hand side of Eq. (12.3.10) by parts yields

$$\frac{p_j^2}{c^2} \int_0^L \phi_i \phi_j \, dx_1 = -[\phi_i \phi_j']_0^L + \int_0^L \phi_i' \phi_j' \, dx_1 \quad (12.3.11)$$

The bracketed term in Eq. (12.3.11) vanishes for any combination of clamped ($\phi_j = 0$) or free ($EA\phi_j' = 0$) end conditions; for these boundary conditions we obtain

$$\frac{p_j^2}{c^2} \int_0^L \phi_i \phi_j \, dx_1 = \int_0^L \phi_i' \phi_j' \, dx_1 \quad (12.3.12)$$

Likewise it can be shown (by interchanging the indices i and j) that

$$\frac{p_i^2}{c^2} \int_0^L \phi_j \phi_i \, dx_1 = \int_0^L \phi_j' \phi_i' \, dx_1 \quad (12.3.13)$$

Subtracting Eq. (12.3.13) from (12.3.12) yields

$$\frac{p_j^2 - p_i^2}{c^2} \int_0^L \phi_i \phi_j \, dx_1 = 0 \quad (12.3.14)$$

For the case $p_i^2 \neq p_j^2$, it follows from Eqs. (12.3.14) and (12.3.12) that

$$\int_0^L \phi_i \phi_j \, dx_1 = 0 \qquad \int_0^L \phi_i' \phi_j' \, dx_1 = 0 \quad (12.3.15)$$

On the other hand if $p_i^2 = p_j^2$ these integrals will, in general, be nonzero. Defining the generalized mass $\mathcal{M}_j$ for the jth mode of vibration by the relation

$$\mathcal{M}_j = \mu \int_0^L \phi_j^2 \, dx_1 \quad (12.3.16)$$

we arrive at the following conditions:

$$\mu \int_0^L \phi_i \phi_j \, dx_1 = \begin{cases} 0 & i \neq j \\ \mathcal{M}_j & i = j \end{cases}$$

$$EA \int_0^L \phi_i' \phi_j' \, dx_1 = \begin{cases} 0 & i \neq j \\ \mathcal{M}_j p_j^2 & i = j \end{cases} \quad (12.3.17)$$

Equations (12.3.17) represent the *orthogonality relations* for a uniform bar having any combination of clamped and free ends. The resemblance of these relations with the orthogonality conditions (11.5.12) for a discrete-mass system should be noted.

Let us now see how an application of the orthogonality conditions simplifies the solution to an initial value problem. From the general free-vibration solution (12.3.8), it is found that

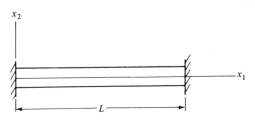

FIGURE 12.3
Clamped bar.

$$u_1(x_1,0) = \sum_{j=1}^{\infty} A_j \phi_j \qquad (12.3.18)$$

$$\dot{u}_1(x_1,0) = \sum_{j=1}^{\infty} p_j B_j \phi_j \qquad (12.3.19)$$

Multiplying Eq. (12.3.18) by ϕ_i and integrating over the length of the bar gives

$$\int_0^L u_1(x_1,0)\phi_i \, dx_1 = \sum_{j=1}^{\infty} A_j \int_0^L \phi_i \phi_j \, dx_1$$

$$= \frac{1}{\mu} A_i \mathcal{M}_i \qquad (12.3.20)$$

where use has been made of the first orthogonality condition (12.3.17). The constants A_j can therefore be expressed in terms of the specified initial displacement $u_1(x_1,0)$ as

$$A_j = \frac{\mu}{\mathcal{M}_j} \int_0^L u_1(x_1,0)\phi_j \, dx_1 \qquad (12.3.21)$$

Similarly it can be shown that the constants B_j are related to the initial velocity $\dot{u}_1(x_1,0)$ by

$$B_j = \frac{\mu}{\mathcal{M}_j p_j} \int_0^L \dot{u}_1(x_1,0)\phi_j \, dx_1 \qquad (12.3.22)$$

Example 12.3   Free, longitudinal vibration of a clamped bar   We wish to compute the natural frequencies and the corresponding mode shapes for the uniform, clamped bar shown in Fig. 12.3. Since the axial displacement at each end of the bar is zero, it follows from Eq. (12.3.5) that

$$\phi(0) = \phi(L) = 0 \qquad (12.3.23)$$

Substitution of the solution

$$\phi(x_1) = C_1 \cos \frac{px_1}{c} + C_2 \sin \frac{px_1}{c} \qquad (12.3.24)$$

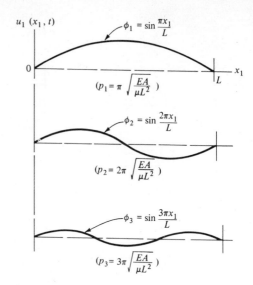

$u_1(x_1, t)$

$\phi_1 = \sin \dfrac{\pi x_1}{L}$

0 x_1

L

$\left(p_1 = \pi \sqrt{\dfrac{EA}{\mu L^2}}\right)$

$\phi_2 = \sin \dfrac{2\pi x_1}{L}$

$\left(p_2 = 2\pi \sqrt{\dfrac{EA}{\mu L^2}}\right)$

$\phi_3 = \sin \dfrac{3\pi x_1}{L}$

FIGURE 12.4
Mode shapes for a clamped bar.

$\left(p_3 = 3\pi \sqrt{\dfrac{EA}{\mu L^2}}\right)$

into the boundary conditions (12.3.23) yields

$$C_1 = 0$$

$$C_1 \cos \frac{pL}{c} + C_2 \sin \frac{pL}{c} = 0 \qquad (12.3.25)$$

For a nontrivial solution the determinant of the coefficients of C_1 and C_2 must vanish. Thus

$$\sin \frac{pL}{c} = 0 \qquad (12.3.26)$$

From this *frequency equation* it is clear that the bar has an infinite number of natural frequencies given by

$$p_j = \frac{j\pi c}{L} = j\pi \sqrt{\frac{EA}{\mu L^2}} \qquad j = 1,2,\ldots,\infty \qquad (12.3.27)$$

Arbitrarily taking C_2 as unity, the eigenfunctions are found to be

$$\phi_j = \sin \frac{p_j x_1}{c} = \sin \frac{j\pi x_1}{L} \qquad j = 1,2,\ldots,\infty \qquad (12.3.28)$$

Fig. 12.4 shows the mode shapes corresponding to the lowest three normal modes ($j = 1,2,3$). The general free-vibration solution (12.3.8) becomes

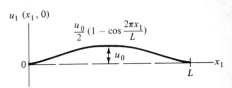

FIGURE 12.5
Prescribed initial displacement.

$$u_1(x_1,t) = \sum_{j=1}^{\infty} \sin \frac{j\pi x_1}{L} \left(A_j \cos j\pi \sqrt{\frac{EA}{\mu L^2}}\, t + B_j \sin j\pi \sqrt{\frac{EA}{\mu L^2}}\, t \right) \quad (12.3.29)$$

Now let us compute the response of the bar to the initial displacement shown in Fig. 12.5. The initial conditions in this case are

$$u_1(x_1,0) = \frac{u_0}{2} \left(1 - \cos \frac{2\pi x_1}{L} \right)$$

$$\dot{u}_1(x_1,0) = 0 \qquad (12.3.30)$$

The generalized mass for the jth mode is

$$\mathcal{M}_j = \mu \int_0^L \phi_j{}^2 \, dx_1 = \mu \int_0^L \sin^2 \frac{j\pi x_1}{L}\, dx_1 = \frac{\mu L}{2} \quad (12.3.31)$$

Substituting Eqs. (12.3.30) and (12.3.31) into (12.3.21) and (12.3.22) gives

$$A_j = \frac{2}{L} \int_0^L \frac{u_0}{2} \left(1 - \cos \frac{2\pi x_1}{L} \right) \sin \frac{j\pi x_1}{L}\, dx_1 = \frac{-4u_0(1 - \cos j\pi)}{\pi j(j-2)(j+2)} \quad j = 1, 2, \ldots, \infty$$

$$B_j = 0 \qquad (12.3.32)$$

Introducing these values into the general solution (12.3.29) and evaluating the first few terms of the series, it is found that

$$u_1(x_1,t) = 0.849 u_0 \sin \frac{\pi x_1}{L} \cos \pi \sqrt{\frac{EA}{\mu L^2}}\, t$$

$$- 0.170 u_0 \sin \frac{3\pi x_1}{L} \cos 3\pi \sqrt{\frac{EA}{\mu L^2}}\, t$$

$$- 0.024 u_0 \sin \frac{5\pi x_1}{L} \cos 5\pi \sqrt{\frac{EA}{\mu L^2}}\, t$$

$$- 0.008 u_0 \sin \frac{7\pi x_1}{L} \cos 7\pi \sqrt{\frac{EA}{\mu L^2}}\, t$$

$$- \cdots \qquad (12.3.33)$$

Note that modes higher than the fifth ($j = 5$) contribute less than 1 percent to the amplitude of the response. ////

12.4 FREE, LATERAL VIBRATION OF A BAR

The partial differential equation governing the flexural vibration of an elastic bar was developed earlier. Exact solutions to this equation can rarely be obtained for bars of variable cross section, and in such problems one must usually resort to an approximate technique such as the Rayleigh-Ritz method. Here we shall consider the free vibration of a uniform bar, in which case the equation of motion (12.2.35) reduces to

$$EIu_2{}^{iv} + \mu \ddot{u}_2 = 0 \qquad (12.4.1)$$

To investigate the normal modes of vibration, we take as a trial solution

$$u_2(x_1,t) = \phi(x_1)\sin(pt + \delta) \qquad (12.4.2)$$

Substitution into Eq. (12.4.1) gives

$$\phi^{iv} - \beta^4 \phi = 0 \qquad (12.4.3)$$

in which

$$\beta^4 = \frac{\mu p^2}{EI} \qquad (12.4.4)$$

The general solution to Eq. (12.4.3) is

$$\phi(x_1) = C_1 \cosh \beta x_1 + C_2 \sinh \beta x_1 + C_3 \cos \beta x_1 + C_4 \sin \beta x_1 \qquad (12.4.5)$$

By introducing this expression into the appropriate boundary conditions and solving the resulting eigenvalue problem, one obtains the beam's natural frequencies p_j and mode shapes ϕ_j ($j = 1, 2, \ldots, \infty$). The general free-vibration solution obtained by superposing the normal mode solutions is then

$$u_2(x_1,t) = \sum_{j=1}^{\infty} \phi_j(x_1)(A_j \cos p_j t + B_j \sin p_j t) \qquad (12.4.6)$$

As in the case of longitudinal vibrations, the constants of integration A_j and B_j can be easily determined if one takes advantage of the orthogonal properties of the eigenfunctions $\phi_j(x_1)$. Following the general procedure used in Sec. 12.3, the orthogonality conditions for a uniform beam with any combination of clamped, free, simply supported, and guided ends (Fig. 8.13) are found to be[1]

---

[1] In Prob. 12.13 the student is asked to derive the orthogonality conditions for a beam of variable cross section.

FIGURE 12.6
Cantilever beam.

$$\mu \int_0^L \phi_i \phi_j \, dx_1 = \begin{cases} 0 & i \neq j \\ \mathcal{M}_j & i = j \end{cases}$$
$$EI \int_0^L \phi_i'' \phi_j'' \, dx_1 = \begin{cases} 0 & i \neq j \\ \mathcal{M}_j p_j^2 & i = j \end{cases} \qquad (12.4.7)$$

where the generalized mass $\mathcal{M}_j$ for the jth mode of vibration is defined as

$$\mathcal{M}_j = \mu \int_0^L \phi_j^2 \, dx_1 \qquad (12.4.8)$$

Using the orthogonality relations (12.4.7), it can be shown that the constants A_j and B_j in Eq. (12.4.6) are related to the initial displacements and velocities by the formulas

$$A_j = \frac{\mu}{\mathcal{M}_j} \int_0^L u_2(x_1,0)\phi_j \, dx_1$$
$$B_j = \frac{\mu}{\mathcal{M}_j p_j} \int_0^L \dot{u}_2(x_1,0)\phi_j \, dx_1 \qquad (12.4.9)$$

Example 12.4   Free vibration of a cantilever beam   Let us examine the free, flexural vibration of the uniform cantilever beam shown in Fig. 12.6. Since the displacement and rotation are zero at the clamped end of the beam, and the bending moment and shear force vanish at the free end, $\phi(x_1)$ must satisfy the conditions

$$\phi(0) = \phi'(0) = \phi''(L) = \phi'''(L) = 0 \quad (12.4.10)$$

Substitution of the expression (12.4.5) into these end conditions gives

$$C_1 + C_3 = 0$$
$$C_2 \beta + C_4 \beta = 0$$
$$C_1 \beta^2 \cosh \beta L + C_2 \beta^2 \sinh \beta L - C_3 \beta^2 \cos \beta L - C_4 \beta^2 \sin \beta L = 0$$
$$C_1 \beta^3 \sinh \beta L + C_2 \beta^3 \cosh \beta L + C_3 \beta^3 \sin \beta L - C_4 \beta^3 \cos \beta L = 0$$
$$(12.4.11)$$

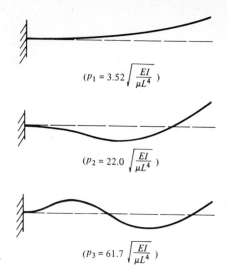

$(p_1 = 3.52\sqrt{\dfrac{EI}{\mu L^4}}\,)$

$(p_2 = 22.0\sqrt{\dfrac{EI}{\mu L^4}}\,)$

FIGURE 12.7
Mode shapes for a cantilever beam.

$(p_3 = 61.7\sqrt{\dfrac{EI}{\mu L^4}}\,)$

For a nontrivial solution, the determinant of the coefficients of the constants C_1, C_2, C_3, and C_4 in Eqs. (12.4.11) must vanish. The evaluation of this determinant leads to the frequency equation

$$\cosh \beta L \cos \beta L + 1 = 0 \quad (12.4.12)$$

This transcendental equation possesses an infinite number of roots β_j. Using a trial-and-error technique, one can show that the first three roots are

$$\beta_1 = \frac{1.88}{L} \qquad \beta_2 = \frac{4.69}{L} \qquad \beta_3 = \frac{7.85}{L} \quad (12.4.13)$$

The corresponding natural circular frequencies are, according to the relation (12.4.4),

$$p_1 = 3.52\sqrt{\frac{EI}{\mu L^4}} \qquad p_2 = 22.0\sqrt{\frac{EI}{\mu L^4}} \qquad p_3 = 61.7\sqrt{\frac{EI}{\mu L^4}} \quad (12.4.14)$$

The shapes of the deflection curves for the normal modes of vibration are given by the eigenfunctions $\phi_j(x_1)$ in Eq. (12.4.5). If one of the undetermined coefficients is chosen arbitrarily, the remaining constants can be computed using Eqs. (12.4.11). Taking C_1 as unity and solving for C_2, C_3, and C_4 yields the jth normal mode shape

$$\phi_j(x_1) = \cosh \beta_j x_1 - \cos \beta_j x_1 - \frac{\cosh \beta_j L + \cos \beta_j L}{\sinh \beta_j L + \sin \beta_j L} (\sinh \beta_j x_1 - \sin \beta_j x_1)$$

$$(12.4.15)$$

The first three mode shapes are plotted in Fig. 12.7.   ////

12.5 RAYLEIGH-RITZ METHOD

For other than the most elementary structures (those having simple geometries, uniform mass and stiffness distributions, and ideal constraints), it is generally impossible to determine the exact dynamic response. Even when the partial differential equations of motion are available, exact solutions to these equations can rarely be found. In such instances the continuous structure can be approximated by a lumped-mass system following the approach outlined in Chap. 11. An alternate method, which also has the effect of reducing the number of degrees of freedom of the system from infinity to a finite value, is the Rayleigh-Ritz method. This method was used to obtain approximate solutions for the displacements and buckling loads of structures in Chaps. 5 and 9. We shall now extend the method to vibration problems.

To illustrate the application of the Rayleigh-Ritz procedure, let us consider the lateral vibration of a beam of variable cross section. The beam's deflection curve $u_2(x_1,t)$ is approximated by the representation

$$u_2(x_1,t) = \sum_{j=1}^{n} \psi_j(x_1)q_j(t) \quad (12.5.1)$$

where $q_j(t)$ $(j = 1, 2, \ldots, n)$ are time-dependent generalized displacements and $\psi_j(x_1)$ are arbitrarily chosen functions which satisfy the structure's kinematic boundary conditions but not necessarily the static conditions. In terms of these assumed functions, the kinetic energy of the beam is

$$T = \int_0^L \frac{\mu}{2} \dot{u}_2{}^2 \, dx_1$$

$$= \int_0^L \frac{\mu}{2} \left(\sum_{i=1}^{n} \psi_i \dot{q}_i \right) \left(\sum_{j=1}^{n} \psi_j \dot{q}_j \right) dx_1$$

$$= \tfrac{1}{2} \sum_{i=1}^{n} \sum_{j=1}^{n} \dot{q}_i \dot{q}_j \int_0^L \mu \psi_i \psi_j \, dx_1 \quad (12.5.2)$$

Defining the generalized masses $\bar{m}_{ij}$ $(i,j = 1, 2, \ldots, n)$ by

$$\bar{m}_{ij} = \int_0^L \mu \psi_i \psi_j \, dx_1 \quad (12.5.3)$$

the kinetic energy expression becomes

$$T = \tfrac{1}{2} \sum_{i=1}^{n} \sum_{j=1}^{n} \bar{m}_{ij} \dot{q}_i \dot{q}_j \quad (12.5.4)$$

Equation (12.5.4) may be written in matrix form as

$$T = \tfrac{1}{2} \{\dot{q}\}^T [\bar{m}] \{\dot{q}\} \quad (12.5.5)$$

in which $[\bar{m}]$ is a symmetric matrix having elements $\bar{m}_{ij}$.

The strain energy resulting from bending of the beam is

$$U = \int_0^L \frac{EI}{2} (u_2'')^2 \, dx_1$$

$$= \int_0^L \frac{EI}{2} \left(\sum_{i=1}^n \psi_i'' q_i \right) \left(\sum_{j=1}^n \psi_j'' q_j \right) dx_1$$

$$= \tfrac{1}{2} \sum_{i=1}^n \sum_{j=1}^n q_i q_j \int_0^L EI \psi_i'' \psi_j'' \, dx_1 \qquad (12.5.6)$$

or

$$U = \tfrac{1}{2} \sum_{i=1}^n \sum_{j=1}^n \bar{k}_{ij} q_i q_j \qquad (12.5.7)$$

where the generalized stiffness coefficients $\bar{k}_{ij}$ are defined by

$$\bar{k}_{ij} = \int_0^L EI \psi_i'' \psi_j'' \, dx_1 \qquad (12.5.8)$$

The strain energy expression (12.5.7) may be written in the matrix form

$$U = \tfrac{1}{2} \{q\}^T [\bar{k}] \{q\} \qquad (12.5.9)$$

in which the elements of $[\bar{k}]$ are the coefficients $\bar{k}_{ij}$.

Following the procedure outlined in Sec. 11.3, the equations of motion for the structure are obtained by substituting the relations for T and U into Lagrange's equations. In terms of the generalized coordinates q_j, the matrix equation governing the free vibration is found to be

$$[\bar{m}]\{\ddot{q}\} + [\bar{k}]\{q\} = \{0\} \qquad (12.5.10)$$

This equation has exactly the same form as the matrix equation of motion (11.4.1) for a discrete-mass structure. Hence the method of solution described in Chap. 11 is applicable here. To investigate the normal modes of vibration, we assume a solution of the form

$$\{q\} = \{\gamma\} \sin(pt + \delta) \qquad (12.5.11)$$

where $\{\gamma\}$ represents the amplitudes of the generalized displacements $\{q\}$. Substitution of the trial solution (12.5.11) into Eq. (12.5.10) gives

$$-p^2 [\bar{m}]\{\gamma\} + [\bar{k}]\{\gamma\} = \{0\} \qquad (12.5.12)$$

The system's natural circular frequencies p_j and the eigenvectors $\{\gamma\}_j$ can therefore be found by solving the eigenvalue problem [see Eq. (11.4.8)]

$$\left([\bar{D}] - \frac{1}{p^2} [I] \right) \{\gamma\} = \{0\} \qquad (12.5.13)$$

where

$$[\bar{D}] = [\bar{k}]^{-1} [\bar{m}] \qquad (12.5.14)$$

The same results can also be obtained by solving the inverse eigenvalue problem
[see Eq. (11.9.7)]

$$([\bar{E}] - p^2[I])\{\gamma\} = \{0\} \qquad (12.5.15)$$

in which

$$[\bar{E}] = [\bar{m}]^{-1}[\bar{k}] \qquad (12.5.16)$$

Once the frequencies and eigenfunctions have been found, the general solution
to Eq. (12.5.10) may be expressed in the form

$$\{q\} = [\gamma]\{A \cos pt + B \sin pt\} \qquad (12.5.17)$$

where the jth column of the matrix $[\gamma]$ is the eigenvector $\{\gamma\}_j$ and $\{A \cos pt + B \sin pt\}$ is the matrix defined by Eq. (11.4.16). In expanded notation, the
generalized displacement q_j is given by

$$q_j = \sum_{i=1}^{n} \gamma_{ji}(A_i \cos p_i t + B_i \sin p_i t) \qquad (12.5.18)$$

Finally, substitution of Eq. (12.5.18) into (12.5.1) yields the approximate free-
vibration solution

$$u_2(x_1, t) = \sum_{j=1}^{n} \sum_{i=1}^{n} \psi_j(x_1)\gamma_{ji}(A_i \cos p_i t + B_i \sin p_i t) \qquad (12.5.19)$$

The displacement may also be written as

$$u_2(x_1, t) = \sum_{i=1}^{n} \phi_i(x_1)(A_i \cos p_i t + B_i \sin p_i t) \qquad (12.5.20)$$

where the approximate mode shapes $\phi_i(x_1)$ for the beam are related to the assumed
functions $\psi_j(x_1)$ by

$$\phi_i(x_1) = \sum_{j=1}^{n} \psi_j(x_1)\gamma_{ji} \qquad i = 1, 2, \ldots, n \qquad (12.5.21)$$

Note that the approximate solution (12.5.20) has the same general form as the
exact free-vibration solution (12.4.6), except that the range of the summation
in Eq. (12.5.20) is finite rather than infinite. In effect, the Rayleigh-Ritz method
reduces the continuous structure to one having a finite number of degrees of
freedom. As pointed out in Sec. 5.10, reducing the degrees of freedom of a structure is equivalent to introducing additional geometric constraints. Owing to
these additional constraints, the idealized system is stiffer than the actual one.
Consequently the frequencies computed by the Rayleigh-Ritz method are larger,
and the amplitudes of vibration are generally smaller than the corresponding exact
values.[1]

---

[1] For a proof that the frequencies predicted by the Rayleigh-Ritz method give upper
bounds to the exact natural frequencies, see: L. Collatz, "Eigenwertproblems,"
Chelsea House Publishers, New York, 1948.

In general, the accuracy of the computed results is best for the fundamental mode and is progressively poorer for the higher modes. The data for the highest mode considered is often extremely inaccurate. In order to obtain a good approximation to the first n modes of vibration, it is therefore advisable to include at least $n + 1$ functions ψ_j in the assumed solution (12.5.1). The decision as to what functions to select should be based upon an intuitive idea of what the normal modes look like. How well each mode shape can be represented by a superposition of the functions ψ_j determines the accuracy of the final results.

Example 12.5   Rayleigh-Ritz method for a cantilever beam   To demonstrate the application of the Rayleigh-Ritz procedure, we shall compute the lowest few natural frequencies and the corresponding mode shapes for the uniform cantilever beam shown in Fig. 12.6. Since the displacement and slope are zero at the left end of the beam, and the bending moment and shear force vanish at the right end, the boundary conditions are

$$u_2(0,t) = u_2'(0,t) = 0 \quad (12.5.22)$$

and

$$u_2''(L,t) = u_2'''(L,t) = 0 \quad (12.5.23)$$

The functions $\psi_j(x_1)$ in our assumed solution must yield displacements which satisfy the kinematic boundary conditions (12.5.22); the displacements need not necessarily satisfy the static conditions (12.5.23). Let us therefore approximate the deflection curve by the three-term series

$$u_2(x_1,t) = \psi_1(x_1)q_1(t) + \psi_2(x_1)q_2(t) + \psi_3(x_1)q_3(t) \quad (12.5.24)$$

where the functions $\psi_j(x_1)$ are taken to be

$$\psi_1 = \left(\frac{x_1}{L}\right)^2 \qquad \psi_2 = \left(\frac{x_1}{L}\right)^3 \qquad \psi_3 = \left(\frac{x_1}{L}\right)^4 \quad (12.5.25)$$

The generalized masses $\bar{m}_{ij}$, computed from Eq. (12.5.3), are

$$\bar{m}_{11} = \frac{\mu L}{5} \qquad \bar{m}_{22} = \frac{\mu L}{7} \qquad \bar{m}_{33} = \frac{\mu L}{9}$$

$$\bar{m}_{12} = \bar{m}_{21} = \frac{\mu L}{6} \qquad \bar{m}_{13} = \bar{m}_{31} = \frac{\mu L}{7} \qquad \bar{m}_{23} = \bar{m}_{32} = \frac{\mu L}{8} \qquad (12.5.26)$$

and the generalized stiffnesses $\bar{k}_{ij}$, computed according to Eq. (12.5.8), are

$$\bar{k}_{11} = \frac{4EI}{L^3} \qquad \bar{k}_{22} = \frac{12EI}{L^3} \qquad \bar{k}_{33} = \frac{144}{5}\frac{EI}{L^3}$$

$$\bar{k}_{12} = \bar{k}_{21} = \frac{6EI}{L^3} \qquad \bar{k}_{13} = \bar{k}_{31} = \frac{8EI}{L^3} \qquad \bar{k}_{23} = \bar{k}_{32} = \frac{18EI}{L^3} \qquad (12.5.27)$$

Therefore the inertia and stiffness matrices are given by

$$[\bar{m}] = \begin{bmatrix} 0.2000 & 0.1667 & 0.1429 \\ 0.1667 & 0.1429 & 0.1250 \\ 0.1429 & 0.1250 & 0.1111 \end{bmatrix} \mu L \qquad [\bar{k}] = \begin{bmatrix} 4.0 & 6.0 & 8.0 \\ 6.0 & 12.0 & 18.0 \\ 8.0 & 18.0 & 28.8 \end{bmatrix} \frac{EI}{L^3} \qquad (12.5.28)$$

Substituting these matrices into either Eq. (12.5.13) or (12.5.15) and solving the resulting eigenvalue problem, yields the natural circular frequencies

$$p_1 = 3.52\sqrt{\frac{EI}{\mu L^4}} \qquad p_2 = 22.2\sqrt{\frac{EI}{\mu L^4}} \qquad p_3 = 371.0\sqrt{\frac{EI}{\mu L^4}} \qquad (12.5.29)$$

and the matrix of eigenvectors

$$[\gamma] = \begin{bmatrix} 1.000 & 1.000 & 1.000 \\ -0.550 & -1.933 & -2.919 \\ 0.103 & 0.843 & 2.009 \end{bmatrix} \qquad (12.5.30)$$

The beam's normal mode shapes, calculated using Eq. (12.5.21) are,

$$\phi_1 = 1.0\left(\frac{x_1}{L}\right)^2 - 0.550\left(\frac{x_1}{L}\right)^3 + 0.103\left(\frac{x_1}{L}\right)^4$$

$$\phi_2 = 1.0\left(\frac{x_1}{L}\right)^2 - 1.933\left(\frac{x_1}{L}\right)^3 + 0.843\left(\frac{x_1}{L}\right)^4 \qquad (12.5.31)$$

$$\phi_3 = 1.0\left(\frac{x_1}{L}\right)^2 - 2.919\left(\frac{x_1}{L}\right)^3 + 2.009\left(\frac{x_1}{L}\right)^4$$

A comparison of the natural frequencies predicted by the Rayleigh-Ritz method with the corresponding exact values is given in Table 12.1. Also shown

Table 12.1 A COMPARISON OF THE APPROXIMATE AND EXACT NATURAL FREQUENCIES OF A UNIFORM CANTILEVER BEAM

Frequency	Rayleigh-Ritz approximation	Lumped-mass approximation (Example 11.3)	Exact solution (Example 12.4)
$\dfrac{p_1}{\sqrt{EI/\mu L^4}}$	3.52	3.36	3.52
$\dfrac{p_2}{\sqrt{EI/\mu L^4}}$	22.2	18.9	22.0
$\dfrac{p_3}{\sqrt{EI/\mu L^4}}$	371.0	47.2	61.7

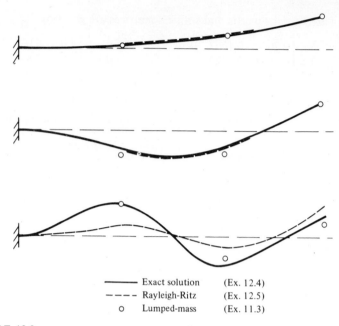

—————— Exact solution	(Ex. 12.4)
— — — Rayleigh-Ritz	(Ex. 12.5)
o Lumped-mass	(Ex. 11.3)

FIGURE 12.8
Comparison of the approximate and exact mode shapes for a uniform cantilever beam.

in the table are the values which were obtained when the beam was approximated by a lumped-mass system (Example 11.3). Notice that the Rayleigh-Ritz results are in excellent agreement with the exact solution for the first two modes, but that the third natural frequency is approximately six times too large. The discrete-mass analysis yields less accurate results than the Rayleigh-Ritz procedure for the first two frequencies but a somewhat closer approximation to the third frequency. The accuracy of the approximations could be improved by considering a larger number of degrees of freedom.

The mode shapes obtained in the three separate analyses are plotted in Fig. 12.8. Each shape has been normalized, such that its maximum amplitude is unity. It is again evident that the accuracy of the approximate solutions becomes progressively poorer for the higher modes.   ////

Example 12.6   Rayleigh-Ritz method for a rod with an attached mass

A uniform elastic rod of mass per unit length μ is fixed at one end and carries a mass $m = \mu L/2$ at its other end, as shown in Fig. 12.9. We wish to obtain an

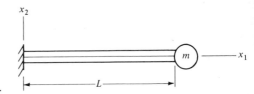

FIGURE 12.9
Bar with an attached mass.

approximate solution for the free, longitudinal vibration of the bar. Let us therefore approximate the displacement $u_1(x_1,t)$ by the finite series

$$u_1(x_1,t) = \sum_{j=1}^{n} \psi_j(x_1)q_j(t) \quad (12.5.32)$$

where, as before, $q_j(t)$ represent time-dependent generalized displacements, and $\psi_j(x_1)$ are functions which satisfy the kinematic boundary conditions. In terms of these assumed functions, the structure's kinetic energy is

$$
\begin{aligned}
T &= \int_0^L \frac{\mu}{2} \dot{u}_1^2 \, dx_1 + \frac{1}{2} m \dot{u}_1(L,t)^2 \\
&= \int_0^L \frac{\mu}{2} \left(\sum_{i=1}^{n} \psi_i \dot{q}_i \right) \left(\sum_{j=1}^{n} \psi_j \dot{q}_j \right) dx_1 + \frac{1}{2} m \left(\sum_{i=1}^{n} \psi_i(L)\dot{q}_i \right) \left(\sum_{j=1}^{n} \psi_j(L)\dot{q}_j \right) \\
&= \tfrac{1}{2} \sum_{i=1}^{n} \sum_{j=1}^{n} \dot{q}_i \dot{q}_j \left[\int_0^L \mu \psi_i \psi_j \, dx_1 + m \psi_i(L)\psi_j(L) \right] \\
&= \tfrac{1}{2} \sum_{i=1}^{n} \sum_{j=1}^{n} \overline{m}_{ij} \dot{q}_i \dot{q}_j \quad (12.5.33)
\end{aligned}
$$

or in matrix form

$$T = \tfrac{1}{2}\{\dot{q}\}^T [\overline{m}]\{\dot{q}\} \quad (12.5.34)$$

where the elements $\overline{m}_{ij}$ of the inertia matrix $[\overline{m}]$ are given by

$$\overline{m}_{ij} = \int_0^L \mu \psi_i \psi_j \, dx_1 + m \psi_i(L)\psi_j(L) \quad (12.5.35)$$

The strain energy resulting from the axial deformation of the bar is

$$
\begin{aligned}
U &= \int_0^L \frac{EA}{2} (u_1')^2 \, dx_1 \\
&= \int_0^L \frac{EA}{2} \left(\sum_{i=1}^{n} \psi_i' q_i \right) \left(\sum_{j=1}^{n} \psi_j' q_j \right) dx_1 \\
&= \tfrac{1}{2} \sum_{i=1}^{n} \sum_{j=1}^{n} q_i q_j \int_0^L EA \psi_i' \psi_j' \, dx_1 \\
&= \tfrac{1}{2} \sum_{i=1}^{n} \sum_{j=1}^{n} \overline{k}_{ij} q_i q_j \quad (12.5.36)
\end{aligned}
$$

or
$$U = \tfrac{1}{2}\{q\}^T[\bar{k}]\{q\} \tag{12.5.37}$$

in which the elements $\bar{k}_{ij}$ of the stiffness matrix $[\bar{k}]$ are defined as

$$\bar{k}_{ij} = \int_0^L EA\psi_i' \psi_j' \, dx_1 \tag{12.5.38}$$

Since the matrix expressions (12.5.34) and (12.5.37) for the kinetic and potential energies of the rod are identical in form to those for the vibrating beam, we may make use of our previous results. That is, the free-vibration solution can be found by solving either the eigenvalue problem (12.5.13) or the inverse problem (12.5.15).

For the problem at hand, let us consider the functions

$$\psi_1 = \frac{x_1}{L} \qquad \psi_2 = \left(\frac{x_1}{L}\right)^2 \qquad \psi_3 = \left(\frac{x_1}{L}\right)^3 \tag{12.5.39}$$

each of which clearly satisfies the rod's kinematic boundary conditions. Using Eq. (12.5.35), the generalized masses $\bar{m}_{ij}$ are found to be

$$\begin{aligned}
&\bar{m}_{11} = \tfrac{5}{6}\mu L && \bar{m}_{22} = \tfrac{7}{10}\mu L && \bar{m}_{33} = \tfrac{9}{14}\mu L \\
&\bar{m}_{12} = \bar{m}_{21} = \tfrac{3}{4}\mu L && \bar{m}_{13} = \bar{m}_{31} = \tfrac{7}{10}\mu L && \bar{m}_{23} = \bar{m}_{32} = \tfrac{2}{3}\mu L
\end{aligned} \tag{12.5.40}$$

and from Eq. (12.5.38) the generalized stiffness coefficients $\bar{k}_{ij}$ are

$$\begin{aligned}
&\bar{k}_{11} = \frac{EA}{L} && \bar{k}_{22} = \frac{4}{3}\frac{EA}{L} && \bar{k}_{33} = \frac{9}{5}\frac{EA}{L} \\
&\bar{k}_{12} = \bar{k}_{21} = \frac{EA}{L} && \bar{k}_{13} = \bar{k}_{31} = \frac{EA}{L} && \bar{k}_{23} = \bar{k}_{32} = \frac{3}{2}\frac{EA}{L}
\end{aligned} \tag{12.5.41}$$

Consequently, the inertia and stiffness matrices become

$$[\bar{m}] = \begin{bmatrix} 0.8333 & 0.7500 & 0.7000 \\ 0.7500 & 0.7000 & 0.6667 \\ 0.7000 & 0.6667 & 0.6429 \end{bmatrix} \mu L \qquad [\bar{k}] = \begin{bmatrix} 1.0 & 1.0 & 1.0 \\ 1.0 & 1.333 & 1.500 \\ 1.0 & 1.500 & 1.800 \end{bmatrix} \frac{EA}{L} \tag{12.5.42}$$

Substituting these matrices into either Eq. (12.5.13) or Eq. (12.5.15) and solving the resulting eigenvalue problem, we obtain the natural circular frequencies

$$p_1 = 1.08\sqrt{\frac{EA}{\mu L^2}} \qquad p_2 = 3.69\sqrt{\frac{EA}{\mu L^2}} \qquad p_3 = 6.85\sqrt{\frac{EA}{\mu L^2}} \tag{12.5.43}$$

and the matrix of eigenvectors

$$[\gamma] = \begin{bmatrix} 1.000 & 1.000 & 1.000 \\ -0.024 & -1.429 & -3.110 \\ -0.162 & 0.332 & 2.142 \end{bmatrix} \tag{12.5.44}$$

The mode shapes for the rod, computed according to Eq. (12.5.21), are

$$\phi_1 = 1.0 \frac{x_1}{L} - 0.024\left(\frac{x_1}{L}\right)^2 - 0.162\left(\frac{x_1}{L}\right)^3$$

$$\phi_2 = 1.0 \frac{x_1}{L} - 1.429\left(\frac{x_1}{L}\right)^2 + 0.332\left(\frac{x_1}{L}\right)^3 \qquad (12.5.45)$$

$$\phi_3 = 1.0 \frac{x_1}{L} - 3.110\left(\frac{x_1}{L}\right)^2 + 2.142\left(\frac{x_1}{L}\right)^3$$

The computed natural frequencies are approximately 0.5, 1.3, and 4.2 percent greater, respectively, than the corresponding exact values (Prob. 12.9). ////

12.6 FORCED, LONGITUDINAL VIBRATION OF A BAR

As we have already seen, the problem of the forced response of a continuous structure is characterized by a nonhomogeneous partial differential equation. Solutions can sometimes be obtained using standard mathematical techniques such as the Laplace or Fourier transform methods.[1] Alternatively, the forced-vibration solution can be determined using the *method of normal modes*. To illustrate the application of the latter procedure, let us now consider the response of a bar of uniform cross section to a distributed axial force $F(x_1,t)$ (Fig. 12.1). Analogous to the approach used for a discrete-mass system, the longitudinal displacement $u_1(x_1,t)$ of the bar is expressed in terms of normal coordinates r_j as

$$u_1(x_1,t) = \sum_{j=1}^{\infty} \phi_j(x_1)r_j(t) \qquad (12.6.1)$$

in which $\phi_j(x_1)$ is the eigenfunction of the jth normal mode of vibration. It will be shown that the expansion (12.6.1) reduces the problem from one involving a partial differential equation of motion to one governed by an infinite number of uncoupled ordinary differential equations.

The kinetic energy of the bar, found by substituting Eq. (12.6.1) into (12.2.18) is

$$T = \int_0^L \frac{\mu}{2} \dot{u}_1{}^2 \, dx_1$$

$$= \int_0^L \frac{\mu}{2} \left(\sum_{i=1}^{\infty} \phi_i \dot{r}_i\right)\left(\sum_{j=1}^{\infty} \phi_j \dot{r}_j\right) dx_1$$

$$= \frac{1}{2} \sum_{i=1}^{\infty} \sum_{j=1}^{\infty} \dot{r}_i \dot{r}_j \int_0^L \mu \phi_i \phi_j \, dx_1 \qquad (12.6.2)$$

[1] R. V. Churchill, "Operational Mathematics," 2d ed., McGraw-Hill, 1958; also Ref. 12.5.

By virtue of the orthogonality relations (12.3.17) for a bar with either clamped or free ends, all cross-product terms in the series (12.6.2) vanish, leaving

$$T = \tfrac{1}{2} \sum_{j=1}^{\infty} \mathcal{M}_j \dot{r}_j^2 \qquad (12.6.3)$$

in which the generalized masses $\mathcal{M}_j$ are defined by

$$\mathcal{M}_j = \mu \int_0^L \phi_j^2 \, dx_1 \qquad (12.6.4)$$

The strain energy resulting from the bar's axial deformation is

$$U = \int_0^L \frac{EA}{2} (u_1')^2 \, dx_1$$

$$= \int_0^L \frac{EA}{2} \left(\sum_{i=1}^{\infty} \phi_i' r_i \right) \left(\sum_{j=1}^{\infty} \phi_j' r_j \right) dx_1$$

$$= \tfrac{1}{2} \sum_{i=1}^{\infty} \sum_{j=1}^{\infty} r_i r_j \int_0^L EA \phi_i' \phi_j' \, dx_1 \qquad (12.6.5)$$

Again using the orthogonality relations (12.3.17), it is seen that all cross-product terms in the series (12.6.5) vanish, so that

$$U = \tfrac{1}{2} \sum_{j=1}^{\infty} \mathcal{M}_j p_j^2 r_j^2 \qquad (12.6.6)$$

The potential of the external loading $F(x_1,t)$ may be written as

$$V_E = - \int_0^L F(x_1,t) u_1(x_1,t) \, dx_1$$

$$= - \int_0^L F(x_1,t) \left(\sum_{j=1}^{\infty} \phi_j r_j \right) dx_1$$

$$= - \sum_{j=1}^{\infty} \left[\int_0^L F(x_1,t) \phi_j \, dx_1 \right] r_j \qquad (12.6.7)$$

or

$$V_E = - \sum_{j=1}^{\infty} (R_j)_a r_j \qquad (12.6.8)$$

where the generalized applied force $(R_j)_a$ associated with the normal coordinate r_j is given by

$$(R_j)_a = \int_0^L F(x_1,t) \phi_j \, dx_1 \qquad (12.6.9)$$

If instead of a distributed load $F(x_1,t)$ the bar is subject to a discrete force $F(t)$ located at $x_1 = d$, then the generalized force is

$$(R_j)_a = F(t) \phi_j(d) \qquad (12.6.10)$$

Substitution of the above expressions for T, U, and $(R_j)_a$ into the Lagrange's equations for r_j, namely

$$\frac{d}{dt}\left(\frac{\partial T}{\partial \dot{r}_j}\right) - \frac{\partial T}{\partial r_j} + \frac{\partial U}{\partial r_j} = (R_j)_a \qquad j = 1,2,\ldots,\infty \quad (12.6.11)$$

yields

$$\mathcal{M}_j\ddot{r}_j + \mathcal{M}_j p_j{}^2 r_j = (R_j)_a \qquad j = 1,2,\ldots,\infty \quad (12.6.12)$$

Hence, the equations of motion are decoupled when the displacement $u_1(x_1,t)$ is expressed in terms of the normal coordinates r_j. Although the number of equations is then infinite, we shall see that, in most cases, the response of a structure can be described adequately by considering just the few lowest modes.

It should be emphasized that the equations of motion (12.6.12) are applicable even if the normal modes have been computed using the Rayleigh-Ritz technique or a similar energy method. In this case the system's eigenfunctions are approximate; nevertheless they are mutually orthogonal, and the corresponding equations of motion are therefore uncoupled.

Various representations of the general solution to an equation of the form (12.6.12) have been given earlier. For instance, the solution for the jth normal mode may be written as

$$r_j = a_j \cos p_j t + b_j \sin p_j t + D_j(t) \quad (12.6.13)$$

where $D_j(t)$ represents the Duhamel's integral

$$D_j(t) = \frac{1}{\mathcal{M}_j p_j} \int_0^t [R_j(\xi)]_a \sin p_j(t - \xi)\, d\xi \quad (12.6.14)$$

The arbitrary constants a_j and b_j are chosen to satisfy the initial conditions. Following the procedure given in Sec. 12.3 it can be shown that

$$a_j = \frac{\mu}{\mathcal{M}_j} \int_0^L u_1(x_1,0)\phi_j\, dx_1$$

$$(12.6.15)$$

$$b_j = \frac{\mu}{\mathcal{M}_j p_j} \int_0^L \dot{u}_1(x_1,0)\phi_j\, dx_1$$

in which $u_1(x_1,0)$ and $\dot{u}_1(x_1,0)$ denote the structure's initial displacement and velocity, respectively.

Example 12.7 Forced, longitudinal vibration of a clamped-free bar   The dynamic reponse of the bar shown in Fig. 12.10 will now be examined. One end of the bar is clamped, while the opposite end is subject to a suddenly applied force of magnitude F_0. We shall assume that the force $F(t)$ acts at a cross section immediately to the left of the end of the bar $x_1 = L$, so that the boundary itself

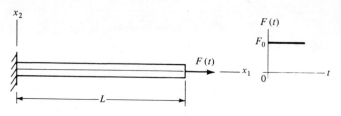

FIGURE 12.10
Clamped-free bar subject to a step force.

is stress-free.[1] In this case the normal modes are those of a clamped-free bar, and the orthogonality relations (12.3.17) are applicable.

The natural circular frequencies and the eigenfunctions for the bar are (see Prob. 12.7)

$$p_j = \frac{(2j-1)\pi}{2} \sqrt{\frac{EA}{\mu L^2}} \qquad j = 1,2,\dots,\infty \quad (12.6.16)$$

and

$$\phi_j = \sin \frac{(2j-1)\pi x_1}{2L} \qquad j = 1,2,\dots,\infty \quad (12.6.17)$$

To find the response of the bar to the applied load $F(t)$, we introduce the normal coordinates defined by Eq. (12.6.1). As shown earlier, this leads to the uncoupled equations of motion

$$\ddot{r}_j + p_j{}^2 r_j = \frac{(R_j)_a}{\mathcal{M}_j} \qquad j = 1,2,\dots,\infty \quad (12.6.18)$$

According to Eq. (12.6.4), the generalized mass $\mathcal{M}_j$ is[2]

$$\mathcal{M}_j = \mu \int_0^L \phi_j{}^2 \, dx_1 = \mu \int_0^L \sin^2 \frac{(2j-1)\pi x_1}{2L} \, dx_1 = \frac{\mu L}{2} \quad (12.6.19)$$

and from Eq. (12.6.10), the generalized force $(R_j)_a$ is

$$(R_j)_a = F(t)\phi_j(L) = F_0 \sin \frac{(2j-1)\pi}{2} \quad (12.6.20)$$

The equation of motion for the jth normal mode then becomes

$$\ddot{r}_j + p_j{}^2 r_j = \frac{F_0}{\mathcal{M}_j} \sin \frac{(2j-1)\pi}{2} \quad (12.6.21)$$

[1]This problem can also be solved by assuming that the force $F(t)$ acts directly at the end of the beam, in which case the boundary condition is $EAu_1{}'(L,t) = F(t)$. The Laplace transform technique provides a convenient means for finding the solution to problems involving nonhomogeneous boundary conditions of this type (see footnote p. 323).
[2]The generalized masses could be normalized such that they all equal some value other than $\mu L/2$. For example, if each eigenfunction were multiplied by $\sqrt{2}$, then $\mathcal{M}_j = \mu L$, for $j = 1, 2,\dots,\infty$.

Assuming that the displacement and velocity of every point in the beam are initially zero [so that $r_j(0) = \dot{r}_j(0) = 0$], the resulting solution for r_j is

$$r_j = \frac{F_0}{\mathscr{M}_j p_j{}^2} \sin \frac{(2j-1)\pi}{2} (1 - \cos p_j t) \quad (12.6.22)$$

Substitution of Eq. (12.6.22) into (12.6.1) yields the displacement curve

$$
\begin{aligned}
u_1(x_1,t) &= \sum_{j=1}^{\infty} \phi_j(x_1) r_j(t) \\
&= \sum_{j=1}^{\infty} \frac{F_0}{\mathscr{M}_j p_j{}^2} \sin \frac{(2j-1)\pi}{2} \sin \frac{(2j-1)\pi x_1}{2L} (1 - \cos p_j t) \\
&= \frac{F_0 L}{EA} \sum_{j=1}^{\infty} \frac{8}{(2j-1)^2 \pi^2} \sin \frac{(2j-1)\pi}{2} \sin \frac{(2j-1)\pi x_1}{2L} (1 - \cos p_j t)
\end{aligned}
$$
$$(12.6.23)$$

The displacement is largest at the free end of the bar, and is maximum at the instant of time when $1 - \cos p_j t = 2$ for all modes. In this case

$$|u_1(L,t)|_{\max} = \frac{F_0 L}{EA} \sum_{j=1}^{\infty} \frac{16}{(2j-1)^2 \pi^2} \sin^2 \frac{(2j-1)\pi}{2} = \frac{F_0 L}{EA} \sum_{j=1}^{\infty} \frac{16}{(2j-1)^2 \pi^2}$$
$$(12.6.24)$$

It can be shown that the series in Eq. (12.6.24) converges to the value 2.[†] Therefore

$$|u_1(L,t)|_{\max} = 2 \frac{F_0 L}{EA} \quad (12.6.25)$$

and we see that the maximum displacement produced by a suddenly applied force is twice as large as that caused by a statically applied force of equal magnitude.

From a consideration of the coefficients in the series solution (12.6.23), it is clear that the higher modes contribute progressively less to the total displacement response of the bar. A relatively accurate approximation for $u_1(x_1,t)$ is obtained if one considers just the first four normal modes ($j = 1, 2, 3,$ and 4).

The axial stress at an arbitrary cross section of the bar is given by

$$
\begin{aligned}
\sigma_{11}(x_1,t) &= E \frac{\partial u_1}{\partial x_1} \\
&= \frac{F_0}{A} \sum_{j=1}^{\infty} \frac{4}{(2j-1)\pi} \sin \frac{(2j-1)\pi}{2} \cos \frac{(2j-1)\pi x_1}{2L} (1 - \cos p_j t) \quad (12.6.26)
\end{aligned}
$$

[†]Series such as the one in Eq. (12.6.24) which result from a Fourier expansion converge to values known as the *Bernoulli numbers*. A tabulation of these numbers may be found in *C.R.C. Standard Mathematical Tables*, 16th edition, Chemical Rubber Co., Cleveland, Ohio, 1968.

The stress is greatest at the clamped end of the bar; an upper bound for the amplitude of $\sigma_{11}(0,t)$ is

$$\left|\sigma_{11}(0,t)\right|_{max} = \frac{F_0}{A} \sum_{j=1}^{\infty} \frac{8}{(2j-1)\pi} \sin \frac{(2j-1)\pi}{2} = 2\frac{F_0}{A} \quad (12.6.27)$$

Hence, in this problem the maximum dynamic stress is twice as large as the corresponding static stress. Note that the rate of convergence of the series in Eq. (12.6.26) is not as rapid as that for the displacement. In order to obtain the same degree of accuracy, one must therefore consider a larger number of normal modes when computing σ_{11} than when computing u_1. ////

12.7 FORCED, LATERAL VIBRATION OF A BAR

The method of normal modes described in the preceding section can be used to obtain the forced response of any continuous structure for which the natural frequencies and eigenfunctions are known. Let us now consider a symmetric, uniform beam acted upon by a distributed transverse force $F(x_1,t)$ (Fig. 12.2). As in the case of longitudinal vibrations, we express the deflection curve in terms of an infinite series of the eigenfunctions $\phi_j(x_1)$ as

$$u_2(x_1,t) = \sum_{j=1}^{\infty} \phi_j(x_1)r_j(t) \quad (12.7.1)$$

where $r_j(t)$ are the normal coordinates for the system.

The kinetic energy of the beam is

$$T = \int_0^L \frac{\mu}{2} \dot{u}_2{}^2 \, dx_1$$

$$= \int_0^L \frac{\mu}{2} \left(\sum_{i=1}^{\infty} \phi_i \dot{r}_i \right) \left(\sum_{j=1}^{\infty} \phi_j \dot{r}_j \right) dx_1$$

$$= \frac{1}{2} \sum_{i=1}^{\infty} \sum_{j=1}^{\infty} \dot{r}_i \dot{r}_j \int_0^L \mu \phi_i \phi_j \, dx_1 \quad (12.7.2)$$

Using the orthogonality relations (12.4.7), for a beam having any combination of clamped, free, simply supported, or guided end conditions, expression (12.7.2) reduces to

$$T = \frac{1}{2} \sum_{j=1}^{\infty} \mathcal{M}_j \dot{r}_j{}^2 \quad (12.7.3)$$

where as before the generalized masses $\mathcal{M}_j$ are given by

$$\mathcal{M}_j = \mu \int_0^L \phi_j{}^2 \, dx_1 \quad (12.7.4)$$

The strain energy of bending may be expressed in terms of the normal coordinates as

$$U = \int_0^L \frac{EI}{2} (u_2'')^2 \, dx_1$$

$$= \int_0^L \frac{EI}{2} \left(\sum_{i=1}^{\infty} \phi_i'' r_i \right) \left(\sum_{j=1}^{\infty} \phi_j'' r_j \right) dx_1$$

$$= \tfrac{1}{2} \sum_{i=1}^{\infty} \sum_{j=1}^{\infty} r_i r_j \int_0^L EI \phi_i'' \phi_j'' \, dx_1 \qquad (12.7.5)$$

Again making use of the orthogonality conditions (12.4.7), we obtain

$$U = \tfrac{1}{2} \sum_{j=1}^{\infty} \mathcal{M}_j p_j^2 r_j^2 \qquad (12.7.6)$$

The potential of the external force $F(x_1,t)$ is

$$V_E = - \int_0^L F(x_1,t) u_2(x_1,t) \, dx_1$$

$$= - \int_0^L F(x_1,t) \left(\sum_{j=1}^{\infty} \phi_j r_j \right) dx_1$$

$$= - \sum_{j=1}^{\infty} \left[\int_0^L F(x_1,t) \phi_j \, dx_1 \right] r_j$$

$$= - \sum_{j=1}^{\infty} (R_j)_a r_j \qquad (12.7.7)$$

where

$$(R_j)_a = \int_0^L F(x_1,t) \phi_j \, dx_1 \qquad (12.7.8)$$

If a discrete force $F(t)$ is applied to the cross section $x_1 = d$ of the beam, then the generalized force $(R_j)_a$ becomes

$$(R_j)_a = F(t) \phi_j(d) \qquad (12.7.9)$$

Substitution of the above expressions for T, U, and $(R_j)_a$ into Lagrange's equations (12.6.11) yields

$$\mathcal{M}_j \ddot{r}_j + \mathcal{M}_j p_j^2 r_j = (R_j)_a \qquad j = 1,2,\ldots,\infty \qquad (12.7.10)$$

The general solution for the jth normal mode can be written as

$$r_j = a_j \cos p_j t + b_j \sin p_j t + D_j(t) \qquad (12.7.11)$$

where

$$a_j = \frac{\mu}{\mathcal{M}_j} \int_0^L u_2(x_1,0) \phi_j \, dx_1$$

$$b_j = \frac{\mu}{\mathcal{M}_j p_j} \int_0^L \dot{u}_2(x_1,0) \phi_j \, dx_1$$

$$(12.7.12)$$

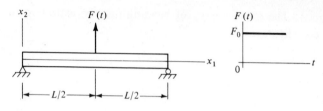

FIGURE 12.11
Simply supported beam subject to a step force.

and
$$D_j(t) = \frac{1}{\mathcal{M}_j p_j} \int_0^t [R_j(\xi)]_a \sin p_j(t - \xi)\, d\xi \quad (12.7.13)$$

Substitution of the normal mode solutions (12.7.11) into equation (12.7.1) yields the desired deflection response $u_2(x_1, t)$.

Example 12.8   Forced vibration of a simply supported beam   The response of the simply supported beam shown in Fig. 12.11 will now be investigated. From the free-vibration solution for the beam (Prob. 12.10), the natural circular frequencies are found to be

$$p_j = j^2 \pi^2 \sqrt{\frac{EI}{\mu L^4}} \quad j = 1, 2, \ldots, \infty \quad (12.7.14)$$

and the corresponding mode shapes are

$$\phi_j = \sin \frac{j\pi x_1}{L} \quad j = 1, 2, \ldots, \infty \quad (12.7.15)$$

In order to compute the response of the beam to the step force $F(t)$ shown, we introduce the normal coordinates r_j defined by Eq. (12.7.1). The equation of motion for the jth mode then becomes

$$\ddot{r}_j + p_j^2 r_j = \frac{(R_j)_a}{\mathcal{M}_j} \quad (12.7.16)$$

where the generalized mass $\mathcal{M}_j$ is

$$\mathcal{M}_j = \mu \int_0^L \phi_j^2\, dx_1 = \mu \int_0^L \sin^2 \frac{j\pi x_1}{L}\, dx_1 = \frac{\mu L}{2} \quad (12.7.17)$$

and the generalized applied force $(R_j)_a$ is

$$(R_j)_a = F(t)\phi_j\left(\frac{L}{2}\right) = F_0 \sin \frac{j\pi}{2} \quad (12.7.18)$$

Equation (12.7.16) may then be written as

$$\ddot{r}_j + p_j^2 r_j = \frac{F_0}{\mathscr{M}_j} \sin \frac{j\pi}{2} \quad (12.7.19)$$

Assuming that the beam is at rest initially, the solution to Eq. (12.7.19) is

$$r_j = \frac{F_0}{\mathscr{M}_j p_j^2} \sin \frac{j\pi}{2} (1 - \cos p_j t) \quad (12.7.20)$$

Substitution of Eq. (12.7.20) into (12.7.1) gives the displacement response

$$u_2(x_1,t) = \sum_{j=1}^{\infty} \phi_j(x_1) r_j(t)$$

$$= \sum_{j=1}^{\infty} \frac{F_0}{\mathscr{M}_j p_j^2} \sin \frac{j\pi}{2} \sin \frac{j\pi x_1}{L} (1 - \cos p_j t)$$

$$= \frac{F_0 L^3}{EI} \sum_{j=1}^{\infty} \frac{2}{j^4 \pi^4} \sin \frac{j\pi}{2} \sin \frac{j\pi x_1}{L} (1 - \cos p_j t) \quad (12.7.21)$$

The maximum displacement occurs at the beam's midpoint at the instant when $1 - \cos p_j t = 2$ for all modes. Then

$$\left| u_2\left(\frac{L}{2},t\right) \right|_{\max} = \frac{F_0 L^3}{EI} \sum_{j=1}^{\infty} \frac{4}{j^4 \pi^4} \sin^2 \frac{j\pi}{2} = \frac{F_0 L^3}{24EI} \quad (12.7.22)$$

which is equal to twice the displacement produced by a statically applied force of magnitude F_0. The series in Eq. (12.7.21) converges very rapidly, owing to the presence of the factor $1/j^4$. A good approximation to the displacement is obtained if one considers, say, the lowest three normal modes ($j = 1, 2,$ and 3).

The bending moment $M(x_1,t)$ at an arbitrary cross section of the beam can be calculated as follows:

$$M(x_1,t) = EI u_2''(x_1,t)$$

$$= -F_0 L \sum_{j=1}^{\infty} \frac{2}{j^2 \pi^2} \sin \frac{j\pi}{2} \sin \frac{j\pi x_1}{L} (1 - \cos p_j t) \quad (12.7.23)$$

The maximum bending moment occurs at the center of the beam and is

$$\left| M\left(\frac{L}{2},t\right) \right|_{\max} = F_0 L \sum_{j=1}^{\infty} \frac{4}{j^2 \pi^2} \sin^2 \frac{j\pi}{2} = \frac{F_0 L}{2} \quad (12.7.24)$$

which is equal to twice the corresponding static value. Note that the convergence of the series for the bending moment is slower than that for the displacement. ////

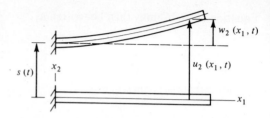

FIGURE 12.12
Cantilever beam subject to a foundation motion $s(t)$.

12.8   RESPONSE TO A FOUNDATION MOTION

In this section we shall investigate the response of a continuous structure to motion of the structure's foundation. Consider, for example, the uniform cantilever beam shown in Fig. 12.12. The displacement $w_2(x_1,t)$ of the beam relative to the foundation is related to the absolute displacement $u_2(x_1,t)$ by

$$w_2(x_1,t) = u_2(x_1,t) - s(t) \qquad (12.8.1)$$

where $s(t)$ represents the movement of the foundation. We shall again make use of the normal coordinates $r_j(t)$ $(j = 1, 2, \ldots, n)$, which in this case are related to the displacement $w_2(x_1,t)$ by

$$w_2(x_1,t) = \sum_{j=1}^{\infty} \phi_j(x_1)r_j(t) \qquad (12.8.2)$$

The kinetic energy of the beam is then

$$
\begin{aligned}
T &= \int_0^L \frac{\mu}{2} \dot{u}_2{}^2 \, dx_1 \\
&= \int_0^L \frac{\mu}{2} (\dot{w}_2 + \dot{s})^2 \, dx_1 \\
&= \int_0^L \frac{\mu}{2} \left[\left(\sum_{i=1}^{\infty} \phi_i \dot{r}_i \right) \left(\sum_{j=1}^{\infty} \phi_j \dot{r}_j \right) + 2\dot{s} \left(\sum_{j=1}^{\infty} \phi_j \dot{r}_j \right) + \dot{s}^2 \right] dx_1 \\
&= \tfrac{1}{2} \sum_{i=1}^{\infty} \sum_{j=1}^{\infty} \dot{r}_i \dot{r}_j \mu \int_0^L \phi_i \phi_j \, dx_1 + \mu \dot{s} \sum_{j=1}^{\infty} \dot{r}_j \int_0^L \phi_j \, dx_1 + \tfrac{1}{2} \mu L \dot{s}^2 \qquad (12.8.3)
\end{aligned}
$$

Using the orthogonality relations (12.4.7) this expression simplifies to

$$T = \tfrac{1}{2} \sum_{j=1}^{\infty} \mathcal{M}_j \dot{r}_j{}^2 + \mu \dot{s} \sum_{j=1}^{\infty} \dot{r}_j \int_0^L \phi_j \, dx_1 + \tfrac{1}{2} \mu L \dot{s}^2 \qquad (12.8.4)$$

where as before the generalized masses are given by

$$\mathcal{M}_j = \mu \int_0^L \phi_j{}^2 \, dx_1 \qquad (12.8.5)$$

The strain energy of bending may be written in terms of the normal coordinates as

$$U = \int_0^L \frac{EI}{2} (w_2'')^2 \, dx_1 = \frac{1}{2} \sum_{j=1}^{\infty} \mathcal{M}_j p_j^2 r_j^2 \qquad (12.8.6)$$

where the computations leading to Eq. (12.8.6) were given in Sec. 12.7.

Substitution of the energy relations (12.8.4) and (12.8.6) into Lagrange's equations (12.6.11) gives

$$\mathcal{M}_j \ddot{r}_j + \mathcal{M}_j p_j^2 r_j = -\mu \ddot{s} \int_0^L \phi_j \, dx_1 \qquad j = 1, 2, \ldots, \infty \qquad (12.8.7)$$

These equations are identical to the Eqs. (12.7.10) governing the forced vibration of a beam, providing we now let

$$(R_j)_a = -\mu \ddot{s} \int_0^L \phi_j \, dx_1 \qquad (12.8.8)$$

Hence, by analogy with our previous result (12.7.11), the solution for r_j is

$$r_j = a_j \cos p_j t + b_j \sin p_j t + D_j(t) \qquad (12.8.9)$$

where

$$a_j = \frac{\mu}{\mathcal{M}_j} \int_0^L w_2(x_1, 0) \phi_j \, dx_1$$

$$b_j = \frac{\mu}{\mathcal{M}_j p_j} \int_0^L \dot{w}_2(x_1, 0) \phi_j \, dx_1 \qquad (12.8.10)$$

and

$$D_j(t) = \frac{-\mu \int_0^L \phi_j \, dx_1}{\mathcal{M}_j p_j} \int_0^t \ddot{s}(\xi) \sin p_j(t - \xi) \, d\xi \qquad (12.8.11)$$

The general solution for $w_2(x_1, t)$ is then found by substituting Eq. (12.8.9) into (12.8.2).

12.9   DAMPED VIBRATION

In general, the analysis of a continuous structure becomes much more complicated when the effects of damping are included. Unlike the undamped problem, it is generally impossible to reduce the partial differential equation of motion to a system of uncoupled ordinary differential equations. There are, however, a few special kinds of damping for which the equations do decouple upon introduction of the normal coordinates defined earlier. Two such cases will be studied in this section.

Consider the flexural vibration of a beam of uniform cross section for which the undamped frequencies p_j and the corresponding eigenfunctions $\phi_j(x_1)$ are known. We again introduce an infinite set of coordinates r_j defined by

$$u_2(x_1,t) = \sum_{j=1}^{\infty} \phi_j(x_1)r_j(t) \qquad (12.9.1)$$

The kinetic and potential energies for the beam were written in terms of the coordinates r_j in Sec. 12.7. Substitution of these relations into Lagrange's equations gives

$$\mathcal{M}_j\ddot{r}_j + \mathcal{M}_j p_j{}^2 r_j = (R_j)_N \qquad j = 1,2,\ldots,\infty \qquad (12.9.2)$$

where the nonconservative generalized forces $(R_j)_N$ now consist of both applied forces $(R_j)_a$ and damping forces $(R_j)_d$; that is,

$$(R_j)_N = (R_j)_a + (R_j)_d \qquad (12.9.3)$$

The generalized forces corresponding to an applied transverse force $F(x_1,t)$ were found to be

$$(R_j)_a = \int_0^L F(x_1,t)\phi_j \, dx_1 \qquad j = 1,2,\ldots,\infty \qquad (12.9.4)$$

In order to determine the quantities $(R_j)_d$, we shall compute the virtual work done by the forces of damping during virtual displacements δr_j. The two types of damping which we shall consider here are *viscous damping* and *strain-rate damping*.

Let us first assume that the beam is subject to a distributed viscous damping force $-c\dot{u}_2(x_1,t)$, where c is a constant called the *coefficient of viscous damping*. The virtual work δW_E done by this force during a distortion δu_2 of the beam is

$$\delta W_E = -\int_0^L c\dot{u}_2(x_1,t)\, \delta u_2 \, dx_1 \qquad (12.9.5)$$

or, in terms of the coordinates r_j,

$$\delta W_E = -\int_0^L c\left(\sum_{k=1}^{\infty} \phi_k \dot{r}_k\right)\left(\sum_{j=1}^{\infty} \phi_j \, \delta r_j\right) dx_1$$

$$= \sum_{j=1}^{\infty} \left(-c\sum_{k=1}^{\infty} \dot{r}_k \int_0^L \phi_k \phi_j \, dx_1\right) \delta r_j \qquad (12.9.6)$$

It is evident from Eq. (12.9.6) that the generalized damping force corresponding to the displacement r_j is

$$(R_j)_d = -c\sum_{k=1}^{\infty} \dot{r}_k \int_0^L \phi_k \phi_j \, dx_1 \qquad (12.9.7)$$

Using the orthogonality conditions (12.4.7), we obtain

$$(R_j)_d = -2\zeta \mathcal{M}_j \dot{r}_j \qquad (12.9.8)$$

where we have defined a new constant ζ by

$$2\zeta = \frac{c}{\mu} \qquad (12.9.9)$$

Substituting Eqs. (12.9.3) and (12.9.8) into (12.9.2) gives

$$\ddot{r}_j + 2\zeta \dot{r}_j + p_j{}^2 r_j = \frac{(R_j)_a}{\mathcal{M}_j} \qquad j = 1, 2, \ldots, \infty \qquad (12.9.10)$$

It is observed that these uncoupled equations are identical to the Eqs. (11.10.8) for a discrete-mass system which has a damping matrix proportional to the inertia matrix. The general solution (11.10.9) and the conclusions stated in Sec. 11.10 are applicable here.

A second type of damping for which the equations of motion (12.9.2) decouple is known as strain-rate damping. Strain-rate damping occurs in materials for which the stress depends upon both the strain and strain rate, i.e., viscoelastic materials. The distributed damping force in this case is equal to $-\eta I \dot{u}_2{}^{iv}$, where η is the *coefficient of strain-rate damping* (see Prob. 12.6). The virtual work performed by the damping force during a distortion δu_2 of the beam is

$$\begin{aligned}
\delta W_E &= -\int_0^L \eta I \dot{u}_2{}^{iv}\, \delta u_2\, dx_1 \\
&= -\int_0^L \eta I \left(\sum_{k=1}^{\infty} \phi_k{}^{iv} \dot{r}_k \right) \left(\sum_{j=1}^{\infty} \phi_j\, \delta r_j \right) dx_1 \\
&= \sum_{j=1}^{\infty} \left(-\eta I \sum_{k=1}^{\infty} \dot{r}_k \int_0^L \phi_k{}^{iv} \phi_j\, dx_1 \right) \delta r_j \qquad (12.9.11)
\end{aligned}$$

from which it is seen that the generalized forces for strain-rate damping are

$$(R_j)_d = -\eta I \sum_{k=1}^{\infty} \dot{r}_k \int_0^L \phi_k{}^{iv} \phi_j\, dx_1 \qquad (12.9.12)$$

Recalling that $\phi_k{}^{iv} = (\mu p_k{}^2/EI)\phi_k$ [see Eq. (12.4.3)], and making use of the orthogonality conditions (12.4.7), we obtain

$$(R_j)_d = -2\gamma \mathcal{M}_j p_j{}^2 \dot{r}_j \qquad (12.9.13)$$

where the constant γ is defined as

$$2\gamma = \frac{\eta}{E} \qquad (12.9.14)$$

The equations of motion (12.9.2) then become

$$\ddot{r}_j + 2\gamma p_j{}^2 \dot{r}_j + p_j{}^2 r_j = \frac{(R_j)_a}{\mathcal{M}_j} \qquad j = 1, 2, \ldots, \infty \qquad (12.9.15)$$

These equations are identical to the Eqs. (11.10.12) for a discrete-mass system which has a viscous damping matrix proportional to the stiffness matrix. The solution (11.10.13) and the conclusions reached in Sec. 11.10 are therefore applicable to the case of strain-rate damping.

PROBLEMS

12.1 The quantity $\mathcal{L} = U - T + V_E$ appearing in Hamilton's principle (12.2.13) is called the *lagrangian*. Show that when the lagrangian is a function of n independent generalized displacements q_j and generalized velocities $\dot{q}_j$, that is, $\mathcal{L}(q_j, \dot{q}_j)$, application of Hamilton's principle yields the system of equations

$$\frac{\partial \mathcal{L}}{\partial q_j} - \frac{d}{dt} \left(\frac{\partial \mathcal{L}}{\partial \dot{q}_j} \right) = 0 \qquad j = 1, 2, \ldots, n$$

Then noting that $U = U(q_j)$, $T = T(q_j, \dot{q}_j)$, and $V_E = -\sum_{j=1}^{\infty} (Q_j)_a q_j$, show that the latter equations represent Lagrange's equations (11.2.19).

12.2 A shaft of length L has a uniform, circular cross section of area A and polar moment of inertia J. The material of the shaft has a mass density ρ and a shear modulus G. A distributed torque per unit length $\mathcal{T}(x_1, t)$ is applied as shown. Use Hamilton's principle to find the partial differential equation of motion for the angle of twist $\theta(x_1, t)$. Also derive the system's natural boundary conditions.

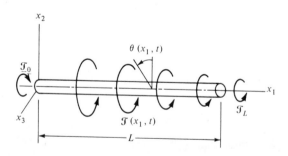

PROBLEM 12.2

12.3 A uniform elastic rod having mass per unit length μ supports a rigid mass m as shown. Using Hamilton's principle, show that: (a) the free longitudinal vibration of the bar is governed by the differential equation of motion $EAu_1'' = \mu \ddot{u}_1$; and (b) that the boundary condition at $x_1 = L$ is $EAu_1'(L, t) = -m\ddot{u}_1(L, t)$.

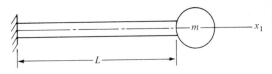

PROBLEM 12.3

12.4   A uniform elastic beam rests on flexible supports of stiffness k as shown. Using Hamilton's principle, verify that (a) the free flexural vibration of the beam is governed by the differential equation $EIu_2{}^{iv} + \mu\ddot{u}_2 = 0$; and (b) that the boundary conditions are $EIu_2''(0,t) = 0$, $EIu_2'''(0,t) = -ku_2(0,t)$, $EIu_2''(L,t) = 0$, and $EIu_2'''(L,t) = ku_2(L,t)$.

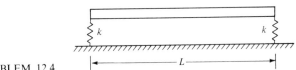

PROBLEM 12.4

12.5   A rod of uniform cross sectional area A and mass per unit length μ is subject to a distributed axial loading $F(x_1,t)$. The rod is constructed of a viscoelastic material having a stress-strain law of the form $\sigma_{11} = Ee_{11} + \eta\dot{e}_{11}$. Using Hamilton's principle in the form of Eq. (12.2.11), verify that the equation of motion for forced, longitudinal vibration is

$$EAu_1'' + \eta A\dot{u}_1'' - \mu\ddot{u}_1 + F(x_1,t) = 0$$

Also derive the admissible boundary conditions for the rod.

12.6   A beam of constant cross-sectional moment of inertia I and mass per unit length μ is subject to a distributed transverse force $F(x_1,t)$. The beam is constructed of a viscoelastic material having a stress-strain law of the form $\sigma_{11} = Ee_{11} + \eta\dot{e}_{11}$. Using Hamilton's principle in the form of Eq. (12.2.11), verify that the flexural vibration is governed by the differential equation

$$EIu_2{}^{iv} + \eta I\dot{u}_2{}^{iv} + \mu\ddot{u}_2 = F(x_1,t)$$

Also derive the admissible boundary conditions for the beam.

12.7   Determine the natural frequencies and the mode shapes for the longitudinal vibration of a uniform elastic bar having one end clamped and the other end free.

12.8   Compute the natural frequencies and the eigenfunctions for the torsional vibration of the shaft considered in Prob. 12.2. Assume that the shaft is clamped at one end and stress free at the other end.

12.9   (a) Verify that the frequency equation for the structure of Prob. 12.3 is $(pL/c)\tan pL/c - \mu L/m = 0$ where $c = \sqrt{EA/\mu}$; (b) compute the lowest three natural frequencies and the corresponding eigenfunctions for the case $m = \mu L/2$; and (c) show that the orthogonality conditions for the rod are:

$$\mu \int_0^L \phi_i \phi_j \, dx_1 + m\phi_i(L)\phi_j(L) = \begin{cases} 0 & i \neq j \\ \mathscr{M}_j & i = j \end{cases}$$

$$EA \int_0^L \phi_i' \phi_j' \, dx_1 = \begin{cases} 0 & i \neq j \\ \mathscr{M}_j p_j^2 & i = j \end{cases}$$

where

$$\mathscr{M}_j = \mu \int_0^L \phi_j^2 \, dx_1 + m\phi_j^2(L)$$

12.10 Determine the natural frequencies and the mode shapes for a uniform, simply supported beam.

12.11 Derive the frequency equation for a uniform beam which is stress-free at each end. Compute the three lowest natural frequencies.

12.12 Compute the fundamental frequency for the structure of Prob. 12.4, for the special case $k = EI/12L^3$.

12.13 Show that the eigenfunctions $\phi_i(x_1)$ for a variable section beam with any combination of clamped, free, simply supported, and guided ends satisfy the following orthogonality conditions:

$$\int_0^L \mu \phi_i \phi_j \, dx_1 = \begin{cases} 0 & i \neq j \\ \mathscr{M}_j & i = j \end{cases}$$

$$\int_0^L EI\phi_i'' \phi_j'' \, dx_1 = \begin{cases} 0 & i \neq j \\ \mathscr{M}_j p_j^2 & i = j \end{cases}$$

where

$$\mathscr{M}_j = \int_0^L \mu \phi_j^2 \, dx_1$$

12.14 The cross-sectional area of an elastic rod varies linearly from A_0 at $x_1 = 0$ to $0.5A_0$ at $x_1 = L$, as shown. Likewise the rod's mass per unit length varies linearly from μ_0 to $0.5\mu_0$. Use the Rayleigh-Ritz method in order to find approximate values of the two lowest natural frequencies of longitudinal vibration.

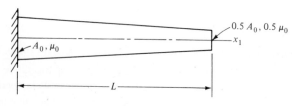

PROBLEM 12.14

12.15 The mass per unit length of the elastic cantilever beam shown is given by $\mu = \mu_0 (1 - x_1/L)$, and the moment of inertia of the cross section is $I = I_0 (1 - x_1/L)$. Apply the Rayleigh-Ritz method to find approximate values for the two lowest natural frequencies and mode shapes.

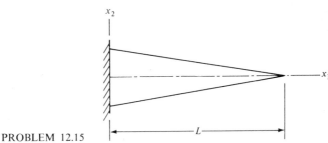

PROBLEM 12.15

12.16 Use the Rayleigh-Ritz method to determine the fundamental frequency and mode shape for the structure of Prob. 12.4. The stiffness of each support is given by $k = EI/12L^3$. *Hint*: Take $\psi_1 = 1$ and $\psi_2 = \sin(\pi x_1/L)$.

12.17 A uniform rod having clamped ends is disturbed from rest by the concentrated force $F(t)$ shown. Derive expressions for the displacement $u_1(x_1,t)$ and the axial stress $\sigma_{11}(x_1,t)$ in the rod.

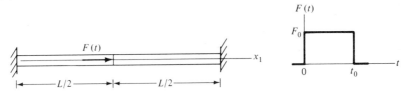

PROBLEM 12.17

12.18 The structure of Prob. 12.3 is excited by a step force $F(t) = F_0$ applied to the mass m in the x_1 direction. Assuming that the system is at rest initially, compute the maximum displacement u_1 and the maximum axial stress σ_{11} which result. Take $m = \mu L/2$, and consider just the first three normal modes (as determined in Prob. 12.9).

12.19 A torque $\mathcal{T}(t) = \mathcal{T}_0 \sin \omega t$ is applied to the free end of a circular shaft as shown. Derive expressions for the angle of twist and the torque in the shaft. Assume the system is at rest initially.

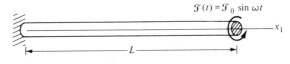

PROBLEM 12.19

12.20   A simply supported beam is subjected to the uniformly distributed load $F(t)$ shown. Assuming that the system is at rest initially, compute the maximum displacement and the maximum bending moment which occur. Compare your results with the corresponding static values.

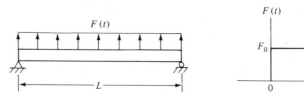

PROBLEM 12.20

12.21   A simply supported beam is subjected to the force $F(x_1,t) = -(Q_0 x_1/L)f(t)$ where $f(t)$ is a unit-step function. Assuming the beam is at rest initially, derive expressions for the displacement $u_2(x_1,t)$ and the bending moment $M(x_1,t)$.

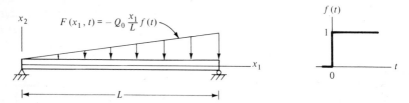

PROBLEM 12.21

12.22   A step force $F(t) = F_0$ is applied to the tip of the tapered cantilever beam of Prob. 12.15. Using the natural frequencies and eigenfunctions found by means of the Rayleigh-Ritz method, compute the displacement $u_2(x_1,t)$ of the beam. Assume the structure is at rest initially.

12.23   A step couple $M(t) = M_0$ acts at the center of a simply supported beam as shown. Assuming that the structure is at rest initially, compute the maximum displacement u_2 and the maximum rotation u_2' which result. Compare your results with the corresponding static values.

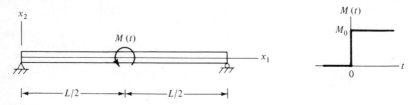

PROBLEM 12.23

12.24 A simply supported beam which is initially at rest experiences the half-sine acceleration-pulse foundation motion shown. (*a*) Derive an expression for the relative displacement $w_2(x_1,t)$. (*b*) Compute the maximum bending which occurs in the beam for the case $t_0 = \tau_1/4$, where $\tau_1 = 2\pi/p_1$ is the fundamental period of the system; assume that the maximum occurs at a time $t > t_0$.

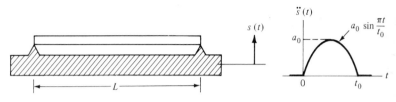

PROBLEM 12.24

12.25 Derive an expression for the relative displacement of the beam in Prob. 12.24 for the case of the " velocity shock " foundation motion $s = v_0 t$. *Hint*: The foundation's acceleration may be expressed as $\ddot{s}(t) = v_0 \, \delta(t)$, where $\delta(t)$ is the *Dirac delta function* defined by

$$\delta(t) = \lim_{\Delta \to 0} \begin{cases} 0 & \text{when} & t < -\dfrac{\Delta}{2} \\[2mm] \dfrac{1}{\Delta} & \text{when} & -\dfrac{\Delta}{2} < t < \dfrac{\Delta}{2} \\[2mm] 0 & \text{when} & t > \dfrac{\Delta}{2} \end{cases}$$

The integral of the delta function is the unit step function, that is,

$$\int_{-\infty}^{t} \delta(\xi) \, d\xi = \begin{cases} 0 & t < 0 \\ 1 & t > 0 \end{cases}$$

12.26 The distributed damping force in a rod exhibiting strain-rate damping is equal to $\eta A \dot{u}_1'$ (see Prob. 12.5). Show that the equations of motion governing the longitudinal, forced vibration of such a rod decouple when normal coordinates are used.

12.27 A uniform, simply supported beam is subject to a force $F(t) = F_0 \sin \omega t$, as shown. Investigate the beam's steady-state response for the case of (*a*) zero damping, (*b*) viscous damping, and (*c*) strain-rate damping. In particular, derive an expression for the amplitude of the steady-state vibration which occurs when ω is equal to one of the beam's natural circular frequencies p_J.

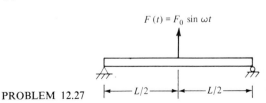

PROBLEM 12.27

REFERENCES

For a more detailed account of Hamilton's principle and its applications in structural dynamics:

12.1 LANGHAAR, H. L.: "Energy Methods in Applied Mechanics," Wiley, New York, 1962.

12.2 FUNG, Y. C.: "Foundations of Solid Mechanics," Prentice-Hall, Englewood Cliffs, N.J., 1965.

For a thorough treatment of the dynamic response of continuous structures, including bars, beams, plates, shells, etc.:

12.3 TIMOSHENKO, S. and D. H. YOUNG; "Vibration Problems in Engineering," 3d ed., Van Nostrand, Princeton, N.J., 1955.

12.4 YOUNG, D.: Continuous Systems, in W. Flügge (ed.), "Handbook of Engineering Mechanics," McGraw-Hill, New York, 1962.

12.5 NOWACKI, W.: "Dynamics of Elastic Systems," Wiley, New York, 1963.

For an account of the propagation of stress waves in elastic solids:

12.6 KOLSKY, H.: "Stress Waves in Solids," Dover, New York, 1963.

For a discussion of the dynamic response of inelastic continuous structures see Ref. 12.5; also:

12.7 ROGERS, G. L.: "Dynamics of Framed Structures," Wiley, New York, 1959.

APPENDIX **A**

CARTESIAN TENSORS

A.1 INTRODUCTION

Vector notation provides a convenient means for describing many physical quantities and laws, as students of particle mechanics are well aware. However, in the study of deformable media there are important quantities of a more complex nature, such as stress and strain, which cannot adequately be described by vectors. *Tensors*, which represent a generalization of vectors, offer a suitable way of representing these quantities. Like vectors, tensors are independent of any particular frame of reference. But also like vectors, a tensor quantity is most conveniently described by specifying its *components* in an appropriate system of coordinates. Here we shall restrict our attention to cartesian coordinate systems and a discussion of *cartesian tensors*. A few simple conventions in notation will be introduced at the outset, in order that the equations governing tensors may be written in a simple and concise fashion.

A.2 INDEX NOTATION

Consider a set of three independent quantities which might, for example, represent the coordinates of a point or the components of a vector. We may denote this set by a single subscripted symbol. That is, we may write the three quantities as (a_1, a_2, a_3), or simply a_i, where it is understood that the *index* or subscript i can take the values 1, 2, and 3.

Similarly, we can denote any set of 9 quantities by a single letter carrying two different subscripts, for example b_{ij}. Since both the indices i and j can take on the values 1, 2, and 3, it is clear that b_{ij} does indeed represent 9 quantities, namely, b_{11}, b_{12}, b_{13}, b_{21}, b_{22}, b_{23}, b_{31}, b_{32}, b_{33}. Likewise it should be clear that a set of 27 quantities can be denoted by a_{ijk}; a set of 81 quantities can be represented by b_{ijkl}; and so forth.

In order to shorten the writing of equations involving subscripted quantities, we adopt the following conventions:

Range convention Whenever a subscript appears only once in a term, the subscript takes all the values 1, 2, and 3.

Summation convention Whenever a subscript appears twice in the same term, the repeated index is summed from 1 to 3. No index may be repeated more than once.

The following examples serve to illustrate these conventions.

$$a_i b_i = \sum_{i=1}^{3} a_i b_i = a_1 b_1 + a_2 b_2 + a_3 b_3$$

$$
\begin{aligned}
a_{ij} b_i c_j = \sum_{i=1}^{3} \sum_{j=1}^{3} a_{ij} b_i c_j = {} & a_{11} b_1 c_1 + a_{12} b_1 c_2 + a_{13} b_1 c_3 \\
& + a_{21} b_2 c_1 + a_{22} b_2 c_2 + a_{23} b_2 c_3 \qquad \text{(A.2.1)} \\
& + a_{31} b_3 c_1 + a_{32} b_3 c_2 + a_{33} b_3 c_3
\end{aligned}
$$

$$a_{ij} b_j = \sum_{j=1}^{3} a_{ij} b_j = a_{i1} b_1 + a_{i2} b_2 + a_{i3} b_3$$

The first two expressions contain only repeated indices, and application of the summation convention thus indicates that the terms $a_i b_i$ and $a_{ij} b_i c_j$ represent single quantities. In the last example, the summation is carried out on the repeated index j, while the unrepeated index i remains free to take the values 1, 2, 3 (by virtue of the range convention). For this reason an unrepeated index is sometimes called a *free* index. The number of free indices determines how many quantities are represented by a symbol. Thus $a_{ij} b_j$ stands for $3^1 = 3$ quantities (1 free index); $a_{ij} b_{jk}$ represents $3^2 = 9$ quantities (2 free indices); $a_{ij} b_k$ stands for $3^3 = 27$ quantities (3 free indices); etc.

Repeated subscripts are sometimes called *dummy* indices, since the particular letter used in the subscript is immaterial. The student can verify that

$$a_i b_i = a_l b_l$$

$$a_{ij} b_{jk} = a_{in} b_{nk} \qquad \text{(A.2.2)}$$
$$\text{etc.}$$

To shorten the writing of expressions involving differentiation with respect to a set of coordinates axes x_i, we introduce the following convention:

Comma convention A subscript comma followed by an index i indicates partial differentiation with respect to each coordinate x_i.

For example,

$$a_{i,j} = \frac{\partial a_i}{\partial x_j}$$

$$b_{ij,k} = \frac{\partial b_{ij}}{\partial x_k} \qquad \text{(A.2.3)}$$

$$c_{ij,kl} = \frac{\partial^2 c_{ij}}{\partial x_k \, \partial x_l}$$

where $a_{i,j}$ represents 9 quantities $(\partial a_1/\partial x_1, \ \partial a_1/\partial x_2, \ \ldots, \ \partial a_3/\partial x_3)$; $b_{ij,k}$ represents 27 quantities $(\partial b_{11}/\partial x_1, \ \partial b_{11}/\partial x_2, \ \partial b_{12}/\partial x_1, \ \ldots, \ \partial b_{33}/\partial x_3)$; and $c_{ij,kl}$ denotes 81 quantities $(\partial^2 c_{11}/\partial x_1{}^2, \ \partial^2 c_{11}/\partial x_1 \, \partial x_2, \ \partial^2 c_{12}/\partial x_1{}^2, \ \ldots, \ \partial^2 c_{33}/\partial x_3{}^2)$. The summation and range conventions apply to indices following the comma as well. Hence

$$a_{i,i} = \frac{\partial a_i}{\partial x_i} = \frac{\partial a_1}{\partial x_1} + \frac{\partial a_2}{\partial x_2} + \frac{\partial a_3}{\partial x_3}$$

$$b_{ij,j} = \frac{\partial b_{ij}}{\partial x_j} = \frac{\partial b_{i1}}{\partial x_1} + \frac{\partial b_{i2}}{\partial x_2} + \frac{\partial b_{i3}}{\partial x_3} \qquad \text{(A.2.4)}$$

$$b_{i,jj} = \frac{\partial^2 b_i}{\partial x_j \, \partial x_j} = \frac{\partial^2 b_i}{\partial x_1{}^2} + \frac{\partial^2 b_i}{\partial x_2{}^2} + \frac{\partial^2 b_i}{\partial x_3{}^2}$$

A.3   TRANSFORMATION OF COORDINATES

Let (x_1, x_2, x_3) and (x'_1, x'_2, x'_3) be two cartesian coordinate systems having a common origin 0, as shown in Fig. A.1. A point P with coordinates x_i in the first system will have the coordinates x'_i in the second system given by the linear transformation

$$x'_1 = \alpha_{11}x_1 + \alpha_{12}x_2 + \alpha_{13}x_3$$

$$x'_2 = \alpha_{21}x_1 + \alpha_{22}x_2 + \alpha_{23}x_3 \qquad \text{(A.3.1)}$$

$$x'_3 = \alpha_{31}x_1 + \alpha_{32}x_2 + \alpha_{33}x_3$$

The nine quantities α_{ij}, called the *direction cosines*, represent the cosines of the angles between the ith primed axis and the jth unprimed axis, or

$$\alpha_{ij} = \cos(x'_i, x_j) \qquad \text{(A.3.2)}$$

Using our index notation, the three transformation equations (A.3.1) can be written in the compact form

$$x'_i = \alpha_{ij}x_j \qquad \text{(A.3.3)}$$

Consider now the position vector $\mathbf{r}$ of the point P shown in Fig. A.1. The components of $\mathbf{r}$ are x_i in the unprimed system and x'_i in the primed system. Eq. (A.3.3) gives the transformation relation between the two sets of components. It can be shown

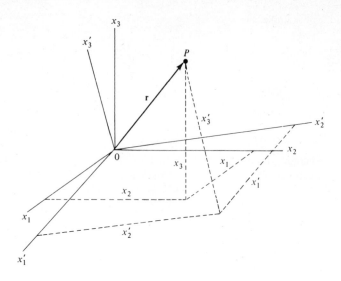

FIGURE A.1

that the components of any other vector will obey the same transformation law. It should be remembered that while a vector quantity is independent of the choice of the coordinate system used, its components are different for different coordinate systems and are related by this transformation rule. We now can generalize the concept of vector transformation to higher order quantities called tensors.

A.4   TENSOR OF FIRST ORDER

We now proceed to define a tensor of first order, or a vector, as follows:

> *A tensor of first order* is a set of $3^1 = 3$ quantities which, when transformed from one set of axes x_i to another set x_i', obeys the following transformation law
>
> $$A_i' = \alpha_{ij} A_j \qquad (A.4.1)$$

Here A_i are the components of the tensor in the x_i coordinate system, A_i' are the components of the same tensor in the x' system, and α_{ij} are the direction cosines defined by Eq. (A.3.2). We have already seen that the components of the position vector $\mathbf{r}$ transform according to this law and hence represent a tensor of first order. Other examples include the components of a force, displacement, or velocity.

A.5   TENSORS OF HIGHER ORDER

By analogy with a tensor of order 1, we define a tensor of second order as follows:

A *tensor of second order* is a set of $3^2 = 9$ quantities A_{ij} referred to coordinate axes x_i which transform to another set of axes x_i' by the law

$$A_{ij}' = \alpha_{ik} \alpha_{jl} A_{kl} \qquad \text{(A.5.1)}$$

Examples of second-order tensors are the components of stress σ_{ij} and strain e_{ij}. In fact, the name tensor comes from its historical association with stress (tension). The well-known Mohr circle construction of these quantities is nothing more than a graphical representation of the transformation rule (A.5.1). The student may verify that the moments and products of inertia I_{ij} also form a second-order tensor. Each of these examples represents a special class of tensors, namely, *symmetric tensors*. A tensor A_{ij} is symmetric if $A_{ij} = A_{ji}$ for all i and j.

Third and higher order tensors may be defined in an analogous way; i.e.,

A *tensor of third order* is a set of $3^3 = 27$ quantities A_{ijk} referred to coordinate axes x_i which transform to another set x_i' by the law

$$A_{ijk}' = \alpha_{il} \alpha_{jm} \alpha_{kn} A_{lmn} \qquad \text{(A.5.2)}$$

A.6   THE KRONECKER DELTA AND THE PERMUTATION SYMBOL

A special second-order tensor, which will prove very useful in dealing with tensors, is the *Kronecker delta* δ_{ij} defined as

$$\delta_{ij} = \begin{cases} 1 & \text{if} \quad i = j \\ 0 & \text{if} \quad i \neq j \end{cases} \qquad \text{(A.6.1)}$$

Using the summation and range conventions, the student may verify that

$$\delta_{ij} A_j = A_i$$
$$\delta_{ij} B_{jk} = B_{ik} \qquad \text{(A.6.2)}$$
$$\delta_{ij} C_{jkl} = C_{ikl}$$
$$\text{etc.}$$

From these expressions it is seen that the application of δ_{ij} to another tensor has the effect of replacing the dummy index j by the free index i in the tensor being operated upon by δ_{ij}. Hence δ_{ij} is sometimes referred to as the substitution tensor.

A useful third-order tensor, called the *permutation symbol*, is defined as follows:

$$\epsilon_{ijk} = \begin{cases} 0 & \text{if any two of } i, j, k \text{ are equal} \\ +1 & \text{if } i, j, k \text{ are an even permutation of 1, 2, 3} \\ -1 & \text{if } i, j, k \text{ are an odd permutation of 1, 2, 3} \end{cases} \qquad \text{(A.6.3)}$$

Hence $\epsilon_{123} = \epsilon_{231} = \epsilon_{312} = +1$ and $\epsilon_{213} = \epsilon_{132} = \epsilon_{321} = -1$. The permutation tensor provides a convenient way for expressing the vector cross product in index notation (see Prob. A.9).

A.7 TENSOR OPERATIONS

It can be shown that the *sum* or *difference* of two tensors of the same order is a tensor, also of the same order. For example, if A_{ij} and B_{ij} are tensors, then

$$C_{ij} = A_{ij} + B_{ij} \qquad (A.7.1)$$

where C_{ij} is a second-order tensor. A tensor equation such as (A.7.1) which is true in one coordinate system is necessarily true in all systems. Note, also, that in order for a tensor equation to be meaningful, each term in it must contain the same free indices.

Next consider the two tensors A_i and B_{jk}. We may define a new set of quantities C_{ijk} by a process called tensor *multiplication*:

$$C_{ijk} = A_i B_{jk} \qquad (A.7.2)$$

where C_{ijk} is a third-order tensor. In general, tensor multiplication yields a new tensor whose order is the sum of the orders of the tensors being multiplied. This rule holds also for multiplication of a tensor by a *scalar*, if we define a scalar as a *tensor of order* 0. That is, if λ is any scalar quantity, the product λA_{ij} is the tensor of order 2 found by multiplying each component of A_{ij} by λ.

Contraction is a process whereby the order of a tensor is reduced by equating two of its indices and invoking the summation convention. This occurs, for example, when we multiply a tensor by the Kronecker delta as follows:

$$\delta_{ij} A_{ijk} = A_{jjk} \qquad (A.7.3)$$

Note that the tensor A_{jjk} resulting from the contraction is a first-order tensor since it has only one free index. Thus we see that contraction reduces the order of a tensor by two.

PROBLEMS

A.1 Verify Eq. (A.2.2).

A.2 Verify Eq. (A.3.1).

A.3 Verify Eq. (A.6.2).

A.4 If $A_i' = \alpha_{ij} A_j$, where α_{ij} is defined by Eq. (A.3.2), verify that $A_i = \alpha_{ji} A_j'$.

A.5 Show that
   (a) $\delta_{ij} \delta_{ij} = 3$
   (b) $\delta_{ij} \delta_{jk} = \delta_{ik}$

(c) $\epsilon_{ijk}\epsilon_{kji} = -6$

(d) $\epsilon_{ijk}A_jA_k = 0$

A.6 Prove that $\alpha_{ik}\alpha_{jk} = \delta_{ij}$. (Verify this equation for the two special cases: $i = j = 1$, and $i = 1, j = 2$; the general result follows naturally.)

A.7 Prove that if we define $\delta_{ij} = \delta'_{ij}$, then δ_{ij} is a second-order tensor.

A.8 Demonstrate that the scalar product of two vectors, $\mathbf{A} \cdot \mathbf{B}$, can be written in index notation as A_iB_i.

Also show that the divergence of a vector $\nabla \cdot \mathbf{A}$ (where ∇ is the differential operator) can be written as $A_{i,i}$.

A.9 Demonstrate that the components of the cross product $\mathbf{C} = \mathbf{A} \times \mathbf{B}$ may be expressed in index notation as $C_i = \epsilon_{ijk}A_jB_k$. Verify that the components of the curl of a vector $\mathbf{D} = \nabla \times \mathbf{A}$ can be written as $D_i = \epsilon_{ijk}A_{k,j}$.

A.10 Show that the *divergence theorem* $\int_{\mathscr{V}} \nabla \cdot \mathbf{A}\, d\mathscr{V} = \int_{\mathscr{S}} \mathbf{n} \cdot \mathbf{A}\, d\mathscr{S}$ (where $\mathbf{n}$ is the unit outward normal to an element of the surface $\mathscr{S}$ which bounds the volume $\mathscr{V}$) may be written in index notation as

$$\int_{\mathscr{V}} A_{i,i}\, d\mathscr{V} = \int_{\mathscr{S}} n_iA_i\, d\mathscr{S}$$

A.11 Prove that $C_{ij} = A_{ij} + B_{ij}$ is a tensor if A_{ij} and B_{ij} are.

A.12 Prove that $C_{ij} = A_iB_j$ is a tensor of order 2 if A_i and B_j are tensors of order 1.

A.13 Prove that $C_{ijk} = A_iB_{jk}$ is a tensor of order 3 if A_i and B_{jk} are tensors of order 1 and 2 respectively.

A.14 Prove that if A_i and B_i are first-order tensors, and $A_i = C_{ij}B_j$, then C_{ij} is a second-order tensor.

A.15 Show that $A_{i,j}$ is a second-order tensor if A_i is a first-order tensor.

REFERENCES

For an elementary account of cartesian tensors and their applications in mechanics:

A.1 MYKLESTAD, N. O.: "Cartesian Tensors," Van Nostrand, Princeton, N.J., 1967.

For a more detailed treatment of general tensor analysis:

A.2 SYNGE, J. L., and A. SCHILD: "Tensor Calculus," University of Toronto Press, 1949.

A.3 SOKOLNIKOFF, I. S.: "Tensor Analysis," 2d ed., Wiley, New York, 1964.

APPENDIX B

MATRICES

B.1 INTRODUCTION

Matrices provide a concise and meaningful way of representing systems of linear algebraic or differential equations. When a matrix formulation is used, the algebra or calculus involved in the solutions of the equations can be performed efficiently on a digital computer. Consequently matrices have assumed an important role in engineering analysis.

A brief introduction to matrix algebra is given in this appendix. Only those definitions and operations which are prerequisite to an understanding of the matrix methods presented in Chaps. 7 and 11 have been included.

B.2 DEFINITIONS AND NOTATIONS

A *matrix* is defined as a rectangular array of elements. The elements are generally either numbers or symbols which represent numbers, although in some instances they are functions of certain variables. The array is often denoted by a single letter within brackets [ ] as follows:

$$[A] = \begin{bmatrix} A_{11} & A_{12} & \cdots & A_{1n} \\ A_{21} & A_{22} & \cdots & A_{2n} \\ \cdots & \cdots & \cdots & \cdots \\ A_{m1} & A_{m2} & \cdots & A_{mn} \end{bmatrix} \qquad (B.2.1)$$

Note that the first and second subscripts applied to an element denote, respectively, the *row* and *column* in which the element appears.

A matrix having m rows and n columns is said to be of *order* (m,n).

The matrix formed by interchanging the rows and columns of a given matrix is called the *transpose* of that matrix. A superscript T will be used to denote transposition. For example, the transpose of the matrix $[A]$ in Eq. (B.2.1) is given by

$$[A]^T = \begin{bmatrix} A_{11} & A_{21} & \cdots & A_{m1} \\ A_{12} & A_{22} & \cdots & A_{m2} \\ \cdots & \cdots & \cdots & \cdots \\ A_{1n} & A_{2n} & \cdots & A_{mn} \end{bmatrix} \qquad (B.2.2)$$

It is evident that the order of the transposed matrix is (n,m) if the original matrix is of order (m,n).

If a matrix has just one row, it is called a *row matrix*. The position of an element in such a matrix can be designated by means of a single subscript. Thus a row matrix of order $(1,n)$ is usually written in the form

$$[B] = [B_1 \quad B_2 \quad \cdots \quad B_n] \qquad (B.2.3)$$

A matrix consisting of a single column is referred to as a *column matrix*. A column matrix of order $(n,1)$ is expressed in the form

$$\{C\} = \begin{bmatrix} C_1 \\ C_2 \\ \cdots \\ C_n \end{bmatrix} \qquad (B.2.4)$$

where braces { } are used to distinguish the column matrix from a row matrix.

When an array has the same number of rows and columns, it is called a *square matrix*. The elements of a square matrix situated along an imaginary line joining the upper left and lower right corners of the array (i.e., the elements $A_{11}, A_{22}, \ldots, A_{nn}$ of the matrix $[A]$) constitute the *principal diagonal* of the matrix.

A square matrix whose elements A_{ij} are symmetric with respect to the principal diagonal, so that $A_{ij} = A_{ji}$, is called a *symmetric matrix*. Note that a symmetric matrix $[A]$ has the property that $[A] = [A]^T$.

If all the " off diagonal " terms of a square matrix are identically zero, the matrix is said to be a *diagonal matrix*. An example of a diagonal matrix of order (n,n) is the following:

$$[D] = \begin{bmatrix} D_{11} & 0 & \cdots & 0 \\ 0 & D_{22} & \cdots & 0 \\ \cdots & \cdots & \cdots & \cdots \\ 0 & 0 & \cdots & D_{nn} \end{bmatrix} \qquad (B.2.5)$$

A diagonal matrix in which all the diagonal elements are equal to unity is called a unit matrix and is denoted by the symbol $[I]$; that is,

$$[I] = \begin{bmatrix} 1 & 0 & \cdots & 0 \\ 0 & 1 & \cdots & 0 \\ \cdots & \cdots & \cdots & \cdots \\ 0 & 0 & \cdots & 1 \end{bmatrix} \qquad \text{(B.2.6)}$$

If *all* the elements of a matrix are identically zero, then the matrix is termed a *zero* or *null* matrix. A null matrix is denoted by [0], or {0} in the case of a column matrix.

B.3   MATRIX OPERATIONS

Two matrices of the same order are defined to be *equal* if their corresponding elements are identical. In other words

$$[A] = [B] \qquad \text{(B.3.1)}$$

if

$$A_{ij} = B_{ij} \qquad \text{(B.3.2)}$$

The *addition* (or *subtraction*) of two or more matrices of the same order is defined as the matrix obtained by adding (or subtracting) the corresponding elements of the original matrices. For example, the matrix expression

$$[D] = [A] + [B] - [C] \qquad \text{(B.3.3)}$$

implies that

$$D_{ij} = A_{ij} + B_{ij} - C_{ij} \qquad \text{(B.3.4)}$$

The *multiplication of a matrix by a scalar* is defined as the operation of multiplying each element of the matrix by the same scalar quantity. In other words the matrix equation.

$$[B] = c[A] \qquad \text{(B.3.5)}$$

implies that

$$B_{ij} = cA_{ij} \qquad \text{(B.3.6)}$$

Two or more matrices can be multiplied together according to an operation known as *matrix multiplication* providing the matrices are *conformable*. Two matrices are said to be conformable if the number of columns of the first matrix is equal to the number of rows of the second matrix. Consider for example a matrix $[A]$ of order (m,n) and a second matrix $[B]$ of order (p,q). The product $[A][B]$ is not defined unless $n = p$. For $n = p$, the product is a matrix $[C]$ of order (m,q) given by

$$[C] = [A][B] \qquad \text{(B.3.7)}$$

in which the elements of $[C]$ are, by definition,

$$C_{ij} = \sum_{k=1}^{p} A_{ik} B_{kj} \qquad \text{(B.3.8)}$$

Referring to Eq. (B.3.7) we say that $[B]$ is *premultiplied* by $[A]$; alternatively we may say that $[A]$ is *postmultiplied* by $[B]$.

Similarly, a matrix $[A]$ of order (m,n) may be postmultiplied by a column matrix $\{B\}$ of order $(n,1)$. The result in this case is a column matrix of order $(m,1)$; that is,

$$\{C\} = [A]\{B\} \qquad (B.3.9)$$

in which

$$C_i = \sum_{j=1}^{n} A_{ij} B_j \qquad (B.3.10)$$

From the above definitions it can be shown that matrix multiplication is *associative*; that is,

$$([A][B])[C] = [A]([B][C]) \qquad (B.3.11)$$

distributive

$$[A]([B] + [C]) = [A][B] + [A][C] \qquad (B.3.12)$$

but in general *not commutative*

$$[A][B] \neq [B][A] \qquad (B.3.13)$$

Another important matrix operation is *inversion*. The *inverse* of a matrix is defined as the matrix which, when multiplied by the original matrix, yields the unit diagonal matrix. Denoting the inverse of $[A]$ by $[A]^{-1}$, we have

$$[A][A]^{-1} = [A]^{-1}[A] = [I] \qquad (B.3.14)$$

It can be proved that the inverse of a matrix exists only if the determinant of the co-efficients A_{ij} of the matrix is nonzero; that is, if $|A| \neq 0$. When $|A| = 0$, the matrix is said to be *singular*.

The inverse of a given matrix $[A]$ can be computed using the relation

$$[A]^{-1} = \frac{1}{|A|} \operatorname{adj}[A] \qquad (B.3.15)$$

where $\operatorname{adj}[A]$ denotes the *adjoint matrix* of $[A]$, defined as the transpose of the matrix of the cofactors of $|A|$.† For example, suppose that we wish to invert the third-order matrix

$$[A] = \begin{bmatrix} A_{11} & A_{12} & A_{13} \\ A_{21} & A_{22} & A_{23} \\ A_{31} & A_{32} & A_{33} \end{bmatrix} \qquad (B.3.16)$$

The adjoint matrix is

$$
\operatorname{adj}[A] = \begin{bmatrix} A_{22}A_{33} - A_{32}A_{23} & -(A_{21}A_{33} - A_{31}A_{23}) & A_{21}A_{32} - A_{31}A_{22} \\ -(A_{12}A_{33} - A_{32}A_{13}) & A_{11}A_{33} - A_{31}A_{13} & -(A_{11}A_{32} - A_{31}A_{12}) \\ A_{12}A_{23} - A_{22}A_{13} & -(A_{11}A_{23} - A_{21}A_{13}) & A_{11}A_{22} - A_{21}A_{12} \end{bmatrix}^T
$$
$$
= \begin{bmatrix} A_{22}A_{33} - A_{32}A_{23} & -(A_{12}A_{33} - A_{32}A_{13}) & A_{12}A_{23} - A_{22}A_{13} \\ -(A_{21}A_{33} - A_{31}A_{23}) & A_{11}A_{33} - A_{31}A_{13} & -(A_{11}A_{23} - A_{21}A_{13}) \\ A_{21}A_{32} - A_{31}A_{22} & -(A_{11}A_{32} - A_{31}A_{12}) & A_{11}A_{22} - A_{21}A_{12} \end{bmatrix}
$$
$$(B.3.17)$$

---

†The *cofactor* C_{ij} of a determinant $|A|$ is related to the *minor* M_{ij} of the determinant by the relation $C_{ij} = (-1)^{i+j}M_{ij}$. The minor M_{ij} is defined as the determinant formed by eliminating the ith row and jth column from the original determinant $|A|$. For a detailed discussion of this and other inversion techniques, the reader is referred to the references listed at the end of this appendix.

Multiplication of this matrix by the reciprocal of the determinant $|A|$ yields the required inverse matrix $[A]^{-1}$.

It follows from the definition of the inverse matrix that the inverse of the product of two or more square, nonsingular matrices is equal to the product of the inverses of the matrices taken in the reverse order. For example

$$([A][B][C])^{-1} = [C]^{-1}[B]^{-1}[A]^{-1} \qquad (B.3.18)$$

A similar rule holds for the transpose of the product of several matrices; namely,

$$([A][B][C])^{T} = [C]^{T}[B]^{T}[A]^{T} \qquad (B.3.19)$$

Equations (B.3.18) and (B.3.19) are known as the *reversal laws* of inversion and transposition, respectively.

Finally, the *derivative of a matrix* whose elements are functions of a variable, say t, is defined as the matrix of the derivatives of the elements. That is,

$$\frac{d}{dt}[A] = \begin{bmatrix} \dfrac{dA_{11}}{dt} & \dfrac{dA_{12}}{dt} & \cdots & \dfrac{dA_{1n}}{dt} \\ \dfrac{dA_{21}}{dt} & \dfrac{dA_{22}}{dt} & \cdots & \dfrac{dA_{2n}}{dt} \\ \cdots & \cdots & \cdots & \cdots \\ \dfrac{dA_{m1}}{dt} & \dfrac{dA_{m2}}{dt} & \cdots & \dfrac{dA_{mn}}{dt} \end{bmatrix} \qquad (B.3.20)$$

From this definition it follows that

$$\frac{d}{dt}([A] + [B]) = \frac{d}{dt}[A] + \frac{d}{dt}[B] \qquad (B.3.21)$$

and

$$\frac{d}{dt}([A][B]) = \frac{d}{dt}[A][B] + [A]\frac{d}{dt}[B] \qquad (B.3.22)$$

B.4 SOLUTIONS OF LINEAR EQUATIONS

Consider the system of linear algebraic equations:

$$A_{11}x_1 + A_{12}x_2 + \cdots + A_{1n}x_n = B_1$$
$$A_{21}x_1 + A_{22}x_2 + \cdots + A_{2n}x_n = B_2 \qquad (B.4.1)$$
$$\vdots$$
$$A_{n1}x_1 + A_{n2}x_2 + \cdots + A_{nn}x_n = B_n$$

in which A_{ij} and B_i represent known quantities and x_i are unknowns to be determined. These equations may be written in matrix form as

$$\begin{bmatrix} A_{11} & A_{12} & \cdots & A_{1n} \\ A_{21} & A_{22} & \cdots & A_{2n} \\ \cdots & \cdots & \cdots & \cdots \\ A_{n1} & A_{n2} & \cdots & A_{nn} \end{bmatrix} \begin{bmatrix} x_1 \\ x_2 \\ \cdots \\ x_n \end{bmatrix} = \begin{bmatrix} B_1 \\ B_2 \\ \cdots \\ B_n \end{bmatrix} \qquad (B.4.2)$$

or simply as

$$[A]\{x\} = \{B\} \qquad \text{(B.4.3)}$$

Providing that the coefficient matrix $[A]$ is nonsingular, Eq. (B.4.3) may be solved for $\{x\}$ by premultiplying each side of the equation by the inverse of $[A]$. That is,

$$[A]^{-1}[A]\{x\} = [A]^{-1}\{B\}$$
$$[I]\{x\} = [A]^{-1}\{B\} \qquad \text{(B.4.4)}$$
$$\{x\} = [A]^{-1}\{B\}$$

When the quantities B_i on the right hand side of Eq. (B.4.1) are zero, the system of equations is said to be *homogeneous*. In this case Eq. (B.4.3) reduces to

$$[A]\{x\} = \{0\} \qquad \text{(B.4.5)}$$

If the matrix $[A]$ is nonsingular so that its inverse $[A]^{-1}$ exists, then

$$\{x\} = [A]^{-1}\{0\} = \{0\} \qquad \text{(B.4.6)}$$

A solution such as (B.4.6) in which all of the unknowns are zero is called a *trivial solution*.

A nontrivial solution to the system of homogeneous equations (B.4.5) occurs only when the matrix of coefficients $[A]$ is singular, that is, when $|A| = 0$. From the theory of determinants it can be shown that there exists an infinite number of solutions in this case.

A matrix formulation is useful in the solution of simultaneous differential as well as algebraic equations. Consider for example the following set of first-order differential equations:

$$A_{11}x_1 + A_{12}x_2 + \cdots + A_{1n}x_n = B_{11}\dot{x}_1 + B_{12}\dot{x}_2 + \cdots + B_{1n}\dot{x}_n$$
$$A_{21}x_1 + A_{22}x_2 + \cdots + A_{2n}x_n = B_{21}\dot{x}_1 + B_{22}\dot{x}_2 + \cdots + B_{2n}\dot{x}_n \qquad \text{(B.4.7)}$$
$$\vdots$$
$$A_{n1}x_1 + A_{n2}x_2 + \cdots + A_{nn}x_n = B_{n1}\dot{x}_1 + B_{n2}\dot{x}_2 + \cdots + B_{nn}\dot{x}_n$$

in which A_{ij} and B_{ij} are known constants, x_i represent unknown functions of some real variable t, and $\dot{x}_i$ denote the derivatives of x_i with respect to t. The equations can be written in matrix form as

$$[A]\{x\} = [B]\{\dot{x}\} \qquad \text{(B.4.8)}$$

To obtain the solution to Eq. (B.4.8) we let

$$\{x\} = \{X\}e^{\lambda t} \qquad \text{(B.4.9)}$$

in which X_i are arbitrary constants and λ is a scalar to be determined. Equation (B.4.8) then yields

$$([A] - \lambda[B])\{X\} = \{0\} \qquad \text{(B.4.10)}$$

If either of the coefficient matrices $[A]$ or $[B]$ is nonsingular, Eq. (B.4.10) can be reduced to a so called "eigenvalue problem." For example, if $[B]^{-1}$ exist, then Eq. (B.4.10) may be premultiplied by $[B]^{-1}$ to obtain

$$([D] - \lambda[I])\{X\} = \{0\} \qquad \text{(B.4.11)}$$

where

$$[D] = [B]^{-1}[A] \qquad \text{(B.4.12)}$$

Equation (B.4.11) constitutes the eigenvalue problem of the matrix $[D]$. The solution to such a problem is discussed in the following section.

B.5 EIGENVALUE PROBLEMS

Many engineering problems lead to a system of equations of the form

$$
\begin{aligned}
D_{11}X_1 + D_{12}X_2 + \cdots + D_{1n}X_n &= \lambda X_1 \\
D_{21}X_1 + D_{22}X_2 + \cdots + D_{2n}X_n &= \lambda X_2 \\
&\vdots \\
D_{n1}X_1 + D_{n2}X_2 + \cdots + D_{nn}X_n &= \lambda X_n
\end{aligned}
\qquad \text{(B.5.1)}
$$

where D_{ij} are known coefficients, λ is a scalar quantity, and X_i are the unknowns. The eigenvalue problem (B.5.1) can be expressed in matrix form as

$$[D]\{X\} = \lambda\{X\} \qquad \text{(B.5.2)}$$

or

$$([D] - \lambda[I])\{X\} = \{0\} \qquad \text{(B.5.3)}$$

According to the discussion in the preceding section, the system of homogeneous equations (B.5.3) will possess a nontrivial solution only if the coefficient matrix $[D] - \lambda[I]$ is singular. Thus

$$|[D] - \lambda[I]| = 0 \qquad \text{(B.5.4)}$$

Expansion of the determinant in Eq. (B.5.4) yields an nth order polynomial in λ. The *characteristic equation* (B.5.4) therefore possesses n roots, $\lambda_1, \lambda_2, \ldots, \lambda_n$. These roots are referred to as the eigenvalues of the matrix $[D]$. Corresponding to each eigenvalue, say λ_j, there exists an eigenvector $\{X\}_j$ which represents the solution to the matrix equation

$$([D] - \lambda_j[I])\{X\}_j = \{0\} \qquad \text{(B.5.5)}$$

Since the set of equations (B.5.5) is homogeneous, the eigenvector $\{X\}_j$ can be determined only to within a constant factor. That is, only the ratios of the elements of the eigenvector $\{X\}_j$ are uniquely determined. Often it is convenient to take one of the elements equal to unity, and then to determine the remaining elements using Eq. (B.5.5).

PROBLEMS

B.1 Given the matrices

$$[A] = \begin{bmatrix} 0 & 5 \\ -1 & 6 \end{bmatrix} \qquad [B] = \begin{bmatrix} 2 & 2 \\ 1 & -1 \end{bmatrix} \qquad \{C\} = \begin{bmatrix} 1 \\ 3 \end{bmatrix}$$

compute:
  (*a*) $[A] - 3[B]$
  (*b*) $[A][B]$
  (*c*) $[B][A]$
  (*d*) $\{C\}^T[A]\{C\}$

B.2 If $[A]$ and $[B]$ are square matrices of equal order, verify that in general:

$$[A][B] \neq [B][A]$$

B.3 If $[A]$ is a symmetric matrix and $[B]$ is a symmetric matrix of the same order, show that $[A][B]$ is, in general, a nonsymmetric matrix.

B.4 If $[B]$ is a symmetric matrix of order (n,n) and $[A]$ is a matrix of order (n,m), verify that the product $[A]^T[B][A]$ is a symmetric matrix.

B.5 If $[B] = [A][A]^T$, show that $[B] = [B]^T$.

B.6 If $[A]$ and $[B]$ are diagonal matrices of equal order, verify that:
  (*a*) $[A][B] = [B][A]$
  (*b*) $[A][I] = [I][A] = [A]$

B.7 Compute the inverse of the following matrix:

$$\begin{bmatrix} 1 & 1 & 1 \\ 1 & 0 & -2 \\ 1 & -1 & 1 \end{bmatrix}$$

B.8 If $[A][B] = [C]$, verify the following *reversal laws:*
  (*a*) $[C]^T = [B]^T[A]^T$
  (*b*) $[C]^{-1} = [B]^{-1}[A]^{-1}$

B.9 Consider the quadratic form

$$F = \tfrac{1}{2}(A_{11}x_1^2 + A_{12}x_1x_2 + \cdots + A_{1n}x_1x_n$$
$$+ A_{21}x_2x_1 + A_{22}x_2^2 + \cdots + A_{2n}x_2x_n$$
$$+ \cdots$$
$$+ A_{n1}x_nx_1 + A_{n2}x_nx_2 + \cdots + A_{nn}x_n^2)$$

(*a*) Verify that F can be expressed in matrix notation as

$$F = \tfrac{1}{2}\{x\}^T[A]\{x\}$$

(*b*) Show that if $[A]$ is a symmetric matrix, then

$$\left\{ \frac{\partial F}{\partial x} \right\} = [A]\{x\}$$

where $\{\partial F/\partial x\}$ is a column matrix of the elements

$$\frac{\partial F}{\partial x_1}, \frac{\partial F}{\partial x_2}, \ldots, \frac{\partial F}{\partial x_n}$$

B.10   Solve the following simultaneous equations by matrix inversion:

$$\begin{aligned} x_1 + 3x_2 - 2x_3 &= -10 \\ 2x_1 - 4x_2 + 5x_3 &= 5 \\ -3x_1 - 2x_2 - 4x_3 &= -6 \end{aligned}$$

B.11   Compute the eigenvalues and eigenvectors of the following matrix:

$$\begin{bmatrix} 1 & -1 & 0 \\ -1 & 2 & -1 \\ 0 & -1 & 1 \end{bmatrix}$$

REFERENCES

For an elementary account of matrix algebra, including numerous illustrative examples:

B.1   GERE, J. M., and W. WEAVER, JR.: " Matrix Algebra for Engineers," Van Nostrand, Princeton, N.J., 1965.

For a rigorous treatment of matrix algebra and calculus, including applications of matrix methods to problems in structural mechanics and vibrations:

B.2   FRAZER, R. A., W. J. DUNCAN, and A. R. COLLAR: " Elementary Matrices," Cambridge, London, 1938.

B.3   PIPES, L. A.: " Matrix Methods for Engineering," Prentice-Hall, Englewood Cliffs, N.J., 1963.

APPENDIX C

CALCULUS OF VARIATIONS

C.1 INTRODUCTION

Many problems in engineering and physics are most naturally formulated in terms of extremum principles. Ordinary extremum problems of the differential calculus involve finding the extreme values (maxima or minima) of a function of one or more independent variables. The student will recall that a function of a single variable $u(x)$ will possess an extremum at a certain point only if the first derivative of the function vanishes at that point $(du/dx = 0)$. Likewise a necessary condition for the existence of an extreme value of a function of n variables, $u(x_1, \ldots, x_n)$, is that all its partial derivatives of first order are zero $(\partial u/\partial x_1 = \cdots = \partial u/\partial x_n = 0)$. We shall now consider a somewhat more complicated type of problem. Namely, we wish to find, among a set of admissible functions, that function which maximizes or minimizes a certain *functional* (i.e., a function of functions). As an example, consider the problem of determining the function $u(x)$ which makes the following integral a minimum:

$$I(u) = \int_a^b F(u,u',x)\, dx \qquad (C.1.1)$$

and which, in addition, satisfies the prescribed end conditions

$$u(a) = u_a \qquad u(b) = u_b \qquad (C.1.2)$$

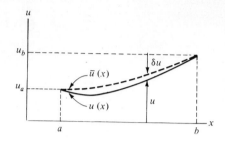

FIGURE C.1

The integrand $F(u,u',x)$ is presumed to be a known function of u, $u' = du/dx$, and x; I is the functional to be minimized. The branch of mathematics concerned with problems of this nature is known as the *calculus of variations*. Some of the basic ideas of the calculus are introduced in this appendix.

C.2 DEFINITIONS AND NOTATIONS

To make certain definitions clear, we shall focus our attention on the problem defined by Eqs. (C.1.1) and (C.1.2). For a given argument function $u(x)$, the integral $I(u)$ yields a specific numerical value. The problem posed here is to find the particular function of all those functions satisfying (C.1.2) which makes the integral a minimum.

Let us hypothesize that the curve $u(x)$ shown in Fig. C.1. is the actual minimizing function. In the calculus of variations we are interested in the behavior of $I(u)$ when the curve $u(x)$ is replaced by a slightly different curve, say $\bar{u}(x)$. A function $\bar{u}(x)$ in the neighborhood of $u(x)$ can be represented in the form

$$\bar{u}(x) = u(x) + \epsilon\eta(x) \qquad \text{(C.2.1)}$$

where ϵ is a small parameter. Since we are only interested in the class of functions which satisfy Eq. (C.1.2), the arbitrary function $\eta(x)$ must satisfy the conditions

$$\eta(a) = \eta(b) = 0 \qquad \text{(C.2.2)}$$

The difference between $\bar{u}(x)$ and $u(x)$ is called the *variation* in $u(x)$ and is denoted by δu; thus

$$\delta u = \bar{u}(x) - u(x) = \epsilon\eta(x) \qquad \text{(C.2.3)}$$

In other words the variation δu is a small arbitrary change in u from that value of u which minimizes the integral $I(u)$; it should not be confused with the differential quantity du which represents an infinitesimal change in u caused by an infinitesimal change dx in the independent variable x.

Similarly, the difference between the slope of the minimizing curve $u(x)$ and the slope of the varied curve $\bar{u}(x)$ is called the variation in the slope and is denoted by $\delta u'$: that is,

$$\delta u' = \bar{u}'(x) - u'(x) \qquad \text{(C.2.4)}$$

Comparing Eq. (C.2.4) with the expression obtained by differentiating (C.2.3), namely,

$$(\delta u)' = \bar{u}'(x) - u'(x) = \epsilon\eta'(x) \qquad \text{(C.2.5)}$$

it is seen that

$$\delta u' = (\delta u)' = \epsilon\eta' \qquad \text{(C.2.6)}$$

Hence the processes of variation and differentiation are permutable.

We next consider the behavior of the function $F(u,u',x)$ in the neighborhood of the minimizing curve $u(x)$. For a fixed value of the independent variable x, F depends upon u and u'; when these quantities are varied, F also varies. The function $F(\bar{u},\bar{u}',x)$ associated with the varied curve $\bar{u}(x)$ differs from the function $F(u,u',x)$ corresponding to the minimizing curve $u(x)$ by the increment

$$\Delta F = F(\bar{u},\bar{u}',x) - F(u,u',x)$$
$$= F(u + \delta u, u' + \delta u', x) - F(u,u',x) \qquad \text{(C.2.7)}$$

We may expand $F(u + \delta u, u' + \delta u', x)$ in a Taylor's series as

$$F(u + \delta u, u' + \delta u', x) = F(u,u',x) + \frac{\partial F}{\partial u}\delta u + \frac{\partial F}{\partial u'}\delta u'$$
$$+ \frac{1}{2!}\left(\frac{\partial^2 F}{\partial u^2}\delta u^2 + 2\frac{\partial^2 F}{\partial u\,\partial u'}\delta u\,\delta u' + \frac{\partial^2 F}{\partial u'^2}\delta u'^2\right) + \cdots \qquad \text{(C.2.8)}$$

in which case

$$\Delta F = \frac{\partial F}{\partial u}\delta u + \frac{\partial F}{\partial u'}\delta u' + \frac{1}{2!}\left(\frac{\partial^2 F}{\partial u^2}\delta u^2 + 2\frac{\partial^2 F}{\partial u\,\partial u'}\delta u\,\delta u' + \frac{\partial^2 F}{\partial u'^2}\delta u'^2\right) + \cdots \qquad \text{(C.2.9)}$$

The first and second variations of F are defined as

$$\delta F = \frac{\partial F}{\partial u}\delta u + \frac{\partial F}{\partial u'}\delta u' \qquad \text{(C.2.10)}$$

and

$$\delta^2 F = \delta(\delta F) = \frac{\partial^2 F}{\partial u^2}\delta u^2 + 2\frac{\partial^2 F}{\partial u\,\partial u'}\delta u\,\delta u' + \frac{\partial^2 F}{\partial u'^2}\delta u'^2 \qquad \text{(C.2.11)}$$

respectively. Higher variations may be defined in an analogous fashion; that is, $\delta^n F = \delta(\delta^{n-1}F)$. Substitution of the definitions (C.2.10) and (C.2.11) into Eq. (C.2.9) yields

$$\Delta F = \delta F + \frac{1}{2!}\delta^2 F + \cdots \qquad \text{(C.2.12)}$$

The difference ΔI between the minimum value of the integral I in Eq. (C.1.1) and the value of I evaluated for the varied curve $\bar{u}(x)$ may now be written as

$$\Delta I = I(\bar{u},\bar{u}',x) - I(u,u',x)$$
$$= \int_a^b F(\bar{u},\bar{u}',x)\,dx - \int_a^b F(u,u',x)\,dx$$
$$= \int_a^b \Delta F\,dx \qquad \text{(C.2.13)}$$

or, by virtue of Eq. (C.2.12),

$$\Delta I = \int_a^b \left(\delta F + \frac{1}{2!} \delta^2 F + \cdots \right) dx \qquad \text{(C.2.14)}$$

The first and second variations of the integral I are defined, respectively, as

$$\delta I = \int_a^b \delta F \, dx \qquad \text{(C.2.15)}$$

and

$$\delta^2 I = \int_a^b \delta^2 F \, dx \qquad \text{(C.2.16)}$$

Similarly the nth variation of I is defined by $\delta^n I = \int_a^b \delta^n F \, dx$.
Introduction of these definitions into Eq. (C.2.14) gives

$$\Delta I = \delta I + \frac{1}{2!} \delta^2 I + \cdots \qquad \text{(C.2.17)}$$

Based upon our hypothesis that $u(x)$ is the function which minimizes I, it follows that $\Delta I \geq 0$. Note that as the parameter ϵ in Eq. (C.2.1) is reduced, the neighboring function $\bar{u}(x)$ approaches $u(x)$, and the increment ΔI tends to zero; when $\bar{u}(x)$ is identical to $u(x)$, ΔI vanishes and I is a minimum.

In order to establish necessary conditions for the existence of a minimum of I, we must examine the relative orders of magnitude of the terms in the series (C.2.17). It can easily be shown that δI is proportional to the infinitesimal parameter ϵ, that $\delta^2 I$ is proportional to the square of ϵ, and that in general $\delta^n I$ is of the order of ϵ^n. To see this, we introduce the relations (C.2.3), (C.2.6), and (C.2.10) into (C.2.15), in which case

$$\delta I = \int_a^b \left(\frac{\partial F}{\partial u} \delta u + \frac{\partial F}{\partial u'} \delta u' \right) dx = \epsilon \int_a^b \left(\frac{\partial F}{\partial u} \eta + \frac{\partial F}{\partial u'} \eta' \right) dx \qquad \text{(C.2.18)}$$

Similarly

$$\delta^2 I = \int_b^b \left(\frac{\partial^2 F}{\partial u^2} \delta u^2 + 2 \frac{\partial^2 F}{\partial u \, \partial u'} \delta u \, \delta u' + \frac{\partial^2 F}{\partial u'^2} \delta u'^2 \right) dx$$

$$= \epsilon^2 \int_a^b \left(\frac{\partial^2 F}{\partial u^2} \eta^2 + 2 \frac{\partial^2 F}{\partial u \, \partial u'} \eta \eta' + \frac{\partial^2 F}{\partial u'^2} \eta'^2 \right) dx \qquad \text{(C.2.19)}$$

Thus, the second and higher variations in the series expression for ΔI are negligible compared with δI when ϵ is sufficiently small. Accordingly, a necessary condition, for I to have a minimum, is that δI vanish identically. To verify that the extremum is indeed a minimum, it is sufficient to show that the first nonvanishing term in the series (C.2.17) is positive definite. For example, if it is possible to prove that $\delta^2 I > 0$ for all admissible nonzero variations, then $\Delta I > 0$, and the extremum is necessarily a minimum.

Using the same arguments as those given above, it can be shown that the condition $\delta I = 0$ also represents a necessary condition for a *maximum* of the integral I in Eq. (C.1.1) under the end conditions (C.1.2). In this case, however, the first nonvanishing term in the series (C.2.17) for ΔI will be negative.

In conclusion, a necessary condition for the integral I to have an extremum is that the first variation δI vanishes. If the second variation $\delta^2 I$ is positive definite, the extremum is a minimum; if $\delta^2 I$ is negative definite, it is a maximum. Quite often the nature of the problem is such that there is no uncertainty as to whether the extremum is a minimum or a maximum, in which case there is no need to investigate the sign of ΔI.

We shall see in the following section that the variational problem governed by the condition $\delta I = 0$ can be reduced to a problem in the differential calculus.

C.3   EULER-LAGRANGE EQUATIONS

As demonstrated above, a necessary condition that the definite integral (C.1.1) have an extremum under the boundary conditions (C.1.2) is

$$\delta I = \int_a^b \left(\frac{\partial F}{\partial u} \delta u + \frac{\partial F}{\partial u'} \delta u' \right) dx = 0 \qquad \text{(C.3.1)}$$

Noting that $\delta u' = d(\delta u)/dx$, the second term in the integrand of (C.3.1) can be integrated by parts as follows:

$$\int_a^b \frac{\partial F}{\partial u'} \delta u' \, dx = \left[\frac{\partial F}{\partial u'} \delta u \right]_a^b - \int_a^b \frac{d}{dx} \left(\frac{\partial F}{\partial u'} \right) \delta u \, dx \qquad \text{(C.3.2)}$$

Therefore
$$\delta I = \int_a^b \left[\frac{\partial F}{\partial u} - \frac{d}{dx} \left(\frac{\partial F}{\partial u'} \right) \right] \delta u \, dx + \left[\frac{\partial F}{\partial u'} \delta u \right]_a^b = 0 \qquad \text{(C.3.3)}$$

Since the variation $\delta u = \bar{u}(x) - u(x)$ vanishes at the end points $x = a$ and $x = b$, the integrated term in Eq. (C.3.3) vanishes. Furthermore it can be proved that since δu is arbitrary in the range $a < x < b$, the bracketed term inside the integral must also vanish independently; thus

$$\frac{\partial F}{\partial u} - \frac{d}{dx} \left(\frac{\partial F}{\partial u'} \right) = 0 \qquad \text{(C.3.4)}$$

This differential equation is called the *Euler-Lagrange equation*. It represents a necessary but not sufficient condition which the function $u(x)$ must satisfy if it is to yield an extremum for $I(u)$.

The Euler-Lagrange equation can be generalized to the case in which the integrand F contains higher order derivatives of u. Suppose for example that we wish to determine an extremum of the integral

$$I = \int_a^b F(u,u',u'',x) \, dx \qquad \text{(C.3.5)}$$

The variation of the integrand F may be written as

$$\delta F = \frac{\partial F}{\partial u} \delta u + \frac{\partial F}{\partial u'} \delta u' + \frac{\partial F}{\partial u''} \delta u'' \qquad \text{(C.3.6)}$$

so that
$$\delta I = \int_a^b \left(\frac{\partial F}{\partial u} \delta u + \frac{\partial F}{\partial u'} \delta u' + \frac{\partial F}{\partial u''} \delta u'' \right) dx = 0 \qquad \text{(C.3.7)}$$

Integration by parts then yields

$$\delta I = \int_a^b \left[\frac{\partial F}{\partial u} - \frac{d}{dx}\left(\frac{\partial F}{\partial u'}\right) + \frac{d^2}{dx^2}\left(\frac{\partial F}{\partial u''}\right) \right] \delta u \, dx$$
$$+ \left[\left[\frac{\partial F}{\partial u'} - \frac{d}{dx}\left(\frac{\partial F}{\partial u''}\right) \right] \delta u \right]_a^b + \left[\frac{\partial F}{\partial u''} \delta u' \right]_a^b = 0 \qquad \text{(C.3.8)}$$

If the value of the function u and its first derivative u' are prescribed at $x = a$ and $x = b$, then $\delta u(a) = \delta u(b) = \delta u'(a) = \delta u'(b) = 0$. In this case both integrated terms in Eq. (C.3.8) are identically zero. Since δu is arbitrary for $a < x < b$, the integral in (C.3.8) will vanish only if

$$\frac{\partial F}{\partial u} - \frac{d}{dx}\left(\frac{\partial F}{\partial u'}\right) + \frac{d^2}{dx^2}\left(\frac{\partial F}{\partial u''}\right) = 0 \qquad \text{(C.3.9)}$$

which is the Euler-Lagrange equation for this variational problem.

The above ideas can also be extended to situations involving several dependent and/or several independent variables. Consider for example the problem of finding the three dependent variables $u_i(x_1, x_2, x_3)$ ($i = 1, 2, 3$) which will yield an extremum for the volume integral

$$I = \int_{\mathcal{V}} F(u_i, u_{j,k}, x_l) \, d\mathcal{V} \qquad i,j,k,l = 1,2,3 \qquad \text{(C.3.10)}$$

Here $u_{j,k}$ denotes the partial derivative $\partial u_j / \partial x_k$, and $d\mathcal{V}$ represents the elemental volume $dx_1 \, dx_2 \, dx_3$. The variation of the integrand F in this case is

$$\delta F = \frac{\partial F}{\partial u_i} \delta u_i + \frac{\partial F}{\partial u_{j,k}} \delta u_{j,k} \qquad \text{(C.3.11)}$$

in which the *summation convention* for repeated subscripts has been used. Again requiring that the first variation of I vanish, we have

$$\delta I = \int_{\mathcal{V}} \left(\frac{\partial F}{\partial u_i} \delta u_i + \frac{\partial F}{\partial u_{j,k}} \delta u_{j,k} \right) d\mathcal{V} = 0 \qquad \text{(C.3.12)}$$

Integration of Eq. (C.3.12) by parts leads to the three Euler-Lagrange equations[1]

$$\frac{\partial F}{\partial u_i} - \frac{\partial}{\partial x_j}\left(\frac{\partial F}{\partial u_{i,j}}\right) = 0 \qquad i = 1,2,3 \qquad \text{(C.3.13)}$$

[1] The details of the derivation of the Euler-Lagrange equations for a function of more than one variable are given in Ref. C.2.

C.4 NATURAL BOUNDARY CONDITIONS

In the preceding sections we postulated that the argument function $u(x)$ had prescribed values at the end points $x = a$ and $x = b$. We shall now see that if the boundary values are left unspecified, then the variational approach leads to a set of boundary conditions which must be satisfied if the functional is to possess an extremum. These boundary conditions are called natural boundary conditions.

Let us again consider the integral

$$I = \int_a^b F(u, u', x)\, dx \qquad \text{(C.4.1)}$$

but no longer impose conditions on the function $u(x)$ at the end points $x = a$ and $x = b$. A necessary condition for I to attain an extremum is (C.3.3)

$$\delta I = \int_a^b \left[\frac{\partial F}{\partial u} - \frac{d}{dx}\left(\frac{\partial F}{\partial u'}\right) \right] \delta u\, dx + \left[\frac{\partial F}{\partial u'} \delta u \right]_a^b = 0 \qquad \text{(C.4.2)}$$

Since δu is now presumed to be arbitrary over the region of integration $a < x < b$, the Euler-Lagrange equation (C.3.4) must still be satisfied. In addition, Eq. (C.4.2) requires that at $x = a$ and $x = b$, either

$$\delta u = 0 \qquad \text{or} \qquad \frac{\partial F}{\partial u'} = 0 \qquad \text{(C.4.3)}$$

Hence if u is not prescribed on the boundary (in which case δu would be zero), then a necessary condition for an extremum is $\partial F/\partial u' = 0$. The first type of boundary condition in (C.4.3) is called a *rigid boundary condition*, whereas the latter is termed a *natural boundary condition*.

Similarly, the necessary condition for an extremum of the integral

$$I = \int_a^b F(u, u', u'', x)\, dx \qquad \text{(C.4.4)}$$

is, according to Eq. (C.3.8),

$$\begin{aligned} \delta I = & \int_a^b \left[\frac{\partial F}{\partial u} - \frac{d}{dx}\left(\frac{\partial F}{\partial u'}\right) + \frac{d^2}{dx^2}\left(\frac{\partial F}{\partial u''}\right) \right] \delta u\, dx \\ & + \left[\left[\frac{\partial F}{\partial u'} - \frac{d}{dx}\left(\frac{\partial F}{\partial u''}\right) \right] \delta u \right]_a^b + \left[\frac{\partial F}{\partial u''} \delta u' \right]_a^b = 0 \end{aligned} \qquad \text{(C.4.5)}$$

In addition to satisfying the Euler-Lagrange equation (C.3.9), the argument function $u(x)$ must satisfy the following conditions at $x = a$ and $x = b$; either

$$\delta u = 0 \qquad \text{or} \qquad \frac{\partial F}{\partial u'} - \frac{d}{dx}\left(\frac{\partial F}{\partial u''}\right) = 0 \qquad \text{(C.4.6)}$$

and also either

$$\delta u' = 0 \qquad \text{or} \qquad \frac{\partial F}{\partial u''} = 0 \qquad \text{(C.4.7)}$$

The second of each pair of conditions represents a natural boundary condition for this problem.

In a similar way, natural boundary conditions can be obtained for variational problems involving several dependent and/or several independent variables.

The concept of natural boundary conditions is of great significance in structural mechanics. One of the major advantages of using a variational approach is that these conditions are obtained in a simple and straightforward way. Applications of this technique of deriving boundary conditions are given in Chaps. 8, 9, and 12.

PROBLEMS

C.1   Consider two functions of the same variables, for instance $F_1(u, u', x)$ and $F_2(u, u', x)$. Prove that

$$\delta(F_1 F_2) = \delta F_1 F_2 + F_1 \, \delta F_2$$

C.2   Verify that Eq. (C.3.8) is a necessary condition for an extremum of the integral (C.3.5); that is, show all the details in the derivation of (C.3.8).

C.3   Derive the Euler-Lagrange equation and the natural boundary conditions for the integral

$$I = \int_{a_2}^{b_2} \int_{a_1}^{b_1} F(u_1, u_2, u_{1,1}, u_{1,2}, u_{2,1}, u_{2,2}, x_1, x_2) \, dx_1 \, dx_2$$

C.4   A prismatic bar is subjected to a distributed axial force per unit length $Q(x_1)$, as shown.

(*a*) Show that by minimizing the total potential energy $U + V_E$, where

$$U = \int_0^L \frac{EA}{2} \left(\frac{\partial u_1}{\partial x_1} \right)^2 dx_1 \qquad \text{and} \qquad V_E = - \int_0^L Q(x_1) u_1 \, dx_1$$

one obtains the Euler-Lagrange equation

$$EA \frac{\partial^2 u_1}{\partial x_1^2} + Q(x_1) = 0$$

(*b*) Show that the natural boundary condition at the free end of the bar is

$$EA \frac{\partial u_1(L)}{\partial x_1} = 0$$

(*c*) Determine the deflection curve $u_1(x_1)$ for the case of the linearly varying load

$$Q(x_1) = Q_0 \frac{x_1}{L}$$

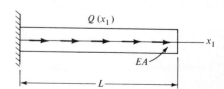

PROBLEM C.4

C.5 A circular shaft is loaded by a distributed torque per unit length $Q(x_1)$, as shown.
(*a*) By minimizing the total potential energy $U + V_E$, where

$$U = \int_0^L \frac{GJ}{2} \left(\frac{\partial \theta}{\partial x_1} \right)^2 dx_1 \quad \text{and} \quad V_E = - \int_0^L Q(x_1)\theta \, dx_1$$

show that the shaft's rotation is governed by the Euler-Lagrange equation

$$GJ \frac{\partial^2 \theta}{\partial x_1{}^2} + Q(x_1) = 0$$

(*b*) Derive the natural boundary condition for the free end of the shaft.

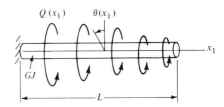

$Q(x_1)$ $\theta(x_1)$

x_1

GJ

PROBLEM C.5 $\longleftarrow \quad L \quad \longrightarrow$

C.6 A cantilever beam of uniform cross section supports a distributed transverse load $Q(x_1)$, as shown.
(*a*) Show that by minimizing the total potential energy $U + V_E$, where

$$U = \int_0^L \frac{EI}{2} \left(\frac{d^2 u_2}{dx_1{}^2} \right)^2 dx_1 \quad \text{and} \quad V_E = - \int_0^L Q(x_1) u_2 \, dx_1$$

one obtains the Euler-Lagrange equation

$$EI \frac{d^4 u_2}{dx_1{}^4} = Q(x_1)$$

(*b*) Show that the natural boundary conditions at the free end of the beam are

$$-EIu_2'''(L) = EIu_2''(L) = 0$$

(*c*) Compute the deflection curve $u_2(x_1)$ for the case of a distributed load of uniform intensity.

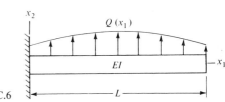

x_2

$Q(x_1)$

EI x_1

PROBLEM C.6 $\longleftarrow \quad L \quad \longrightarrow$

REFERENCES

For a rigorous treatment of the calculus of variations:

C.1 COURANT, R., and D. HILBERT: " Methods of Mathematical Physics," vol. I, Interscience, New York, 1953.

For a variety of applications of variational principles in classical mechanics and in the mechanics of deformable solids:

C.2 LANGHAAR, H. L.: "Energy Methods in Applied Mechanics," Wiley, New York, 1962.

C.3 LANCZOS, C.: "The Variational Principles of Mechanics," 3d ed., University of Toronto Press, 1966.

C.4 WASHIZU, K.: "Variational Methods in Elasticity and Plasticity," Pergamon, Oxford, 1968.

ANSWERS TO SELECTED PROBLEMS

1.4 $\quad |N| = P \cos \phi \qquad |V| = P \sin \phi \qquad |M| = PR \cos \phi$

1.6 $\quad |u_2|_{\max} = 0.00652 \dfrac{w_0 L^4}{EI}$

2.1 $\quad T_n = 2{,}667 \text{ lb}_f/\text{in.}^2 \qquad T_s = 6{,}650 \text{ lb}_f/\text{in.}^2$

2.4 $\quad \sigma^{(1)} = 10{,}270 \text{ lb}_f/\text{in.}^2 \qquad (0.991, 0.132, 0.017)$
$\qquad \sigma^{(11)} = -3{,}146 \text{ lb}_f/\text{in.}^2 \qquad (-0.103, 0.676, 0.730)$
$\qquad \sigma^{(111)} = -7{,}120 \text{ lb}_f/\text{in.}^2 \qquad (-0.085, 0.725, -0.684)$

3.2 $\quad e'_{11} = 0.002 \qquad\qquad e'_{22} = 0.001 \qquad\qquad e'_{33} = -0.001$
$\qquad e'_{12} = 0.0005\sqrt{3} \qquad e'_{23} = -0.001\sqrt{3} \qquad e'_{31} = -0.0005$

3.4 $\quad \Delta^{(AB)} = 0.002 \text{ in.} \qquad \Delta^{(BC)} = 0.002 \text{ in.} \qquad \Delta^{(BD)} = 0.0072\sqrt{5} \text{ in.}$

5.6 $\quad U = \dfrac{1}{2EI} \left[\dfrac{Q_1{}^2 L^5}{20} + \dfrac{Q_2{}^2 L^3}{3} + Q_3{}^2 L + \dfrac{Q_1 Q_2 L^4}{4} \right.$

$$\left. + \dfrac{Q_1 Q_3 L^3}{3} + Q_2 Q_3 L^2 \right] + \dfrac{Q_4^2 L}{2EA}$$

5.9 $\quad Q_1 = (189 q_1 - 48 q_2) \dfrac{EA}{125L} \qquad Q_2 = (-48 q_1 + 36 q_2) \dfrac{EA}{125L}$

5.11 $\Delta = \dfrac{Q_1 L^4}{128EI} + \dfrac{Q_2 L^3}{8EI}$

5.12 $\Delta = \dfrac{Q_1 L^3}{45EI}$

5.13 $\Delta = -\left(1 + \dfrac{\pi}{4}\right)\dfrac{Q_1 R^3}{EI} + \dfrac{Q_2 R^2}{EI}$

5.14 $\Delta = \dfrac{25 Q_1{}^2 L^4}{2 b^2 h^5 E^2}$

5.24 $\Delta = \dfrac{Q_1 L^3}{2EI}$

5.26 $[k] = \begin{bmatrix} 1 & -1 \\ -1 & 3 \end{bmatrix} \dfrac{GJ_0}{L}$

5.27 $u_2(L) = 0.329 \dfrac{Q_1 L^3}{EI}$

6.1 $P^{(AD)} = 0.720 Q_1$ $P^{(CD)} = -0.530 Q_1$
    $P^{(BD)} = -0.114 Q_1$ $P^{(AB)} = 0$ $P^{(BC)} = 0.424 Q_1$

6.3 $q_1 = 7.59 \dfrac{Q_1 L}{EA}$

6.4 (c) $q_1 = \dfrac{31 Q_1 L}{24EI}$

6.6 $\Delta_{max} = \dfrac{Q_1 L^4}{384EI}$

6.10 $q_1 = 0.0191 \dfrac{Q_1 R^3}{EI}$

6.12 $|\sigma_{11}|_{max} = \dfrac{32 Q_1 R}{\pi^2 d^3}$

6.13 $q_1 = 0.259 \dfrac{Q_1 L^3}{EAh^2}$

7.1 $[k] = \begin{bmatrix} 0.766 & 0.0126 \\ 0.0126 & 0.0616 \end{bmatrix} \dfrac{EA}{L}$

7.4 $[c] = \begin{bmatrix} 0.458 \dfrac{L^3}{EI_0} & 0.625 \dfrac{L^2}{EI_0} \\ 0.625 \dfrac{L^2}{EI_0} & 1.125 \dfrac{L}{EI_0} \end{bmatrix}$

7.8 $[k] = \begin{bmatrix} 0.0189 & 0 & 0.00344 \\ 0 & 0.0223 & 0 \\ 0.00344 & 0 & 0.0726 \end{bmatrix} EA$ lb/in.

7.11 $[c] = \begin{bmatrix} 27.8 & 7.22 \\ 7.22 & 5.01 \end{bmatrix} 10^{-6}$ in./lb

8.1 $\sigma_{11} = -\alpha E \Theta_0 \left(\dfrac{1}{3} - \dfrac{4x_2{}^2}{h^2} \right)$

8.3 $\sigma_{11} = \dfrac{Q_1 x_2}{60 LI} (10x_1{}^3 - 9L^2 x_1 + 2L^3) - \dfrac{\alpha E \Theta_0 x_1 x_2}{Lh}$

8.4 $\sigma_{11} = \alpha E \Theta_0 \left[-\dfrac{4}{\pi^2} + \left(\dfrac{2}{\pi} - \cos \dfrac{\pi x_2}{h} \right) \sin \dfrac{\pi x_1}{L} \right]$

8.5 $\Delta = \dfrac{Q_1 L^3}{45 EI} - \dfrac{\alpha L \Theta_0}{3h}$

8.9 $q_1 = \dfrac{Q_1 L}{EA} + \dfrac{4Q_2 L}{3EA} + \dfrac{\alpha L \Theta_0}{2}$ $q_2 = \dfrac{4Q_1 L}{3EA} + \dfrac{21 Q_2 L}{4EA} - \dfrac{3\alpha L \Theta_0}{8}$

8.16 $Q_1 = \dfrac{EA}{L} (0.211 q_1 - 0.642 \alpha L \Theta_0)$

9.2 $(Q_0)_{\text{critical}} = 1.45 \dfrac{k}{L},$ $3.22 \dfrac{k}{L}$

9.5 $(Q_0)_{\text{critical}} = \dfrac{\pi^2 EI}{L^2} \left(n^2 + \dfrac{kL^4}{n^2 \pi^4 EI} \right)$ $n = 1, 2, \dots$

9.7 $(Q_0)_{\text{critical}} = \dfrac{20.2 EI}{L^2}$

9.8 $(Q_0)_{\text{critical}} = \dfrac{4\pi^2 EI}{L^2}$

9.10 $|M|_{\text{max}} = \dfrac{M_0}{\cos \sqrt{Q_0 L^2 / EI}}$

9.11 (a) $|M|_{\text{max}} = 0.168 Q_1 L^2$ (b) $|M|_{\text{max}} = 0.125 Q_1 L^2$
      (c) $|M|_{\text{max}} = 0.099 Q_1 L^2$

10.1 (c) $|u|_{\text{max}} = 0.0635$ in. $|M|_{\text{max}} = 183,000$ in-lb

10.5 $u = \dfrac{F_0}{k} \left(\dfrac{t}{t_0} - \dfrac{1}{2\pi t_0 / \tau} \sin \dfrac{2\pi t}{\tau} \right)$ $t < t_0$

$u = \dfrac{F_0}{k} \left[1 + \dfrac{1}{2\pi t_0 / \tau} \left(\sin \dfrac{2\pi (t - t_0)}{\tau} - \sin \dfrac{2\pi t}{\tau} \right) \right]$ $t > t_0$

10.6 (b) $\quad u = \dfrac{F_0/k}{1 - (2/pt_0)(n/p) + 1/p^2 t_0^2} \left\{ e^{-t/t_0} - e^{-nt} \left[\cos \bar{p}t + \left(\dfrac{n}{\bar{p}} - \dfrac{1}{\bar{p}t_0} \right) \sin \bar{p}t \right] \right\}$

10.10 $\quad \theta = \dfrac{8\mathcal{T}_0 L}{\pi G D^4} \left(1 - \cos \sqrt{\dfrac{\pi G D^4}{8LI_m}} \, t \right) \qquad \tau_{max} = \dfrac{16\mathcal{T}_0}{\pi D^3}$

10.11 (a) $\quad w = \dfrac{-v_0}{\sqrt{4Ebh^3/mL^3}} \sin \sqrt{\dfrac{4Ebh^3}{mL^3}} \, t \qquad \left| \sigma_{11} \right|_{max} = 3v_0 \sqrt{\dfrac{Em}{bhL}}$

11.4 $\quad m_1 \begin{bmatrix} 1 & 0 \\ 0 & 2 \end{bmatrix} \begin{bmatrix} \ddot{u}_1 \\ \ddot{u}_2 \end{bmatrix} + \dfrac{EI}{L^3} \begin{bmatrix} 259.2 & -226.8 \\ -226.8 & 259.2 \end{bmatrix} \begin{bmatrix} u_1 \\ u_2 \end{bmatrix} = \begin{bmatrix} F_1(t) \\ F_2(t) \end{bmatrix}$

11.6 $\quad \begin{bmatrix} I_{m_1} & 0 & 0 \\ 0 & I_{m_2} & 0 \\ 0 & 0 & I_{m_3} \end{bmatrix} \begin{bmatrix} \ddot{\theta}_1 \\ \ddot{\theta}_2 \\ \ddot{\theta}_3 \end{bmatrix} + \dfrac{GJ}{L} \begin{bmatrix} 1 & -1 & 0 \\ -1 & 2 & -1 \\ 0 & -1 & 1 \end{bmatrix} \begin{bmatrix} \theta_1 \\ \theta_2 \\ \theta_3 \end{bmatrix} = \begin{bmatrix} \mathcal{T}_1(t) \\ \mathcal{T}_2(t) \\ \mathcal{T}_3(t) \end{bmatrix}$

11.8 (a) $\quad p_1 = 4.63 \sqrt{\dfrac{EI}{m_1 L^3}} \qquad p_2 = 19.2 \sqrt{\dfrac{EI}{m_1 L^3}} \qquad [\phi] = \begin{bmatrix} 1.0 & 1.0 \\ 1.048 & -0.477 \end{bmatrix}$

(b) $\quad u_1 = (0.639 \cos p_1 t + 0.361 \cos p_2 t)u_0$

$\qquad u_2 = (0.671 \cos p_1 t - 0.171 \cos p_2 t)u_0$

11.14 $\quad \left| u_1 \right|_{max} = 0.0906 \dfrac{F_0 L^3}{EI} \qquad \left| u_2 \right|_{max} = 0.0948 \dfrac{F_0 L^3}{EI} \qquad \left| M_2 \right|_{max} = 1.12 F_0 L$

11.16 $\quad \left| M_2 \right|_{max} = 1.12 a_0 m_1 L$

11.18 $\quad p_1 = 0 \qquad p_2 = 1.20 \sqrt{\dfrac{GJ}{I_{m_1} L}} \qquad p_3 = 2.36 \sqrt{\dfrac{GJ}{I_{m_1} L}}$

$\qquad \theta_1 = \dfrac{\mathcal{T}_0 t^2}{4I_{m_1}} + [0.300(1 - \cos p_2 t) + 0.012(1 - \cos p_3 t)] \dfrac{\mathcal{T}_0 L}{GJ}$

11.21 $\quad p_1 = 0.581 \sqrt{\dfrac{EI}{mL^3}} \qquad p_2 = 3.04 \sqrt{\dfrac{EI}{mL^3}}$

$\qquad p_3 = 3.83 \sqrt{\dfrac{EI}{mL^3}} \qquad p_4 = 8.37 \sqrt{\dfrac{EI}{mL^3}}$

11.22 (a) $\quad p_1 = 0.432 \sqrt{\dfrac{EI}{mL^3}} \qquad p_2 = 1.39 \sqrt{\dfrac{EI}{mL^3}}$

$\qquad p_3 = 3.41 \sqrt{\dfrac{EI}{mL^3}} \qquad p_4 = 5.85 \sqrt{\dfrac{EI}{mL^3}}$

(b) $\quad \theta_2 = (0.067 \cos p_1 t - 0.616 \cos p_2 t - 0.469 \cos p_3 t + 0.018 \cos p_4 t)\theta_0$

(c) $\quad \theta_2 = [0.826(1 - \cos p_1 t) + 0.124(1 - \cos p_2 t)$

$\qquad\qquad + 0.050(1 - \cos p_3 t) - 0.006(1 - \cos p_4 t)] \dfrac{M_0 L}{EI}$

11.25 (a) $p_1 = p_2 = 0$ $\qquad$ $p_3 = 0.828 \sqrt{\dfrac{EI}{mL^3}}$

$\qquad$ $p_4 = 2.45 \sqrt{\dfrac{EI}{mL^3}}$ $\qquad$ $p_5 = 3.52 \sqrt{\dfrac{EI}{mL^3}}$

12.7 $\quad p_j = \dfrac{(2j-1)\pi}{2} \sqrt{\dfrac{EA}{\mu L^2}}$ $\qquad$ $\phi_j = \sin \dfrac{(2j-1)\pi x_1}{2L}$

12.9 (b) $\quad p_1 = 1.075 \sqrt{\dfrac{EA}{\mu L^2}}$ $\qquad$ $p_2 = 3.642 \sqrt{\dfrac{EA}{\mu L^2}}$ $\qquad$ $p_3 = 6.575 \sqrt{\dfrac{EA}{\mu L^2}}$

$\qquad \phi_1 = \sin \dfrac{1.075 x_1}{L}$ $\qquad$ $\phi_2 = \sin \dfrac{3.642 x_1}{L}$ $\qquad$ $\phi_3 = \sin \dfrac{6.575 x_1}{L}$

12.10 $\quad p_j = j^2 \pi^2 \sqrt{\dfrac{EI}{\mu L^4}}$ $\qquad$ $\phi_j = \sin \dfrac{j \pi x_1}{L}$

12.11 $\quad p_1 = 22.4 \sqrt{\dfrac{EI}{\mu L^4}}$ $\qquad$ $p_2 = 61.7 \sqrt{\dfrac{EI}{\mu L^4}}$ $\qquad$ $p_3 = 121.0 \sqrt{\dfrac{EI}{\mu L^4}}$

12.12 $\quad p_1 = 0.395 \sqrt{\dfrac{EI}{\mu L^4}}$

12.16 $\quad p_1 = 0.408 \sqrt{\dfrac{EI}{\mu L^4}}$

12.20 $\quad |u_2|_{max} = \dfrac{5 F_0 L^4}{192 EI}$ $\qquad$ $|M|_{max} = \dfrac{F_0 L^2}{4}$

12.25 $\quad w_2 = \dfrac{-4 v_0}{\pi} \displaystyle\sum_{j=1,3,..}^{\infty} \dfrac{1}{j p_j} \sin \dfrac{j \pi x_1}{L} \sin p_j t$

B.7 $\quad \dfrac{1}{6} \begin{bmatrix} 2 & 2 & 2 \\ 3 & 0 & -3 \\ 1 & -2 & 1 \end{bmatrix}$

B.10 $\quad x_1 = -6$ $\qquad$ $x_2 = 2$ $\qquad$ $x_3 = 5$

B.11 $\quad \lambda_1 = 0 \qquad \{1, 1, 1\}$
$\qquad \lambda_2 = 1 \qquad \{1, 0, -1\}$
$\qquad \lambda_3 = 3 \qquad \{1, -2, 1\}$

C.4 $\quad u_1 = \dfrac{Q_0}{EA} \left(\dfrac{x_1 L}{2} - \dfrac{x_1^3}{6L} \right)$

C.6 $\quad u_2 = \dfrac{Q_0}{24 EI} (x_1^4 - 4 x_1^3 L + 6 x_1^2 L^2)$

HO, C.C. 219–300